T0395437

# MACHINE LEARNING FOR SMALL BODIES IN THE SOLAR SYSTEM

# MACHINE LEARNING FOR SMALL BODIES IN THE SOLAR SYSTEM

Edited by

**VALERIO CARRUBA**

**EVGENY SMIRNOV**

**DAGMARA OSZKIEWICZ**

ELSEVIER

Elsevier
Radarweg 29, PO Box 211, 1000 AE Amsterdam, Netherlands
125 London Wall, London EC2Y 5AS, United Kingdom
50 Hampshire Street, 5th Floor, Cambridge, MA 02139, United States

**Notices**

Knowledge and best practice in this field are constantly changing. As new research and experience broaden our understanding, changes in research methods, professional practices, or medical treatment may become necessary.

Practitioners and researchers must always rely on their own experience and knowledge in evaluating and using any information, methods, compounds, or experiments described herein. In using such information or methods they should be mindful of their own safety and the safety of others, including parties for whom they have a professional responsibility.

To the fullest extent of the law, neither the Publisher nor the authors, contributors, or editors, assume any liability for any injury and/or damage to persons or property as a matter of products liability, negligence or otherwise, or from any use or operation of any methods, products, instructions, or ideas contained in the material herein.

ISBN: 978-0-443-24770-5

For information on all Elsevier publications
visit our website at https://www.elsevier.com/books-and-journals

Publisher: Candice Janco
Acquisitions Editor: Peter Llewellyn
Editorial Project Manager: Aleksandra Packowska
Production Project Manager: Sruthi Satheesh
Cover Designer: Mark Rogers

Front cover image created by Laura May

Typeset by VTeX

Working together
to grow libraries in
developing countries

www.elsevier.com • www.bookaid.org

*To the students, researchers, and general public interested in exploring new fields. "Here's to the ones who dream, foolish as they may seem" (Auditions, La La Land).*

# Contents

## 11. Conclusions and future developments 295

Valerio Carruba, Evgeny Smirnov, and Dagmara Oszkiewicz

Supplementary materials, including jupyter notebooks and examples of the software discussed in Chapters 2, 4, 6, 7, and 9 are available at these links:

The website's URL is: https://solar-system-ml.github.io/book/
The Github's repository URL is: https://github.com/solar-system-ml/book

# Contributors

**Safwan Aljbaae**
National Institute for Space and Research (INPE), Division of Graduate Studies, São José dos Campos, SP, Brazil
Make The Way, São Paulo, SP, Brazil

**Abreuçon Atanasio Alves**
São Paulo State University (UNESP), Department of Mathematics, Guaratinguetá, SP, Brazil

**Bryce T. Bolin**
Goddard Space Flight Center, Greenbelt, MD, United States

**Gabriel Caritá**
National Institute for Space and Research (INPE), Division of Graduate Studies, São José dos Campos, SP, Brazil

**Valerio Carruba**
São Paulo State University (UNESP), Department of Mathematics, Guaratinguetá, SP, Brazil

**Michael W. Coughlin**
School of Physics and Astronomy, University of Minnesota, Twin Cities, MN, United States

**R.C. Domingos**
São Paulo State University (UNESP), School of Engineering, Department of Electronic and Telecommunications Engineering, São João da Boa Vista, SP, Brazil

**Wesley C. Fraser**
National Research Council of Canada, Herzberg Astronomy and Astrophysics Research Centre, Victoria, BC, Canada

**M. Huaman**
Universidad tecnológica del Perú (UTP), Cercado de Lima, Peru

**Hanna Klimczak-Plucińska**
Astronomical Observatory Institute, Faculty of Physics and Astronomy, Adam Mickiewicz University, Poznań, Poland

**M.V.F. Lourenço**
São Paulo State University (UNESP), School of Engineering and Sciences, Department of Mathematics, Guaratinguetá, SP, Brazil

**Renu Malhotra**
The University of Arizona, Tucson, AZ, United States

**Dagmara Oszkiewicz**
Astronomical Observatory Institute, Faculty of Physics and Astronomy, Adam Mickiewicz University, Poznań, Poland

**Antti Penttilä**
Department of Physics, University of Helsinki, Helsinki, Finland

**Evgeny Smirnov**
Belgrade Astronomical Observatory, Belgrade, Serbia

**Kathryn Volk**
Planetary Science Institute, Tucson, AZ, United States

# Foreword

Since my PhD in the late 2000s, I have felt the data analysis methods used to study the small bodies of our Solar System had remained at Stone Age (pun intended) compared to other fields of astronomy. The amount of available data was indeed rather limited, and in most cases still manageable at human scale. There is a historical gap between communities, with roughly a century between the availability of a catalog for a million stars and for a million small bodies.

The large sky surveys of the past decades have, however, provided a tremendous amount of observations. Among millions of images, small bodies have to be identified, and their astrometry and photometry and appearance measured. These quantities are then turned into catalogs of relevant properties (such as orbits and colors) and analyzed to understand how, where, when these bodies accreted; the latter processes help to unveil the history of our planetary system.

This wealth of data requires powerful methods of data analysis, as each of the required steps listed above may be tedious, or even impossible, to perform "manually." Machine learning methods have thus ramped up in the community over the last ten years, with great results. There is likely even much more to come with more surveys starting, and a growing interest from the community, as highlighted by the publication of the present book.

In my opinion, machine-learning methods will play a major role in exploring large datasets for planetary sciences. It is, however, not a purpose by itself but a powerful tool. Mimicking Gargantua,[1] I like to think that "Artificial intelligence without scientific intelligence is but the ruin of the soul"; I regard the thought as a reminder that learning more about our Solar system is the motivation for using machine learning methods.

B. Carry
Nice
August 14th, 2024

---

[1] "Science sans conscience n'est que ruine de l'âme". F. Rabelais.

# Foreword

The benefit of reading this book is twofold, as it presents two general themes: namely, classical machine learning (classical ML) and neural networks (NN), both in various astronomical applications.

Great advantages of the considered methods are substantiated in the both cases. ML allows one to avoid solving complicated algorithmic problems; for example, to identify an asteroid orbiting in a mean-motion resonance with a planet, it is necessary to reveal librations of the corresponding resonant phase. This task is often not at all easy due to the presence of short-period perturbations from planets. Supervised learning, where a "train dataset" is provided, which the model is trained on (where the result is known, for example, for several hundreds objects), allows one to solve the identification problem quickly and reliably.

By applying ML, it is possible to find regularities in astronomical images or diagrams (e.g., lightcurves) when there are no theoretical prerequisites and/or any visible patterns, and one has only data. Moreover, any tasks related to primary data processing (for example, searching for outliers or anomalies when analyzing astronomical images) are also easily solvable. This is especially important in operations with data from large sky surveys.

A sharp economy in computing power is appealing. A tuned model is capable of making predictions within milliseconds, based only on input data. Of course, this is only possible in cases with an observed (or theoretically existing) regularity, such as the case of asteroid families or resonant groups. Although, theoretically, the training of the model itself can be rather time-consuming, in practice it is usually short, and the process of finding optimal model parameters is quite well regularized by various methods, including genetic algorithms.

During the past recent years, large language models (LLM) became very popular in various fields. This book provides some insights of how one can use them in astronomy. For example, ChatGPT answers user questions stemming both from general knowledge and from specified documents. Technically, this may already be useful for astronomy, for example, in the context of searching for references to specific results in astronomical articles and databases. When an exact reference to some fact, or a summary of data, is needed, such systems can be effectively used. From the perspective of efficiency, especially where a large volume of data is involved, automation of routine human actions can be even more interesting and promising.

Why is this book important? Because a number of actual issues, such as considered above, are raised here. The dramatic changes, which are already happening, cannot be ignored. Great opportunities have opened up, something that machine learning specialists have been working towards for the past decades.

It is about 65 years ago, in 1959, that the term "machine learning" was coined (Samuel, 1959). Since then, astronomy has experienced several remarkable reforms, initiated by the introduction of computing technology, the beginning of the space age, the launches of space telescopes, the advent of computer algebra (which revolutionized methods of celestial mechanics), the Internet, expeditions to the Solar System bodies, the discovery of exoplanetary systems. The rise of machine learning (and, generally, artificial intelligence methods) in astronomy definitely adds to this list. As Carl Murray and Stanley Dermott wrote in the preface to their book "Solar System Dynamics" (Murray and Dermott, 1999), "We are living in a new age of discovery." Undoubtedly, the ML techniques pave the way for new discoveries.

## References

Samuel, A., 1959. Some studies in machine learning using the game of checkers. IBM Journal of Research and Development 3 (3), 210–229.
Murray, C.D., Dermott, S.F., 1999. Solar System Dynamics. Cambridge Univ. Press, Cambridge.

Ivan I. Shevchenko
Saint Petersburg State University, Saint Petersburg
September 6th, 2024

# Preface

As editors, it was a pleasure and a privilege to be responsible for the first release of "Machine Learning for small bodies in the Solar System." The astronomical field is entering the big data science age as the number and complexity of astronomical datasets grow rapidly. The enormous size of contemporary astronomy datasets necessitates the use of procedures other than human researcher eye assessment. Machine learning (ML) is the study and development of algorithms that learn from data. The term artificial intelligence (AI) refers to the replication of human intelligence in computers, which are programmed to think and learn like humans.

The main goal of this book was to collect what have been the efforts of several separate independent researchers and to combine them into a coherent picture. The book covers a range of fields, going from applications to asteroids, asteroid families, small bodies interacting with mean-motion and secular resonances, to comets, TNOs, and detection of moving objects. Apart from reviewing the state-of-the-art in the various fields covered by this book, we provide links for GitHub repositories with Jupyter notebooks carrying examples of codes used in areas such as identification of asteroids interacting with mean-motion or secular resonances, and detection and characterization of moving objects with ML.

We believe that the authors have done an outstanding job in presenting the latest information in their respective fields and hope that this book can be useful for students and researchers interested in exploring the new field of machine learning and artificial intelligence applications to Solar System small bodies.

Valerio Carruba
UNESP

Evgeny Smirnov
Belgrade Observatory

Dagmara Oszkiewicz
Adam Mickiewicz University

# CHAPTER ONE

# Artificial intelligence and machine learning methods in celestial mechanics

**Valerio Carruba**[a], **Evgeny Smirnov**[b], **Gabriel Caritá**[c], **and Dagmara Oszkiewicz**[d]

[a]São Paulo State University (UNESP), Department of Mathematics, Guaratinguetá, SP, Brazil
[b]Belgrade Astronomical Observatory, Belgrade, Serbia
[c]National Institute for Space and Research (INPE), Division of Graduate Studies, São José dos Campos, SP, Brazil
[d]Astronomical Observatory Institute, Faculty of Physics and Astronomy, Adam Mickiewicz University, Poznań, Poland

## 1.1. Introduction

The amount and complexity of astronomical datasets are expanding quickly, bringing in the big data science age in astronomy. Baron (2019) discusses how the sheer size of modern astronomical datasets necessitates the use of tools other than human researcher eye examination. There are two important terms in this field. The first one, artificial intelligence (*AI*, McCarthy et al. (1955)) refers to the simulation of human intelligence in machines that are programmed to think and learn like humans. It is a broad field of computer science, which encompasses the development of intelligent machines capable of performing tasks that typically require human intelligence. Machine learning (*ML*, (Mitchell, 1997)) is the study and creation of algorithms that can learn from data, and deep learning is a branch of machine learning that focuses on training artificial neural networks with numerous layers to learn and make predictions or judgments (Goodfellow et al., 2016). It aims to automatically discover hierarchical representations of data by employing algorithms that iteratively learn features at different levels of abstraction.

Many new *AI* and *ML* techniques have lately been used by an increasing number of astronomical subfields to meet the problems of modern astronomy. With much easier-to-write software, clustering techniques to find collections of objects with similar properties—such as variable stars or galaxies—can now be carried out quickly. It is feasible to automatically allocate newly potential members of established groups in sizable databases

*Machine Learning for Small Bodies in the Solar System*
https://doi.org/10.1016/B978-0-44-324770-5.00006-4

to the class that best fits them. These days, anomaly detection algorithms require little human observer training to discover outliers and novel classes of celestial objects.

Despite the recent boom in applications of these techniques in other astronomical subfields, the number of studies using machine learning techniques applied to Solar System small bodies has been low. Thus this book's primary objective is to review current uses of *ML* in the field of Solar System small bodies.

We will discuss applications of *AI* and *ML* to problems, such as the identification of asteroid family members (Chapter 2), asteroids in mean-motion resonances (Chapter 3), asteroid families interacting with secular resonances (Chapter 4), orbital dynamics around asteroids (Chapter 5), asteroid spectro-photometric classification (Chapter 6), Kuiper Belt objects (Chapter 7), identification and localization of cometary activity in Solar System objects (Chapter 8), detection and characterization of moving objects (Chapter 9), and chaotic dynamics (Chapter 10).

Please note that the goal of this chapter is not to provide a rigorous mathematical foundation for machine learning and deep learning algorithms applied to studies of small Solar System bodies. Rather, we wish to provide a light introduction and references to some of the most commonly used methods, with an emphasis on the practical application of such approaches. We start by revising machine learning algorithms.

## 1.2. Machine learning

There are two primary groups of machine learning techniques applied for Solar System small bodies:

- Supervised learning: according to Russell and Norvig (2010), "the computer is presented with the task of learning a function that maps an input to an output based on example input-output pairs."
- Unsupervised learning: the learning process is left unlabeled, allowing it to locate the structure at its entry on its own (Wang, 2001).

We won't go into further detail about reinforcement learning, another ML category that hasn't been applied to the work's focus yet. Regression challenges arise when ML algorithms are used to predict a quantity instead of a label in classification tasks. Our examination will begin with a discussion of supervised training methods. For *AI* & *ML* tasks, the programming language *Python* is one of the most widely used, followed by *R* and *Julia*. Most software packages used for applications to Solar System small bodies

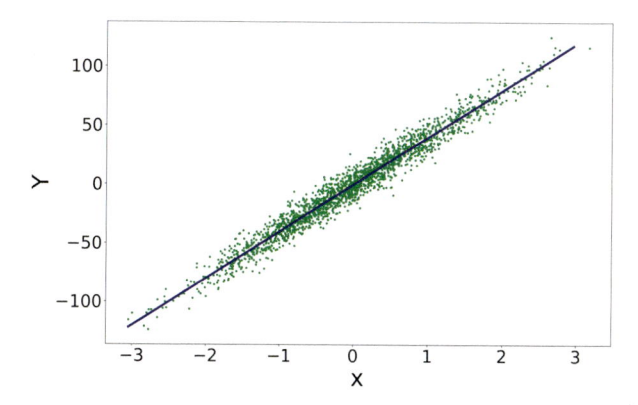

**Figure 1.1** An application of linear regression to data created at random. The data was generated using the *make_regression* function from the *sklearn.datasets* library. Adapted from figure (1) of Carruba et al. (2022) and reproduced with permission from the authors and CMDA (©CMDA).

are written in *Python*, and for this reason, here we will focus on algorithms written in this language.

## 1.2.1 Supervised learning

A number of methods for supervised learning have been created in the *Python* programming language. The scientific community can freely access them using the scikit-learn software (Pedregosa et al., 2012). Supervised learning uses labeled datasets to train algorithms that correctly categorize data or predict outcomes. Approaches built on a single algorithm are known as standalone techniques. In astronomy, decision trees, logistic regression, and linear regression are a few of the most used stand-alone techniques.

The goal of *linear regression* is to find a linear relationship between an input, or independent variable, and an output, or dependent variable. The technique adjusts a linear model to minimize the residual sum of squares difference between observed and forecasted data, as shown in Fig. 1.1. With this method, it is possible to compel the coefficients to be positive, which may be useful in particular physical scenarios.

*Logistic regression* is a statistical approach used when input data must be categorized into distinct categories (Dalpiaz et al., 2021). It is a subset of linear regression, in which the output variable has a single value. Logistic regression uses probability theory to predict whether or not something belongs to a group. This is referred to as a "binary classification problem" since an item is classified as either belonging to a group or not. In this

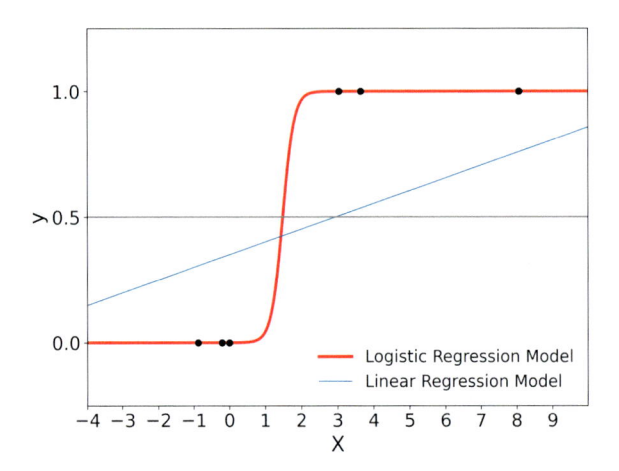

**Figure 1.2** One *logit* function, superimposed with a case of linear regression. The gray line shows the mean value. Adapted from figure (2) of Carruba et al. (2022) and reproduced with permission from the authors and CMDA (©CMDA).

situation, there are two possible outcomes: "yes" or "no." A positive result usually indicates the presence of some identity distinct from those of its members, whereas a negative result indicates the absence of it.

As a result, one must forecast the likelihood that identity exists. In practice, the binary variable answers are coded with 1 and 0 for "yes" and "no" replies, respectively.

To do this, we fit the function *logit* to the set of training data to transfer any actual value to a value between 0 and 1 based on a linear combination of dependent predictor variables as follows:

$$logit P_i = log\left(\frac{P_i}{1 - P_i}\right) = \beta_0 + \beta_1 x_{1,i} + \beta_2 x_{2,i} + ... + \beta_k x_{k,i}, \qquad (1.1)$$

where $P_i$ is the event i term's probability and the coefficients $\beta_i (i = 1, 2, ..., k)$ are model parameters, which are often estimated using maximum likelihood estimation (MLE); $x_i (i = 1, 2, ..., k)$ is a vector of discrete or continuous values (Cramer, 2004; Gudivada et al., 2016). Fig. 1.2 depicts an example of the *logit function*.

A *decision tree* is a model defined by a top-to-bottom tree-like graph. It may be used for classification as well as regression. Decision tree algorithms make choices in the form of successive nodes in a tree. The tree's lowest nodes are known as *leaves* or *terminal nodes*. They are not related among themselves by a condition, but rather with a path inside the tree.

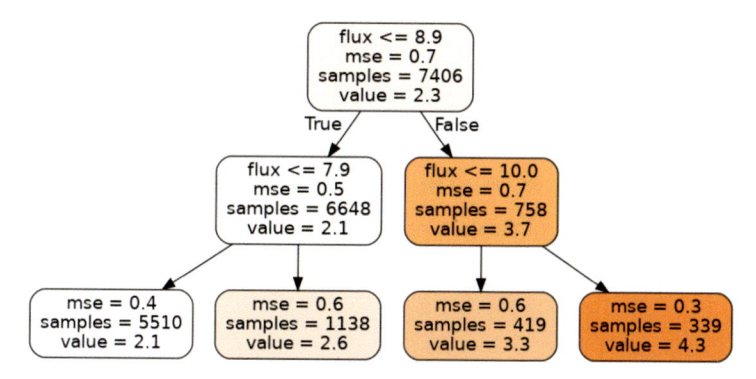

**Figure 1.3** 7406 objects from the Gaia Data Release 2 (Gaia Collaboration et al., 2018) are used to train a decision tree method to find connections between relative flux uncertainties and diameter measurements of the object (de Souza et al., 2021). Adapted from figure (3) of Carruba et al. (2022) and reproduced with permission from the authors and CMDA (©CMDA).

The number of decision nodes, or the maximum depth, "*max_depth*," is a model-free parameter. Fig. 1.3 shows an example of a tree diagram for an *decision tree*, where the authors illustrate the route to categorize a sample of asteroids from the Gaia Data Release 2 (Gaia Collaboration et al., 2018), as utilized by (de Souza et al., 2021). In the first node of this regressor, any objects with relative flux uncertainty $\leq 9$ will follow the True arrow, whereas the remainder will follow the False arrow. Other choices will be made as indicated.

Ensemble approaches provide a solution by combining many standalone algorithms, such as the aforementioned *linear regression, logistic regression,* and *decision tree*. The two types of ensemble approaches are the following:

1. *Bootstrap aggregation (bagging)*: the training data is broken into numerous samples known as bootstrap samples, which are created by randomly picking a sample with the replacement of the training set. Each model is assigned the same weight.

2. *Boosting:* prioritizes training situations, in which previous models produced fewer classification errors, as well as autonomous models that performed better.

*Bagging classifiers* include *random forest* and *extremely randomized trees.* During training, a Random Forest generates a large number of decision trees. Its output for classification problems is the class picked by the majority of trees (Ho, 1995, 1998). For regression tasks, the mean or average forecast of the individual trees is returned. The approach for *Extremely randomized trees,*

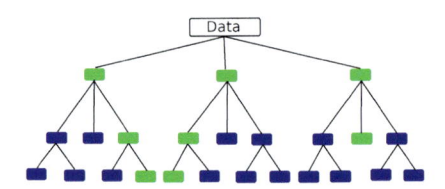

**Figure 1.4** An illustration of the decision-making technique utilized by *Random Forest*. Adapted from figure (4) of Carruba et al. (2022) and reproduced with permission from the authors and CMDA (©CMDA).

or *ExtraTree*, is identical, except that the bagging process is conducted without replacements. Fig. 1.4 depicts an example of the technique employed by *random forest* to achieve an output.

*Boosting* techniques include *adaptive boosting (AdaBoost), gradient boosting (GBoost)*, and *eXtreme gradient boosting (XGBoost)*. The reinforcement algorithms in the *AdaBoost* method (Freund and Schapire, 1995) monitor the models that delivered the least accurate forecast and give them lower weights. As illustrated in Fig. 1.5, the ultimate outcome is a weighted average of the results of each autonomous model. *GBoost* differs from *AdaBoost* in that it uses a gradient descent technique to provide weights to the outcomes of standalone models (Boehmke and Greenwell, 2019). Chen and Guestrin (2016) regularized and improved the Gradient algorithm in *XG-Boost*.

An additional class of ensemble tree-based methods are Bayesian additive regression trees (*BART*) (Chipman et al., 2010; Hill et al., 2020). In the *BART* model, fitting and inference are carried out through an iterative process of Bayesian back-fitting that generates samples a posterior. Each tree in the model is limited by a regularization before it becomes a weak learner. Although *Python scikit-learn* does not contain *BART*, it can be accessed in Python via the *GitHub* repository "https://github.com/JakeColtman/bartpy."

*Naïve Bayes* is a method for developing classifiers. These are models that provide class labels to order situations expressed as vectors of feature values, with the class labels chosen from a small set. For these classifiers, there is not a single training methodology; instead, a variety of approaches are founded on the same idea: all *naïve Bayes* classifiers assume that, given the class variable, the value of one feature is independent of the value of any other feature. For example, a fruit is referred to as an apple if it is spherical, red, and around 10 cm in diameter. Regardless of any correlations between the color, roundness, and diameter variables, a robust *naïve Bayes* classifier

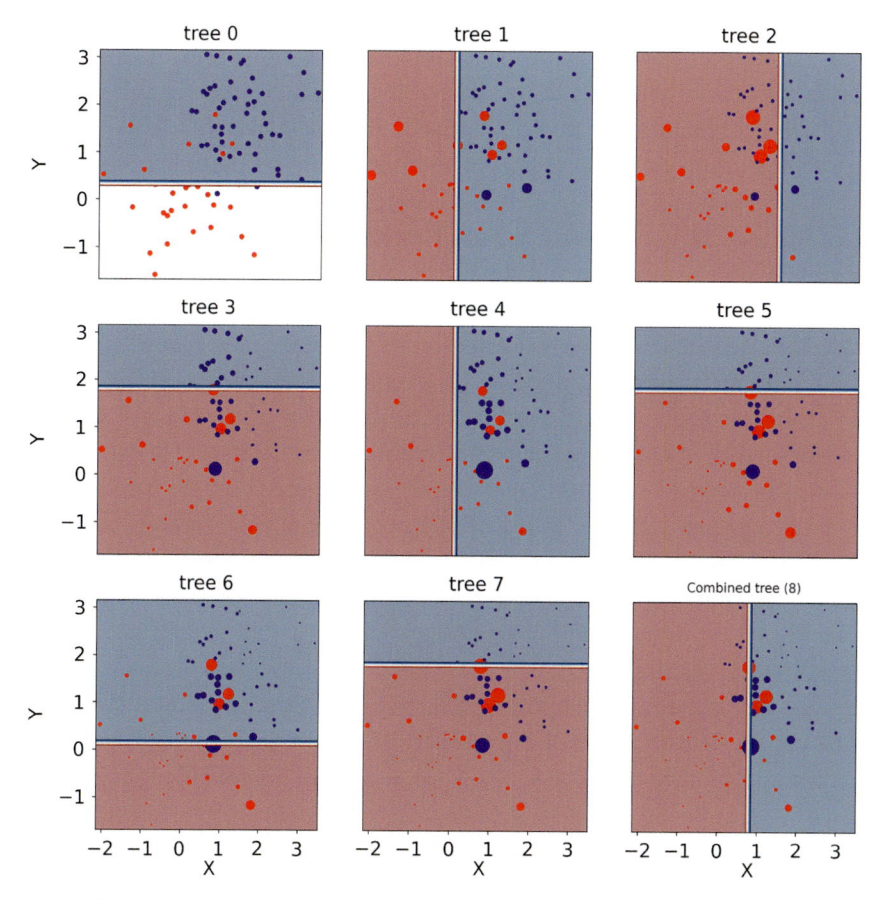

**Figure 1.5** A visual representation of how *AdaBoost* makes judgments. The algorithm combines weak learners, assigns weights to values, and employs sequential tree building, while taking previous mistakes into account. Adapted from figure (5) of Carruba et al. (2022) and reproduced with permission from the authors and CMDA (©CMDA).

assesses each one separately to determine the likelihood that this fruit is an apple (Piryonesi and El-Diraby, 2020). The continuous values linked to each class are presumed to have a normal distribution in the *Gaussian naive Bayes* method.

Lastly, *support-vector machines (SVM)* have been applied recently to problems related to regression and classification (Cortes and Vapnik, 2009). Assuming that some data points fall into one of two categories, the objective is to identify which category a new data point will fall into. We want to know if we can use a $(p-1)$-dimensional hyperplane to segregate data points that are represented as a $p$-dimensional vector (a list of $p$ integers) in support-vector machines. A large number of hyperplanes might be utilized to classify the data. The optimal hyperplane is the one that reflects the largest separation, or margin, between the two classes. *SVMs* are based on an algorithm that automatically determines the optimum hyperplane.

Supervised learning approaches, such as those described in this section, have had several applications on Solar System small bodies, such as identification of objects in mean-motion and secular resonances, identification of asteroid families members, and taxonomic classification of asteroids. Interested readers can find more information in Chapters 2, 3, 4, and 6.

An important part of supervised learning involves metrics and methods to evaluate the algorithms' performance. Examples of metrics are *accuracy, precision, recall,* and *mean squared error.* In binary classification, a true positive (TP) occurs when the model predicts a positive class and the actual value is indeed positive. A false positive (FP) occurs when the model predicts a positive class, but the actual value is negative. False negatives (FN) occur when the model predicts a negative class, but the actual value is positive. Finally, a true negative (TN) happens when both the prediction and the actual values are negative.

*Accuracy* is a measure of how well a classifier correctly predicts the labels of a dataset. It is the ratio of the number of correct predictions to the total number of predictions. The formula for accuracy is as follows:

$$Accuracy = \frac{(Number\ of\ Correct\ Predictions)}{(Total\ Number\ of\ Predictions)} = \frac{TP + TN}{TP + TN + FP + FN}. \quad (1.2)$$

*Accuracy* provides an overall measure of the classifier's performance, indicating the percentage of correct predictions. However, it can be misleading if the dataset is imbalanced, meaning one class dominates the others.

*Precision* is a metric that measures the proportion of correctly predicted positive instances out of all instances predicted as positive. It focuses on the true positives and indicates the classifier's ability to avoid false positives. The formula for precision is as follows:

$$Precision = \frac{(True\ Positives)}{(True\ Positives + False\ Positives)}. \quad (1.3)$$

*Precision* is particularly useful when the cost of false positives is high, as it reflects the classifier's ability to make accurate positive predictions.

*Recall,* also known as sensitivity or true positive rate, measures the proportion of correctly predicted positive instances out of all actual positive instances. It focuses on identifying all positive instances and avoiding false negatives. The formula for recall is as follows:

$$Recall = \frac{(True\ Positives)}{(True\ Positives + FalseNegatives)}. \tag{1.4}$$

*Recall* is important when the cost of false negatives is high, as it reflects the classifier's ability to capture all positive instances.

*Precision* and *recall* are often used together and can be combined into a single metric called the F1 score or F-score. The F1 score is the harmonic mean of precision and recall, providing a balanced measure of a classifier's performance.

The mean squared error (MSE) is a commonly used metric in machine learning to evaluate the performance of a regression model. It measures the average squared difference between the predicted and actual values of a target variable. To calculate the MSE, one takes the difference between the predicted and actual values for each data point, squares the differences, and then takes the average of all the squared differences. The formula for MSE is as follows:

$$MSE = (\frac{1}{n}) \times \sum (y_i - \hat{y})^2, \tag{1.5}$$

where $n$ is the total number of data points, $y_1$ represents the actual values, and $\hat{y}$ represents the predicted values. The MSE metric gives more weight to larger errors because of the squaring operation. It is a nonnegative value, and a lower MSE indicates better model performance. MSE is widely used, because it is differentiable and convex, making it suitable for optimization algorithms.

When one builds a machine learning model, one has to assess its performance. There is a special method used for such a case: cross–validation. Cross-validation is a technique used in supervised learning to assess the performance and generalization ability of a machine learning model. It helps to estimate how well the model will perform on unseen data. The basic idea behind cross-validation is to divide the available labeled dataset into multiple subsets, or folds. The model is then trained and evaluated multiple times, with each fold serving as both a training set and a validation set. The most common type of cross-validation is called k-fold cross-validation, where the dataset is divided into k subsets or folds of approximately equal size.

The cross-validation process typically follows these steps:

1.  Partitioning: The dataset is divided into $k$ subsets of roughly equal size. Each subset is called a fold.
2.  Training and evaluation: The model is trained on $k-1$ folds and evaluated on the remaining fold. This process is repeated $k$ times, with each fold serving as the validation set once.
3.  Performance metrics: The performance metrics, such as accuracy, precision, recall, or mean squared error, are computed for each iteration of the training and evaluation process.
4.  Aggregation: The performance metrics obtained from each iteration are averaged or combined in some way to provide an overall estimate of the model's performance.

The main advantage of cross-validation is that it provides a more robust estimate of the model's performance by reducing the impact of the specific partitioning of the dataset into training and validation sets. It helps to detect overfitting (when a model performs well on the training data but poorly on unseen data) and assesses the model's ability to generalize to new, unseen examples.

Once the model has been evaluated using cross-validation, the performance metrics can guide the selection of hyperparameters, model architectures, or feature representations to improve the model's performance before it is applied to new, unseen data.

## 1.2.2  Unsupervised learning

Unsupervised learning is a general term that includes many different statistical techniques applied to data exploration. We will mostly focus on algorithms for two tasks: *clustering*, which is the problem of organizing a set of items so that objects in the same group (called a cluster) are more comparable (in some sense) to those in other groups (clusters), and *anomaly detection*, a data mining step that finds observations, events, or data points that deviate from a dataset's most observed behavior. In terms of *clustering*, *K-nearest-neighbors (KNN)*, *KMEANS*, *DBSCAN*, and the *hierarchical clustering method (HCM)* were the most often utilized approaches in recent research, whereas *isolation forests* were employed in recently published works that dealt with it anomaly detection. Finally, methods such as Fisher's linear discriminant analysis (LDA) are classification techniques often used in machine learning and pattern recognition,

All of the methods described in this section are available in the scikit-learn (Pedregosa et al., 2012) Python package.

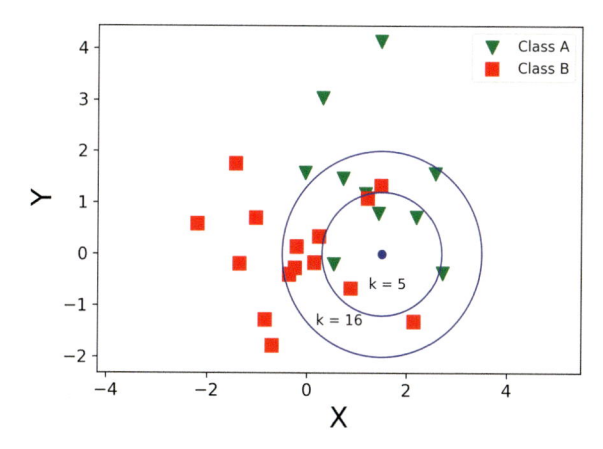

**Figure 1.6** An example of a bi-dimensional use of the KNN for categorization into two classes. Adapted from Figure (6) of Carruba et al. (2022) and reproduced with permission from the authors and CMDA (©CMDA).

The primary principle behind *KNN* is to forecast a point's condition based on its closest neighbors. If the majority of the closest neighbors belong to the same class, the point is classed as well. The number of neighbors utilized for classification is a free parameter of this approach. Fig. 1.6 depicts a scenario in which, with three neighbors, the center point is categorized as belonging to class B, however with a larger sample of seven neighbors, the point is classified as belonging to class A. *KNN* is suitable for both supervised and unsupervised learning.

*K-means* clustering, a centroid-based clustering methodology, is one of the most widely used clustering methods (MacQueen, 1967). The first stage in the *K-means* is the distance assignment between the items in the sample. Although the Euclidean metric is the default, additional metrics that are more appropriate to the dataset at hand might be used. After that, the approach chooses *k* random objects (where *k* is an external free parameter) to act as the initial centroids of the dataset. Each object in the dataset is then assigned to the nearest *k* centroids. Next, new cluster centroids are constructed by averaging the locations of the elements in the chosen cluster. Until convergence is achieved, these two processes—reassigning items to clusters based on their distance from the centroid and recalculating cluster centroids—are repeated. One can characterize convergence in a number of ways, such as when the cluster centroids converge to a single location or when the vast majority of objects (90 percent or more) are no longer reassigned to separate centroids. The cluster centroids and a link between each individual item and each cluster are the algorithm's outputs.

*DBSCAN* (density-based spatial clustering of applications with noise (Ester et al., 1996)) is a prominent density-based clustering technique used in machine learning and data mining. It can find clusters of any shape in a dataset while simultaneously recognizing outliers and noise points. The primary idea underlying *DBSCAN* is to group together data points in the feature space that are close to each other and have a suitable density of neighboring points. Because of this method, it is resistant to noise and can handle datasets with nonuniform density. *DBSCAN* provides a number of advantages. It does not require a predefined number of clusters, can handle clusters of any shape, and is resistant to noise and outliers. However, it may struggle with datasets with considerable density variations or clusters with varied densities. Because of its versatility and capacity to locate clusters in complicated datasets, *DBSCAN* is extensively utilized in a variety of applications, including image processing, geographical data analysis, and anomaly detection.

The *Hierarchical clustering method (HCM, (Ward Jr, 1963))* is a well-known unsupervised learning technique that is used in machine learning and data mining. It is generally used for clustering comparable data points based on their similarity or dissimilarity. The ultimate result is a hierarchical representation of the data, which is frequently displayed as a dendrogram. The HCM method begins by treating each data point as a separate cluster. Then, it combines the two closest clusters repeatedly based on a similarity or dissimilarity measure until all of the data points are merged into a single cluster or a termination condition is satisfied.

Hierarchical clustering algorithms are classified into two types: agglomerative and divisive. Agglomerative clustering is based on hierarchies: it is a bottom-up strategy, in which each data point begins as an independent cluster, and the two closest clusters are merged based on a predefined similarity measure at each iteration. This method is repeated until all of the data points have been combined into a single cluster. To describe the merging process, the program creates a dendrogram, with each node representing a cluster, as shown in Fig. 1.7. Divisive hierarchical clustering is a top-down strategy that begins with a single cluster that contains all of the data points and recursively separates the clusters into smaller clusters until each data point is in its own distinct cluster. The algorithm creates a dendrogram as well, however, the splitting process is depicted in reverse order as compared to the agglomerative technique.

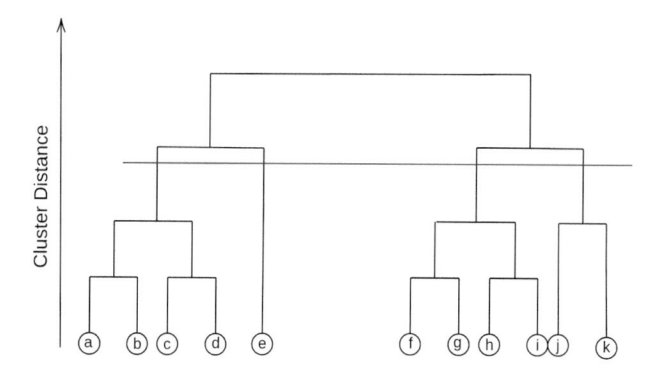

**Figure 1.7** An example of a agglomerative application of *HCM*.

In hierarchical clustering, the choice of similarity or dissimilarity metric is critical. Among the most commonly used distance measurements are Euclidean distance, Manhattan distance, and cosine similarity. Euclidean distance is a commonly used metric for measuring the distance between two points in an Euclidean space. It is named after the ancient Greek mathematician Euclid, who is known for his work on geometry. In a two-dimensional space, the Euclidean distance between two points, denoted as $(x_1, y_1)$ and $(x_2, y_2)$, can be calculated using the Pythagorean theorem:

$$distance = \sqrt{(x_2 - x_1)^2 + (y_2 - y_1)^2}. \tag{1.6}$$

In general, the Euclidean distance between two points in an n-dimensional space is calculated as

$$distance = \sqrt{(x_2 - x_1)^2 + (y_2 - y_1)^2 ... + (z_2 - z_1)^2}, \tag{1.7}$$

where $(x_1, y_1, ..., z_1)$ and $(x_2, y_2, ..., z_2)$ represent the coordinates of the two points in the n-dimensional space. The Manhattan distance, also known as the L1 distance or city block distance, is a metric for measuring the distance between two points in a grid-like space. It is named after the grid-like layout of streets in Manhattan. In a two-dimensional space, the Manhattan distance between two points is calculated as the sum of the absolute differences between their coordinates:

$$distance_{Manhattan} = |x_2 - x_1| + |y_2 - y_1|. \tag{1.8}$$

Unlike the Euclidean distance, the Manhattan distance only considers straight-line movements along the axes of the grid-like space. It does not consider diagonal movements. This makes it more suitable for situations where movement is constrained to a grid-like structure or when diagonal movements are not allowed.

Finally, cosine similarity is a measure of similarity between two vectors in a high-dimensional space. It calculates the cosine of the angle between the two vectors, which represents their orientation or direction relative to each other. Cosine similarity is commonly used in information retrieval, text mining, and recommendation systems. Given two vectors, $\vec{A}$ and $\vec{B}$, the cosine similarity between them is calculated as the dot product of the vectors divided by the product of their magnitudes:

$$cosine\_similarity = \frac{(\vec{A} \cdot \vec{B})}{(||\vec{A}|| \times ||\vec{B}||)}, \tag{1.9}$$

where $\vec{A} \cdot \vec{B}$ represents the dot product of vectors $\vec{A}$ and $\vec{B}$, and $||\vec{A}||$ and $||\vec{B}||$ represent their magnitudes or norms. The cosine similarity ranges between –1 and 1, with values closer to 1 indicating a higher degree of similarity and values closer to –1 indicating a higher degree of dissimilarity or negative similarity. A cosine similarity of 1 means that the vectors have the same direction, whereas a cosine similarity of –1 means they have opposite directions. A cosine similarity of 0 indicates that the vectors are orthogonal (perpendicular) to each other. One of the advantages of cosine similarity is that it is invariant to the magnitude or length of the vectors, focusing solely on the direction. This property makes it useful in situations where the absolute magnitudes of the vectors are not important, such as in text analysis, where the frequency or presence of words in documents is more relevant than their absolute counts.

The linking criterion, which governs how the distance between clusters is determined, also has an impact on the clustering results. Some of the most popular linking criteria include single linkage, full linkage, and average linkage. In a single linkage, the distance between two clusters is defined as the shortest distance between any two points belonging to different clusters. In other words, it considers the nearest neighbor between the two clusters. This means that the proximity of two clusters is determined by the minimum distance between any pair of points, one from each cluster. In full linkage, the distance between two clusters is defined as the maximum distance between any pair of points, one from each cluster. This means that it emphasizes the furthest neighbor between the two clusters. Finally, in average linkage, the distance between two clusters is defined as the average distance between all pairs of points, one from each cluster. It considers the average proximity between the data points in different clusters.

There are various advantages to hierarchical clustering. It does not require a predefined number of clusters, because the cluster hierarchy can be sliced at any desired level. It also includes a readable dendrogram that depicts the clustering process. Hierarchical clustering, on the other hand, can be computationally costly, particularly for big datasets, and it may not scale well to high-dimensional data. Overall, hierarchical clustering is an effective approach for examining and comprehending data structure by generating layered groups based on similarity or dissimilarity metrics. Examples of applications in astronomy involve the identification of asteroid families in domains of proper elements $(a, e, \sin(i))$, to be discussed in Chapter 2.

Unsupervised learning uses anomaly detection to find data points or patterns that differ considerably from the norm or anticipated behavior. These outliers are sometimes referred to as anomalies, outliers, or novelties. Anomaly detection is critical in many fields, including fraud detection, network intrusion detection, industrial quality control, and health monitoring. These kinds of algorithms in unsupervised learning often work on unlabeled data, which means they do not rely on past knowledge or labeled samples of normal and anomalous cases during the training phase. Instead, they seek to capture the data's fundamental structure or distribution and identify examples that deviate considerably from this learned representation.

*Isolation forest* was the name of the first anomaly detection system to use isolation to find abnormalities (Liu et al., 2008). The most widely used methods for identifying anomalies state that anomalies are those occurrences in the dataset that deviate from the typical profile. *Isolation forest* employs an alternative methodology; it specifically separates anomalous points within the dataset, as opposed to endeavoring to construct a model of typical examples. The main benefit of this approach is that sampling techniques may be used in ways that profile-based approaches cannot, leading to a quick algorithm with little memory requirements.

Currently, there are various strategies often employed in unsupervised anomaly detection:

1. Statistical procedures: Statistical approaches are based on the assumption that the majority of the data follows a known statistical distribution, such as the Gaussian (normal) distribution. Anomalies are then detected

as data points with a low probability or that fall outside of a specific range depending on the anticipated distribution.

2. Clustering techniques gather together comparable data points based on various similarity criteria. Anomalies are data points that do not belong to any cluster or that are isolated within their clusters.

3. Methods based on density: These approaches calculate the density of the data points and detect anomalies as cases with a considerably lower density than their neighbors.

4. Proximity-based approaches: These approaches analyze the distance or dissimilarity between data points and detect anomalies as examples that are far apart or have odd proximity connections. The k-nearest neighbors algorithm is a popular proximity-based method.

5. Autoencoders are neural network models (see Section 1.3) that have been trained to recreate the input data. Anomalies are found by measuring the reconstruction error, with large reconstruction errors deemed abnormal.

Interested readers can find more information on the subject in Chandola et al. (2009). Unsupervised learning algorithms have been recently used for identifying members of asteroid families (Chapter 4), Kuiper belt objects (Chapter 7), among other things. Interested readers can find more information in these chapters.

Finally, Fisher's linear discriminant analysis (Fisher (1936), LDA), also known as Fisher's discriminant analysis or simply linear discriminant analysis, is a dimensionality reduction and classification technique used in machine learning and pattern recognition. It aims to find a linear projection of the input data that maximizes the separability between different classes.

The main objective of LDA is to reduce the dimensionality of the input data, while preserving the discriminatory information between classes. It achieves this by finding a projection that maximizes the ratio of between-class scatter to within-class scatter. The LDS algorithm works as follows:

1. Given a labeled dataset with input samples $X$ and corresponding class labels $Y$, calculate the means of each class, as well as the overall mean.

2. Compute the within-class scatter matrix $S_w$ and the between-class scatter matrix $S_b$. These matrices capture the variance within and between classes, respectively. They are defined as follows:

$$S_w = \sum_c \sum_{x_i \in X_c} (x_i - \mu_c)(x_i - \mu_c)^T, \qquad (1.10)$$

$$S_b = \sum_c n_c (x_i - \mu)(x_i - \mu)^T. \qquad (1.11)$$

Here, c represents the class index; $n_c$ is the number of samples in class $c$; $x_i$ is a sample in class $c$; $\mu_c$ is the mean of class $c$, and $\mu$ is the overall mean.

3. Compute the generalized eigenvalue problem $S_w\vec{v} = \lambda S_b\vec{v}$ to obtain the discriminant directions $\vec{v}$ that maximize the Fisher criterion. The eigenvectors corresponding to the largest eigenvalues represent the optimal discriminant directions.

4. Project the input data onto the discriminant directions to obtain the transformed feature vectors.

5. Optionally, apply a classification algorithm (e.g., k–nearest neighbors or linear classifiers) on the reduced feature space to perform classification tasks.

The resulting projection obtained through LDA can be used for various purposes, such as visualization, feature extraction, or classification. By maximizing the between-class scatter and minimizing the within–class scatter, LDA seeks to find a subspace where the classes are well-separated. It assumes that the data is normally distributed, and that the classes have equal covariance matrices.

LDA has been widely used in various applications, such as face recognition, document classification, and bioinformatics. However, it is important to note that LDA assumes linearity in the data distribution and may not perform well in cases where the relationship between classes is nonlinear. In such cases, nonlinear extensions of LDA, such as the quadratic discriminant analysis (Fukunaga (1990), QDA), or the kernel Fisher discriminant analysis (Mika et al. (1999), KFDA), can be employed.

## 1.2.3 Genetic algorithms

A genetic algorithm is a type of optimization algorithm that is inspired by the process of natural selection and evolution. It is commonly used to optimize machine learning models, particularly in cases where traditional optimization methods may not be effective. The basic idea behind a genetic algorithm is to generate a population of potential solutions, which are represented as "genomes" or sets of parameters. The genomes are then evaluated based on their fitness, or how well they perform. The fittest genomes are chosen to breed and generate kids who inherit some of their parents' features. These progeny are then genetically modified in some way to introduce new variants into the population. For many generations, the process of selection, breeding, and mutation is repeated with the expectation that

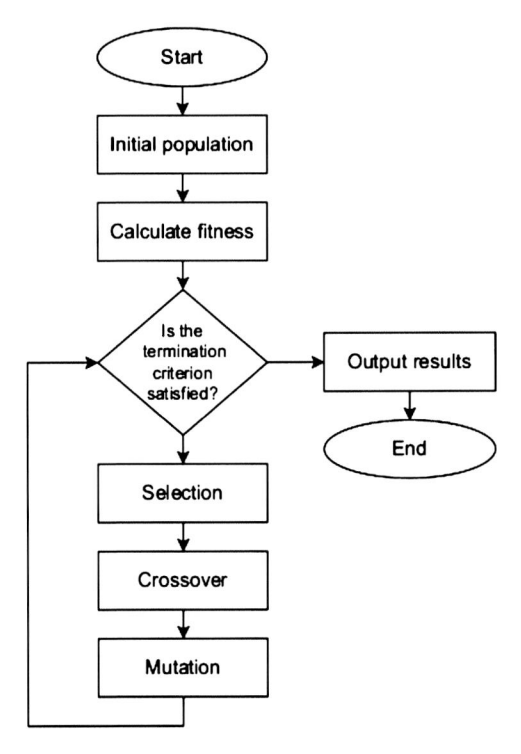

**Figure 1.8** A flowchart describing the procedure used by genetic algorithms to optimize machine learning methods.

the population would progress toward better and better solutions. Fig. 1.8 shows a flowchart that reports this procedure.

Genetic algorithms may be used to improve a wide range of models in machine learning, including neural networks, decision trees, and support vector machines. The characteristics of the model and the optimization issue will determine how a genetic algorithm for machine learning is implemented. For example, the genome may be a neural network's weights and biases, and the fitness function could be the validation accuracy on a certain dataset. There are several libraries that implement genetic algorithm optimization. One of the most commonly used is the *Tpot* library (Chen P.W. et al., 2004). Procedures used to implement genetic algorithms for dynamical problems were discussed in Carruba et al. (2021). Genetic algorithms have been used for the purpose of optimizing methods to identify asteroid families (Chapter 2) and for taxonomic classification of asteroids (Chapter 6).

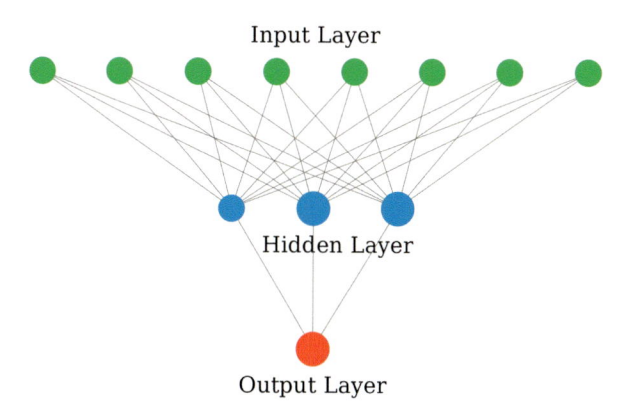

**Figure 1.9** A basic architecture of a neural network. Adapted from figure (9) of Carruba et al. (2022) and reproduced with permission from the authors and CMDA (©CMDA).

## 1.3. Deep learning

Deep learning is a subset of artificial neural network-based machine learning technologies. Deep learning methods use three forms of learning: supervised, semi-supervised, and unsupervised. Artificial neural networks (*ANNs*) were inspired by the neuron networks of biological systems. As demonstrated in Fig. 1.9, a typical basic ANN contains multiple layers between the input and output layers. Neurons, weights, biases, and activation functions are the essential components of all neural networks, which come in a variety of forms and sizes. These components work similarly to the human brain and may be trained in the same manner as any other traditional *ML* approach. Although the terms artificial neural networks and deep learning are occasionally used synonymously in some papers, there is a significant difference between the two. Feature extraction and transformation, which aim to establish a connection between inputs and the associated brain responses, are related to deep learning. Neurons are used in neural networks to transfer data across connections in the form of input and output values. Here, we shall refer to deep learning in its broadest sense, which includes basic neural network applications, for the sake of clarity.

The most commonly used neural networks applied for problems involving Solar System small bodies are the following:

**1.** Feed forward neural networks (*FNN*).
**2.** Multilayer perceptron (*MLP*).
**3.** Recurrent neural networks (*RNN*).
**4.** Long short-term memory (*LTSM*).

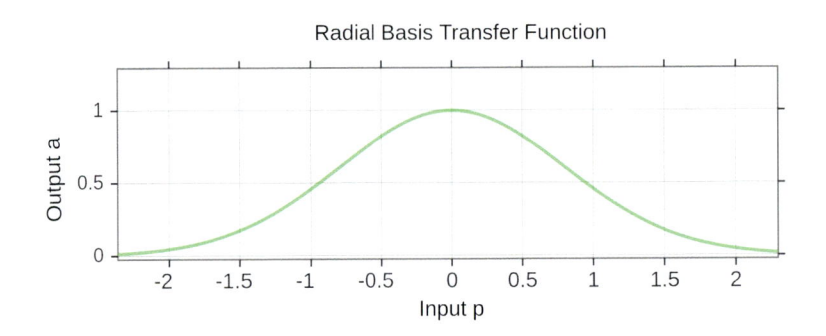

**Figure 1.10** An example of a simple radial basis function (*RBF*).

5. Convolutionary neural networks (*CNN*).
6. Bidirectional feature pyramid network.
7. EfficientDet D0-based network.

An artificial neural network that only allows information to move from the input layer to the output layer is called a feedforward neural network (*FNN*, (McCulloch and Pitts, 1943)). Put differently, there are no connections or feedback loops between nodes in the same layer; instead, the output of one layer becomes the input of the subsequent layer. Three layers usually make up the architecture of a feedforward neural network: an input layer, one or more hidden layers, and an output layer. After the input layer receives the data, the hidden layers process it, and the output layer generates the final result. After completing a weighted sum of its inputs, each node in the output layer and hidden layers uses an activation function. Backpropagation and other algorithms are employed in the training process to teach the nodes their weights and biases. Frequently utilized for tasks, such as classification and regression, feedforward neural networks have been implemented in several domains, for example, image identification, speech recognition, and natural language processing.

Among feedforward neural networks, radial basis function (*RBF* (Broomhead and Lowe, 1988)) neural networks are often used. Unlike traditional activation functions, such as sigmoid or *ReLU*, which are based on the signal value (*ReLU* will let only positive values of the signal to pass, for instance), *RBF*s are based on the concept of radial symmetry, and they take their name from the fact that its activation is determined by the distance between the input data point and a center point in a multidimensional space. The activation value of an *RBF* neuron is high when the input data point is close to the center and decreases as the distance increases (see Fig. 1.10)

An *RBF* network's fundamental design is made up of three layers: an input, a hidden, and an output layer. The data is fed into the hidden layer after being received by the input layer. The radial basis functions that make up the hidden layer individually calculate the distance between a center point and the input data, and then apply a nonlinear function to that distance. The final output is subsequently produced by the output layer using the weighted sum of the hidden layer's outputs. Radial basis function (*RBF*) is a type of activation function used in the hidden layer of the network. It is common practice to select the centers of the radial basis functions using clustering methods, such as $K - Means$ clustering.

The radial basis functions' widths can be manually adjusted or acquired through training via techniques, such as gradient descent. Gradient descent is an optimization algorithm commonly used in machine learning and deep learning to find the optimal values of the parameters for a given model. Its goal is to minimize the cost or loss function associated with the model. The basic idea behind gradient descent is to iteratively update the parameters of the model in the direction of the steepest descent of the cost function. The "gradient" refers to the vector of partial derivatives of the cost function with respect to each parameter. By computing the gradient, we can determine the direction and magnitude of the steepest ascent or descent in the parameter space. *RBF NN* are often used in clustering problems. One advantage of the *RBF* network is that it is able to approximate any continuous function to arbitrary accuracy, given enough hidden units. Additionally, *RBF* networks are less prone to overfitting, i.e., learning the fine features of the training set, and underperformed for other databases, than other types of neural networks, as they have a smaller number of parameters to learn. However, *RBF* networks can be difficult to train, especially when the number of hidden units is large.

A multilayer perceptron (*MLP*, (Widrow and Hoff, 1960)) is a type of feedforward neural network that consists of multiple layers of interconnected perceptrons, or neurons. A *perceptron* is a fundamental building block of artificial neural networks and is a type of binary classifier. It is a simplified model of a biological neuron, designed to mimic the way a single neuron in the brain makes simple decisions. In its basic form, a perceptron takes a set of input values, applies weights to those inputs, and computes a weighted sum. It then applies an activation function to the sum to produce an output. The activation function traditionally used in perceptrons is the step function, which outputs a binary value (e.g., 0 or 1) based on a threshold. The input layer of an MLP receives the input data, and the output layer generates the output of the network. The hidden layers, which calculate intermediate representations of the input data, are the layers that sit between the input and output layers. An MLP's perceptrons are computational units that generate outputs by applying an activation function to the weighted sum of their inputs. Using a backpropagation technique that modifies the weights to minimize the error between the network's output and the expected output, the weights and biases of the perceptrons are learned during training.

MLPs are commonly used for supervised learning tasks, such as classification and regression. For classification tasks, the output layer typically uses a softmax activation function to produce a probability distribution over the possible classes (Bridle, 1990). The softmax function takes a vector of real-valued inputs and transforms them into a vector of probabilities that sum up to 1. It essentially converts the outputs of the previous layer into probabilities representing the likelihood of each class. For regression tasks, the output layer typically uses a linear activation function, such as *ReLU*, to produce a continuous output. MLPs have been shown to be effective for a wide range of tasks, including image classification, natural language processing, and speech recognition. However, they can be computationally expensive to train, especially for large datasets or networks with many layers.

Recurrent neural networks (*RNN*), whose concept was first introduced by Lenz (1920), are a type of artificial neural network that is designed for processing sequential data, such as time series data or natural language text. In contrast to feedforward neural networks, which only evaluate input data once, *RNN* can retain a "memory" of prior inputs, allowing them to consider the context of the current input. The fundamental characteristic of *RNN*s is the loops in their construction, which enable data to be transferred from one stage of the sequence to the next. An input vector and a hidden state vector are fed into the *RNN* at each time step, and it generates an output vector and a fresh hidden state vector as output. Based on the received input and the hidden state from the previous time step, the hidden state vector is updated at each time step. As a result, the *RNN* is able to record input dependencies across time.

One difficulty with ordinary *RNN*s is the "vanishing gradient" problem, which happens when the gradients used to train the network get very small and cause the network to converge slowly or not at all (Rumelhart et al., 1986). Several *RNN* variations have been created to solve this issue. For example, gated recurrent unit (*GRU*) networks and long short-term memory (*LSTM*) networks employ specific techniques to more effectively update and store concealed state data. Natural language processing (including language modeling, machine translation, and sentiment analysis), audio recognition, picture captioning, and time series prediction are just a few of the tasks that *RNNs* have been effectively used for.

Long short-term memory (*LSTM* (Hochreiter and Schmidhuber, 1997)) is a type of recurrent neural network (*RNN*) that is designed to handle the problem of vanishing gradients in traditional *RNNs*. The vanishing gradient problem arises when the gradients that are used to update the network's weights are extremely tiny, which makes it challenging to train the model efficiently. Since its first introduction by Hochreiter and Schmidhuber in 1997, *LSTM* networks have gained popularity as a solution for applications involving the processing of sequential data, including machine translation, speech recognition, and language modeling. Introducing a memory cell that can retain information for a long time and using gating mechanisms to regulate the flow of information into and out of the cell is the fundamental concept underlying *LSTMs*. In essence, the memory cell is a collection of neurons that function similarly to a conveyor belt, allowing information to enter at one end, be stored for a while, and then be retrieved at the other. The input gate, forget gate, and output gate are the three gates that make up an *LSTM* network's gating mechanisms. The memory cell's input, forget, and output gates regulate the amount of fresh information that can enter, the amount of old information that can linger, and the amount of information that can exit. These gating mechanisms allow *LSTM* networks to process sequences of varied lengths without experiencing the vanishing gradient problem, as well as to selectively recall or forget information from the past.

Convolutional neural networks, or *CNNs* (LeCun et al., 1998), are a kind of neural networks that are frequently used for segmentation, classification, and image and video identification. Convolutional layers are the main component of *CNNs*, which enable automatic feature spatial hierarchies to be learned from an image's raw pixel values.

A *CNN* creates a set of feature maps by applying a set of filters to the raw input image after it has been processed through several convolutional layers. Every filter generates a new pixel value in the output feature map by convolving the input image with a tiny matrix of weights. The input image can be subjected to several filters so that the *CNN* can learn to identify features and patterns at various spatial scales and orientations (see Fig. 1.11). To add nonlinearity to the network, the output feature maps are usually

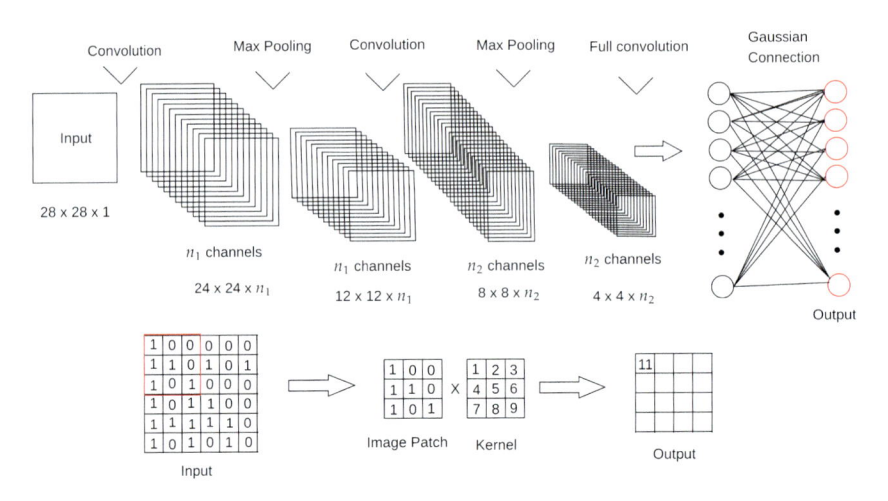

**Figure 1.11** An example of applications of convolutional neural networks (*CNN*). Convolution layers apply a filter by multiplying a filter matrix to the data. Max pooling layers select the maximum values in a $n \times n$ matrix. See the text for further details on this subject.

sent via a nonlinear activation function, such as *ReLU*, after the convolutional layers. The generated feature maps are then downsampled using a pooling layer—max pooling, for example—to preserve the most significant characteristics, while lowering the feature maps' spatial resolution. The network is better able to withstand slight changes in the input image, such as translations or rotations, thanks to this downsampling. To get a final prediction for the input image, the output feature maps are usually routed through one or more fully linked layers. For regression tasks, this prediction can take the form of a set of continuous values or a probability distribution over a set of classes. Many computer vision tasks, such as object detection, semantic segmentation, image classification, and more, have shown great performance using *CNNs*. They have also been applied in other fields, for example, speech recognition and natural language processing.

For image classification problems, the visual geometry group (*VGG*, Simonyan and Zisserman (2014, 2015)) (*VGG*) model is a widely used *CNN* architecture. It was created by Oxford University academics, and it came in second place in the 2014 ImageNet Large Scale Visual Recognition Challenge (Russakovsky et al., 2014). The architecture of the *VGG* model is fairly straightforward and consistent; it consists of a stack of convolutional layers and a stack of fully linked layers. Throughout, tiny $3 \times 3$ convolutional filters are used. Depending on the network's depth, there are various

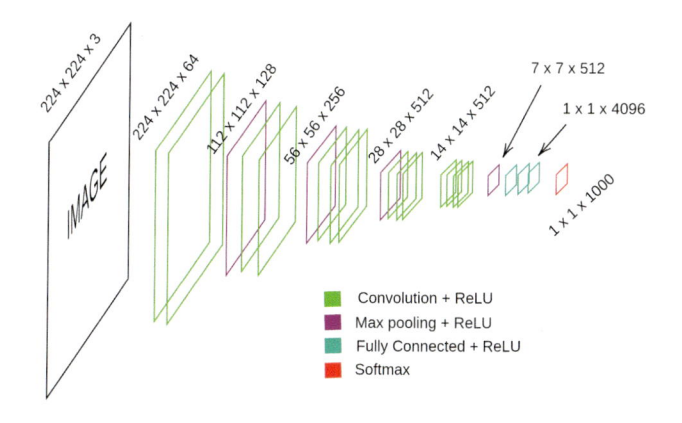

**Figure 1.12** An example of applications of a *VGG* model.

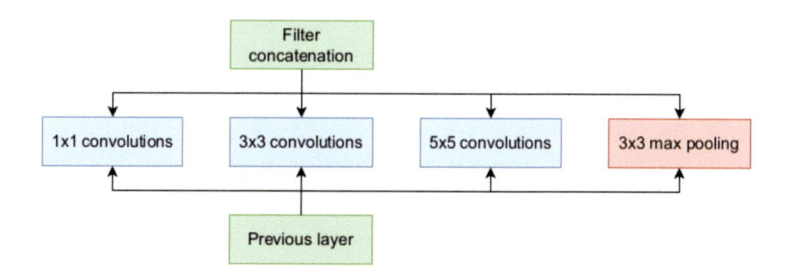

**Figure 1.13** An example of applications of a *Inception* model.

versions of the *VGG* architecture. With 16 layers—13 convolutional layers and 3 fully linked layers—*VGG*16 is the most often used version (see Fig. 1.12). Whereas the deeper layers of the network learn more complicated properties like object pieces and textures, the first few layers learn low-level elements, such as edges and corners. The *VGG* architecture has a drawback in that it requires a lot of memory and is computationally expensive to train due to its huge number of parameters. On a number of image classification tasks, it has still been demonstrated to achieve state-of-the-art performance, and it has been extensively utilized as a baseline architecture for comparison with more recent *CNN* models. In conclusion, the *VGG CNN* model is a well-liked and successful architecture for image classification applications. It is distinguished by its deep stacks of convolutional and fully connected layers, compact convolutional filters, and simple, uniform structure (see Fig. 1.13).

A deep learning architecture known as the *Inception CNN* was initially presented by Google researchers in a paper titled "Going Deeper with Convolutions" in 2014 (Szegedy et al., 2015, 2016, 2017). The Inception CNN was created to decrease the number of parameters and network operations needed, thereby increasing the accuracy and efficiency of image classification jobs. The usage of "Inception modules," which are made up of several convolutional layers with varying sizes and pooling layers, characterizes the *Inception CNN* architecture. These layers are intended to capture various scales and orientations of the input image's features. The network may learn ever more complicated aspects of the image by piling these Inception modules on top of one another. The usage of "bottleneck layers," which lower the number of input channels to the next convolutional layers, is one of the primary advances of the *Inception CNN* design. In doing so, the network's parameter count is lowered, but retains key functionality. Another innovation of the *Inception CNN* is the use of $1 \times 1$ convolutions, which can be used to reduce the dimensionality of the input data. This can help to reduce the computational cost of the network, while still maintaining high accuracy. When it was first introduced, the *Inception CNN* architecture produced state-of-the-art results in the ImageNet Large-Scale Visual Recognition Challenge, among other picture classification challenges. The *Inception CNN*'s accuracy and efficiency for large-scale image classification applications have been substantially enhanced with the introduction of the *Inception* v2, v3, v4, and *Inception − ResNet* designs. All things considered, the *Inception CNN* is a potent deep learning architecture that has proven successful in picture classification tasks, especially those requiring huge, intricate datasets.

Microsoft researchers created the *CNN* architecture known as *ResNet*, short for "Residual Network," in 2015 (He et al., 2016a,b,c). It was made to solve the issue of vanishing gradients, which can arise in very deep neural networks and make it challenging to train the network efficiently. A residual block is the fundamental unit of a *ResNet* (see Fig. 1.14). A batch normalization layer, a *ReLU* activation function, and two convolutional layers make up a residual block. An additional *ReLU* activation function is applied to the sum of the input of the residual block and the output of the second convolutional layer. The "skip connection" facilitates easier gradient passage inside the network, which in turn facilitates deeper network training. *ResNets* can be as deep as desired, with an increase in the amount of residual blocks occurring at a given depth. Many stages, each with a number of residual blocks, make up a typical *ResNet* architecture. Usually,

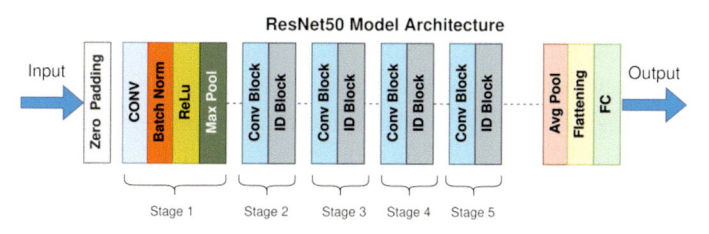

**Figure 1.14** An example of applications of a *ResNet* model. This file is licensed under the Creative Commons Attribution-Share Alike 4.0 International license.

there is only one convolutional layer in the first stage and several blocks in the latter stages. A pooling layer is used to downsample each stage's output, enabling the network to collect information at various scales. On a variety of computer vision tasks, such as semantic segmentation, object detection, and image classification, *ResNet*s have demonstrated state-of-the-art performance.

Of particular interest for problems of object recognition are bidirectional feature pyramid neural networks (*BiFPN*, (Lin et al., 2017; Tan et al., 2019)). This is a type of neural network architecture that is commonly used in object detection tasks, particularly in the field of computer vision. It is an extension of the feature pyramid network (FPN) architecture, which was originally designed to address the challenge of detecting objects at multiple scales.

The main idea behind *BiFPN* is to enhance the feature representation of an input image by fusing multiscale features from different levels of a *CNN*. This is achieved by introducing bidirectional connections between adjacent levels of the network. The *BiFPN* architecture consists of several stages, and each stage contains multiple *BiFPN* blocks. In each block, there are four main operations:

1. Top-down pathway: The higher-level features are downsampled and propagated down to the lower-level features. This is done through a sequence of $1 \times 1$ convolutions to reduce the dimensionality and align the feature sizes.

2. Bottom-up pathway: The lower-level features are upsampled and propagated up to the higher-level features. Similarly, $1 \times 1$ convolutions are applied to align the feature sizes.

3. Lateral connections: These connections serve to fuse the features from the top-down and bottom-up pathways. The feature maps from adjacent levels are combined using element-wise addition or concatenation, allowing information to flow bi-directionally.

4. Weighted feature fusion: To control the importance of the features from different levels, a set of learnable weights is applied to the fused features. These weights are learned during the training process and help to balance the contribution of features at different scales.

The iterative application of these operations in multiple *BiFPN* blocks helps to refine the feature representation and capture objects at various scales. The resulting features are then used for object detection or other downstream tasks, typically by connecting them to additional layers, such as classification or regression heads. *BiFPN* has been shown to improve the performance of object detection models, particularly in scenarios where objects have significant scale variations. It allows the network to effectively integrate multiscale contextual information, leading to more accurate and robust object detection results.

*EfficientDet* (Tan et al., 2020), is a state-of-the-art object detection neural network architecture, which is designed to address the trade-off between accuracy and efficiency in object detection tasks. It achieves this by leveraging a compound scaling method that optimizes the network architecture and scaling parameters. The architecture is based on a backbone network. The backbone network is a high-speed infrastructure that forms the primary communication pathway of a large-scale computer network. It serves as the central conduit for transmitting data, voice, video, and other types of network traffic between different network segments or nodes. Typically, it is a variation of the *EfficientNet* architecture. It is combined with feature pyramid networks (FPN) and a modified version of the *BiFPN* (bidirectional feature pyramid network). The *EfficientDet* architecture demonstrates strong performance across different object detection benchmarks and is computationally efficient.

*EfficientDet* introduces a compound scaling method that scales the depth, width, and resolution of the network in a systematic manner. This method allows the architecture to achieve a good balance between accuracy and efficiency across different resource constraints. *EfficientDet* has several variants, such as *EfficientDet* − D0, *EfficientDet* − D1, *EfficientDet* − D2, and so on, with increasing model sizes and performance. The variant choice depends on the specific requirements of the application, considering factors such as accuracy, speed, and resource constraints. Overall, *EfficientDet* is a highly effective and efficient neural network architecture for object detection tasks, and it has gained significant popularity and adoption in the computer vision community.

Deep learning and neural networks have been applied to various problems for Solar System small bodies, like the identification of asteroids interacting with secular resonances (Chapter 4), orbital dynamics around asteroids (Chapter 5), the identification of cometary objects (Chapter 8), and for chaotic dynamics (Chapter 10). We refer interested readers to these chapters for more information on these topics.

## 1.4. Conclusions

In this chapter we revised basic concepts of *AI*, *ML*, and deep learning that have been recently applied to the field of Solar System small bodies. Algorithms of *ML* were revised in Section 1.2, where we discussed the difference between supervised and unsupervised methods. Standalone and ensemble methods were treated in subsection 1.2.1, whereas clustering and anomaly detection approaches for unsupervised methods were discussed in subsection 1.2.2. Optimization methods based on genetic algorithms were treated in subsection 1.2.3, whereas Section 1.3 dealt with the deep learning methods used in *ML*. This chapter is intended as an introductory material and as a support for Chapters 2 to 10, where detailed applications of *ML* and *AI* methods for Solar System small bodies will be discussed in more detail. More specific details on the methods used in such chapters will be available therein. Interested readers will be able to find more details in the referred papers.

## Acknowledgments

VC acknowledges support from the Brazilian National Research Council (CNPq, grant 304168/2021-1).

## References

Baron, D., 2019. Machine Learning in Astronomy: a practical overview. arXiv e-prints, arXiv:1904.07248.

Boehmke, B., Greenwell, B., 2019. Gradient Boosting. Hands-On Machine Learning with R. Chapman & Hall, London.

Bridle, J.S., 1990. Probabilistic interpretation of feedforward classification network outputs, with relationships to statistical pattern recognition. Connection Science 2, 313–325.

Broomhead, D.S., Lowe, D., 1988. Radial basis functions, multi-variable functional interpolation and adaptive networks. Royal Society of London Proceedings Series A 413, 1–16.

Carruba, V., Aljbaae, S., Domingos, R.C., 2021. Identification of asteroid groups in the $z_1$ and $z_2$ nonlinear secular resonances through genetic algorithms. Celestial Mechanics & Dynamical Astronomy 133, 24. https://doi.org/10.1007/s10569-021-10021-z.

Carruba, V., Aljbaae, S., Domingos, R.C., Huaman, M., Barletta, W., 2022. Machine learning applied to asteroid dynamics. Celestial Mechanics & Dynamical Astronomy 134, 36. https://doi.org/10.1007/s10569-022-10088-2. arXiv:2110.06611.

Chandola, V., Banerjee, A., Kumar, V., 2009. Anomaly detection: a survey. ACM Computing Surveys (CSUR) 41, 1–58.

Chen, T., Guestrin, C., 2016. Xgboost. In: Proceedings of the 22nd ACM SIGKDD International Conference on Knowledge Discovery and Data Mining URL. https://doi.org/10.1145/2939672.2939785.

Chen, P.W., Wang, J.Y., Lee, H., 2004. Model selection of SVMS using GA approach. In: 2004 IEEE International Joint Conference on Neural Networks (IEEE Cat No04CH37541), pp. 2035–2040.

Chipman, H.A., George, E.I., McCulloch, R.E., 2010. Bart: Bayesian additive regression trees. Annals of Applied Statistics 4. https://doi.org/10.1214/09-AOAS285.

Cortes, C., Vapnik, V., 2009. Support-vector networks. Chemical Biology & Drug Design 297, 273–297. https://doi.org/10.1007/%2FBF00994018.

Cramer, J., 2004. The early origins of the logit model. Studies in History and Philosophy of Science Part C: Studies in History and Philosophy of Biological and Biomedical Sciences 35, 613–626. https://doi.org/10.1016/j.shpsc.2004.09.003. https://www.sciencedirect.com/science/article/pii/S1369848604000676.

Dalpiaz, et al., 2021. Applied Statistics with R, STAT, vol. 420. University of Illinois at Urbana-Champaign. https://daviddalpiaz.github.io/appliedstats/.

de Souza, R.S., Krone-Martins, A., Carruba, V., de Cassia Domingos, R., Ishida, E.E.O., Alijbaae, S., Huaman Espinoza, M., Barletta, W., 2021. Probabilistic modeling of asteroid diameters from Gaia DR2 errors. Research Notes of the American Astronomical Society 5, 199. https://doi.org/10.3847/2515-5172/ac205e. arXiv:2108.11814.

Ester, M., Kriegel, H.P., Sander, J., Xu, X., 1996. A density-based algorithm for discovering clusters in large spatial databases with noise. In: Proceedings of the 2nd International Conference on Knowledge Discovery and Data Mining (KDD-96). https://cdn.aaai.org/KDD/1996/KDD96-037.pdf.

Fisher, R.A., 1936. The use of multiple measurements in taxonomic problems. Annals of Eugenics 7, 179–188.

Freund, Y., Schapire, R.E., 1995. A decision-theoretic generalization of on-line learning and an application to boosting.

Fukunaga, K., 1990. Introduction to statistical pattern recognition.

Gaia Collaboration, Spoto, F., Tanga, P., et al., 2018. Gaia Data Release 2. Observations of solar system objects. Astronomy & Astrophysics 616, A13. https://doi.org/10.1051/0004-6361/201832900. arXiv:1804.09379.

Goodfellow, I., Bengio, Y., Courville, A., 2016. Deep learning. Nature 521, 436–444.

Gudivada, V., Irfan, M., Fathi, E., Rao, D., 2016. Cognitive analytics: going beyond big data analytics and machine learning. In: Gudivada, V.N., Raghavan, V.V., Govindaraju, V., Rao, C. (Eds.), Handbook of Statistics. In: Cognitive Computing: Theory and Applications, vol. 35. Elsevier, pp. 169–205. Chapter 5. https://doi.org/10.1016/bs.host.2016.07.010. https://www.sciencedirect.com/science/article/pii/S0169716116300517.

He, K., Zhang, X., Ren, S., Sun, J., 2016a. Deep residual learning for image recognition. In: Proceedings of the IEEE Conference on Computer Vision and Pattern Recognition (CVPR).

He, K., Zhang, X., Ren, S., Sun, J., 2016b. Deep residual learning for image recognition. IEEE Transactions on Pattern Analysis and Machine Intelligence (PAMI).

He, K., Zhang, X., Ren, S., Sun, J., 2016c. Identity mappings in deep residual networks. In: Proceedings of the European Conference on Computer Vision (ECCV).

Hill, J., Linero, A., Murray, J., 2020. Bayesian additive regression trees: a review and look forward. Annual Review of Statistics and Its Application 7, 251–278. https://doi.org/10.1146/annurev-statistics-031219-041110.

Ho, T.K., 1995. Random decision forests. In: Proceedings of the Third International Conference on Document Analysis and Recognition (Volume 1) - Volume 1. IEEE Computer Society, M, pp. 278–282.

Ho, T.K., 1998. The random subspace method for constructing decision forests. IEEE Transactions on Pattern Analysis and Machine Intelligence 20, 832–844. https://doi.org/10.1109/34.709601.

Hochreiter, S., Schmidhuber, J., 1997. Long short-term memory. Neural Computation 9, 1735–1780. https://doi.org/10.1162/neco.1997.9.8.1735.

LeCun, Y., Bottou, L., Bengio, Y., Haffner, P., 1998. Gradient-based learning applied to document recognition. Proceedings of the IEEE 86, 2278–2324.

Lenz, W., 1920. Beiträge zum verständnis der magnetischen eigenschaften in festen körpern. Physikalische Zeitschrift 21, 613–615.

Lin, T.Y., Dollár, P., Girshick, R., He, K., Hariharan, B., Belongie, S., 2017. Feature pyramid networks for object detection. In: Proceedings of the IEEE Conference on Computer Vision and Pattern Recognition (CVPR).

Liu, F.T., Ting, K.M., Zhou, Z.H., 2008. Isolation forest. In: 2008 Eighth IEEE International Conference on Data Mining, pp. 413–422. https://doi.org/10.1109/ICDM.2008.17.

MacQueen, J.B., 1967. Some methods for classification and analysis of multivariate observations. In: Cam, L.M.L., Neyman, J. (Eds.), Proc. of the Fifth Berkeley Symposium on Mathematical Statistics and Probability. University of California Press, pp. 281–297.

McCarthy, J., Minsky, M.L., Rochester, N., Shannon, C.E., 1955. A proposal for the Dartmouth summer research project on artificial intelligence. Dartmouth College.

McCulloch, W., Pitts, W., 1943. A logical calculus of ideas immanent in nervous activity. The Bulletin of Mathematical Biophysics 5, 115–133.

Mika, S., Schölkopf, B., Smola, A.J., Müller, K.R., Scholz, M., Rätsch, G., 1999. Fisher discriminant analysis with kernels. In: Neural Networks for Signal Processing IX: Proceedings of the 1999 IEEE Signal Processing Society Workshop, pp. 41–48.

Mitchell, T.M., 1997. Machine Learning. McGraw Hill.

Pedregosa, F., Varoquaux, G., Gramfort, A., Michel, V., Thirion, B., Grisel, O., Blondel, M., Müller, A., Nothman, J., Louppe, G., Prettenhofer, P., Weiss, R., Dubourg, V., Vanderplas, J., Passos, A., Cournapeau, D., Brucher, M., Perrot, M., Duchesnay, É., 2012. Scikit-learn: Machine Learning in Python. arXiv e-prints. arXiv:1201.0490.

Piryonesi, S.M., El-Diraby, T., 2020. Data analytics in asset management: cost-effective prediction of the pavement condition. Journal of Infrastructure Systems 26. https://doi.org/10.1061/(ASCE)IS.1943-555X.0000512.

Rumelhart, D.E., Hinton, G.E., Williams, R.J., 1986. Learning representations by back-propagating errors. Nature 323, 533–536.

Russakovsky, O., Deng, J., Su, H., Krause, J., Satheesh, S., Ma, S., Huang, Z., Karpathy, A., Khosla, A., Bernstein, M., et al., 2014. 2014 imagenet large scale visual recognition challenge. arXiv preprint. arXiv:1409.0575.

Russell, S., Norvig, P., 2010. Artificial Intelligence: A Modern Approach, 3rd ed. Prentice Hall.

Simonyan, K., Zisserman, A., 2014. Two-stream convolutional networks for action recognition in videos. In: Proceedings of the Conference on Neural Information Processing Systems (NIPS), pp. 568–576.

Simonyan, K., Zisserman, A., 2015. Very deep convolutional networks for large-scale image recognition. In: Proceedings of the International Conference on Learning Representations (ICLR).

Szegedy, C., Ioffe, S., Vanhoucke, V., Alemi, A., 2017. Inception-v4, inception-resnet and the impact of residual connections on learning. In: Proceedings of the AAAI Conference on Artificial Intelligence, pp. 4278–4284.

Szegedy, C., Liu, W., Jia, Y., Sermanet, P., Reed, S., Anguelov, D., et al., Rabinovich, A., 2015. Going deeper with convolutions. In: Proceedings of the IEEE Conference on Computer Vision and Pattern Recognition, pp. 1–9.

Szegedy, C., Vanhoucke, V., Ioffe, S., Shlens, J., Wojna, Z., 2016. Rethinking the inception architecture for computer vision. In: Proceedings of the IEEE Conference on Computer Vision and Pattern Recognition, pp. 2818–2826.

Tan, M., Pang, R., Le, Q.V., 2019. Bifpn: towards efficient and accurate object detection via bidirectional feature pyramid network. arXiv preprint. arXiv:1911.09070.

Tan, M., Pang, R., Le, Q.V., 2020. Efficientdet: scalable and efficient object detection. In: Proceedings of the IEEE/CVF Conference on Computer Vision and Pattern Recognition, pp. 10781–10790.

Wang, D., 2001. Unsupervised learning: foundations of neural computation. AI Magazine 22, 101. https://doi.org/10.1609/aimag.v22i2.1565.

Ward Jr, J.H., 1963. Hierarchical grouping to optimize an objective function. Journal of the American Statistical Association 58, 236–244. https://doi.org/10.1080/01621459.1963.10500845.

Widrow, B., Hoff, M., 1960. Adaptive switching circuits. IRE WESCON Convention Record 4, 96–104.

# CHAPTER TWO

# Identification of asteroid families' members

## R.C. Domingos[a], M. Huaman[b], and M.V.F. Lourenço[c]

[a]São Paulo State University (UNESP), School of Engineering, Department of Electronic and Telecommunications Engineering, São João da Boa Vista, SP, Brazil
[b]Universidad tecnológica del Perú (UTP), Cercado de Lima, Peru
[c]São Paulo State University (UNESP), School of Engineering and Sciences, Department of Mathematics, Guaratinguetá, SP, Brazil

## 2.1. Introduction

The generation of various asteroid families is attributed to factors such as collision events, rotation faults, characteristic ejection speeds, and the consequence of the dynamic region in which they survive. The hierarchical clustering method (HCM, Zappalà et al., 1990, 1995) is the best-known standard technique for the identification of families according to Nesvorný et al. (2015) and Radović et al. (2017). The domain of proper elements $(a, e, \sin(i))$ is used for its constant movement and stability over long periods under non-gravitational influences (unlike the osculating elements).

The method focuses on locating asteroids related to a parent body (primary body). If the distance (defined by a metric) between its proper elements is less than a threshold number known as the limit $d_{cutoff}$, it is included in the family list. Then, this asteroid is used as a new parent, and the method is repeated until no new members are located.

In the 1990s, when this technique was introduced, there were about 10,000 asteroids with proper elements. Currently, according to Spacewatch and LINEAR (robotic surveys on asteroids), the number of reliable proper elements is around 750,000, which is gradually increasing. In regions with a high density of bodies, the conventional HCM approach is not applied effectively: families of asteroids close to space of proper elements may overlap and become unrecognizable as independent entities. Milani et al. (2014) refers to this problem as "chaining" and can be overcome by applying HCM alternatives. These alternatives can be expensive from a computational point of view since they involve determining individual solutions for each specific family in the orbital region where the group is located.

*Machine Learning for Small Bodies in the Solar System*
https://doi.org/10.1016/B978-0-44-324770-5.00007-6

In recent years, different algorithms have been implemented in machine learning for the Python programming language, which is publicly available to solve problems about chaining and categorization. Carruba et al. (2020) uses supervised and unsupervised machine learning techniques to detect new members of recognized asteroid families mentioned in the Asteroid Families Portal (*AFP*[1]; Radović et al., 2017; Novaković et al., 2022). The authors used training ML methods to identify patterns in the distribution of previously established family members, allowing new asteroids with comparable distributions to join these groups without the chaining problem.

We aim to discuss various algorithms using standalone or ensemble methods to determine which method may be most accurate for the problem. For this purpose, we will compare the recently detected asteroids with those located using traditional HCM and study the criteria to evaluate the efficiency of the different machine-learning algorithms.

This chapter contains the following structure: Section 2.2 "Asteroid families" describes the selection criteria for choosing the research test families. In Sections 2.3 and 2.4, the results of the machine learning methods for each asteroid family tested are presented and analyzed. After determining the optimal classification algorithm, the membership of an asteroid family was updated to find new family members. Section 2.5 describes a tool based on a genetic algorithm used by Lourenço and Carruba (2022) to optimize the study to identify the ideal classification algorithm. Our final comments are then presented in the "Final Remarks" section.

## 2.2. Asteroid families

To understand the study presented here, it is essential to clearly understand how to identify asteroid family members using machine learning algorithm training. Some confusion may persist, unless the reader fully understands both the theory involved and the methodology followed.

The literature reports that, whereas it is theoretically possible to do so in the frequency space (Carruba and Michtchenko, 2007, 2009), the detection of asteroid families and the identification of their members are typically performed in the space of the proper elements: major semiaxis $a$, eccentricity

---

[1] http://asteroids.matf.bg.ac.rs/fam/properelements.php

$e$, and inclination $i$ (Milani et al., 2014; Nesvorný et al., 2015). The asteroid families database utilized in Carruba et al. (2020) may be found in *AFP* (accessed on December 5, 2019; Radović et al., 2017; Novaković et al., 2022). Using typical HCM processes, the web page-based algorithms may automatically obtain asteroid families from an asteroid dataset with synthetic self-elements.

Some selection criteria were established to choose the target families for this study. The first criterion was that the families chosen were represented in Radovic's catalog (2017), located in the main belt's inner, central, or outer zones at low inclinations. These are the regions where we find the most numerous families. We suggest the work of Carruba et al. (2013) to define these zones better. The second criterion was that the family had at least ten members with H < 14. According to the literature (Zappalà et al., 1990), a family is not statistically significant if it has less than ten asteroids. To have a more significant number of families located in the inner belt, despite having fewer members (H < 14), the Massalia family was included in the study.

By applying these selection criteria, twenty-one families were selected for the study. The date and location of the chosen families can be seen in Table 2.1. The columns are the location of the family in the main belt, the family ID, the total number of members, and the number of members with $H < 14$. For the study, the total number of family members was divided into training and test samples.

The models used the training sample as an initial basis to learn complex patterns and relationships in the data. The models identified and categorized additional family members, thanks to these patterns. All of the individuals in the training sample have $H < 14$. The work of Milani et al. (2014) motivated the selection of this cutoff magnitude value. It was concluded in that work that dynamic family chaining achieved with hierarchical clustering approaches does not affect families estimated for asteroids with $H < 14$. There should be little to no overlap between these categories and dynamic, local families.

The test sample comprised the rest of the data, which is the dataset not seen by the model during its training phase. The test sample results assess new data the model sees and how well the models can predict.

In the next section, we discuss the implementation of machine learning algorithms to classify new asteroid family members.

**Table 2.1** The table lists the asteroid families localized in the main belt and found on the Asteroid Families Portal. It also includes the number of members in each family and the number of members with $H < 14$.

| Main belt | Family Id. | Number of members | Number of members $H < 14$ |
|---|---|---|---|
| Inner | 20 Massalia | 6400 | 4 |
| | 163 Erigone | 2979 | 16 |
| Central | 15 Eunomia | 6076 | 1072 |
| | 128 Nemesis | 1449 | 19 |
| | 170 Maria | 4183 | 392 |
| | 363 Padua | 807 | 39 |
| | 410 Chloris | 449 | 45 |
| | 569 Misa | 723 | 11 |
| | 668 Dora | 1677 | 128 |
| | 808 Merxia | 1624 | 28 |
| | 847 Agnia | 4432 | 98 |
| | 1726 Hoffmeister | 2396 | 42 |
| Outer | 10 Hygiea | 6224 | 453 |
| | 24 Themis | 5949 | 1029 |
| | 158 Koronis | 7294 | 790 |
| | 221 Eos | 14661 | 2043 |
| | 283 Emma | 610 | 37 |
| | 375 Ursula | 2214 | 285 |
| | 490 Veritas | 1805 | 133 |
| | 845 Naema | 418 | 18 |
| | 1040 Klumpkea | 2840 | 195 |

## 2.3. Machine learning methods

Several supervised machine learning methods for solving classification issues have recently been introduced in the PYTHON programming language (Pedregosa et al., 2011). Given an existing population, these techniques can leverage known data to forecast whether new data belongs to a given group. Without finding a solution for every family in the area, Carruba et al. (2020) examined whether these methods could automatically identify potential new members of a given asteroid family. In that paper, the authors introduced parameters to evaluate the algorithms' performance and examined how the newly found asteroids compared to those identified by the standard HCM. The approach that was most likely to be effective for the given situation was determined. Below, we describe the research carried out by Carruba et al. (2020).

To predict the membership of the test sample population, machine learning techniques are used on the suitable distribution ($a$, $e$, $\sin(i)$) of the training sample. To assess the algorithms' correctness, variables that are simple to understand—such as those associated with specific categories of misclassification errors or accurate classifications—must be created. Additionally, any correlation between these factors must be examined. Thus as a result of applying the algorithms, following the notation from Chapter 1, the accuracy of the algorithms was analyzed using three classification coefficients of the members of the asteroid families. Some asteroids are identified as *true positives* ($T_{Pos}$), which are those asteroids identified as members of the family by the HCM methods and the machine learning algorithm. Another quantity we call *false positives* ($F_{Pos}$), whose asteroids were identified as belonging to the family only by the machine learning algorithm. The third quantity comprises *false negatives* ($F_{Neg}$), which are those asteroids not identified as family members by the machine learning algorithms alone.

Some classification coefficients of the members of the asteroid families were applied to analyze the effectiveness of the algorithms in recovering members of the asteroid families. As seen in Chapter 1, the percentage of family members retrieved by the machine learning method relative to the entire original population is known as the *recall* coefficient, and it is provided by

$$Recall = \frac{T_{Pos}}{T_{Pos} + F_{Neg}} \tag{2.1}$$

On the other hand, the *Precision* coefficient indicates how well the model avoids making incorrect predictions about the data. It is given by

$$Precision = \frac{T_{Pos}}{T_{Pos} + F_{Pos}} = \frac{T_{Pos}}{N_{Retr}}, \tag{2.2}$$

where the number of asteroids retrieved within the family is denoted by $N_{Retr}$.

By examining the formulas for (2.1) and (2.2), we can observe that a high *Recall* value implies that the algorithm must have recovered a significant portion of the initial population. However, a closer inspection does not appear to indicate that this is a favorable outcome. Consider a scenario where a family of 100 objects exists, and the algorithm has recovered 1000 objects, each containing 100 original family members. In this instance, *recall* would equal 1, even though the recovered family is not a good approximation of the original family. As a result, the value of *recall* by itself consistently identified new asteroid family members and trained the method's accuracy; a family with few members may have a high *Precision* coefficient value. An original family of 100 members and a recovered family with 10% of the members categorized as $T_{Pos}$ could serve as examples. Even though both coefficients provide helpful information, more information is required to determine effective family recovery. A final parameter coefficient was created to reach a compromise by merging the data from the *Recall* and *Precision* coefficients. It is given by

$$F_P = \frac{1}{\sqrt{2}}\sqrt{(Recall)^2 + (Precision)^2}. \tag{2.3}$$

Next, we discuss implementing machine learning algorithms to classify new family members. The algorithms were trained, and new members of the asteroid family were reliably found using the *AFP* database.

### 2.3.1 Machine learning algorithms

This section describes the methods used by Carruba et al. (2020) for identifying new asteroid family members. We then describe the concepts involved in the algorithms and the importance of their hyperparameters.

Carruba et al. (2020) investigated stand-alone and ensemble machine-learning methods. Approaches that employ a single algorithm for the classification process are known as stand-alone methods. K-nearest neighbors (*KNN*) and decision tree algorithms were investigated in that study.

Ensemble approaches combine many learning algorithms to get more accurate answers than a single model could provide. Bootstrap aggregating and boosting methods are ensemble methods that use several autonomous algorithms. Many weak classifiers are used in bootstrap aggregating, or bagging, to aggregate the forecasts and choose the best one. One type of bagging method is random forest (RF), which combines the results of multiple random decision trees to attain high classification accuracy. Boosting is a machine learning strategy incorporating numerous weak learners to increase a model's forecast accuracy. It highlights the training examples incorrectly identified by earlier classifiers and the stand-alone classifiers with superior performance. It builds upon the advantages of several models to provide a more potent and precise prediction. Carruba et al. (2020) studied three of the most commonly used boosting algorithms: adaptive boosting (*AdaBoost*), gradient boosting (*GBoost*), and eXtreme gradient boosting

(*XGBoost*). When the problem in the training data is not overfitting, boosting might yield better accuracy than bagging. The reader can find more information on this approach in Swamynathan (2017).

Three types of bootstrap aggregations will be analyzed here: bagging classifiers, random forests, and highly randomized trees (*ExtraTree*). For every procedure, one or more free parameters need to be optimized. These parameters, sometimes called hyperparameters, must be investigated for each family. Carruba et al. (2020) determined the hyperparameter values using the *GridSearchCV* approach of the scikit-learn package (Pedregosa et al., 2011). Using the *GridSearchCV* approach, the algorithm of interest is applied for an entire grid of hyperparameter values. Subsequently, the best-fit value is applied for each family that is being studied. For each of the 21 asteroid families, Carruba et al. (2020) included tables with the mean values of the $F_P$ coefficients and the hyperparameters for nine different machine-learning techniques, along with their errors.

For illustration, histograms of the hyperparameter value (left-hand panel of Figs. 2.1 to 2.8 displays a histogram of the "probability cutoff" hyperparameter) and the $F_P$ coefficient (right-hand panel of Figs. 2.1 to 2.8) were produced for each class of algorithms presented here. The $F_P$ coefficient's mean value can be used to compare the effectiveness of various methods; the greater the value, the more successful the algorithm is in locating relatives. The technique can be used for families not included by utilizing the average hyperparameter value discovered in that study.

## 2.3.2 Stand-alone methods

### 2.3.2.1 KNN algorithm

The main idea is to take a dataset of asteroids that have previously been classified into families and use the k closest asteroids to determine whether an asteroid not in the set is a member of the family or not and check which is the predominant family of the neighboring asteroids. For example, imagine that an asteroid has five neighbors, three of which are the closest to the asteroid and belong to a specific family. Since three of the nearest neighbors are family members, the asteroid would also be labeled as such. Each neighboring asteroid votes for its family, and the asteroid is considered to belong to the family with the most votes. The number of nearby asteroids $K$ is the primary determining element for this inquiry. Therefore the ideal "number of neighbors" to consider is a hyperparameter of this approach.

As previously said, each family must have comprehensively examined its hyperparameter. The results are summarized in Fig. 2.1, which displays

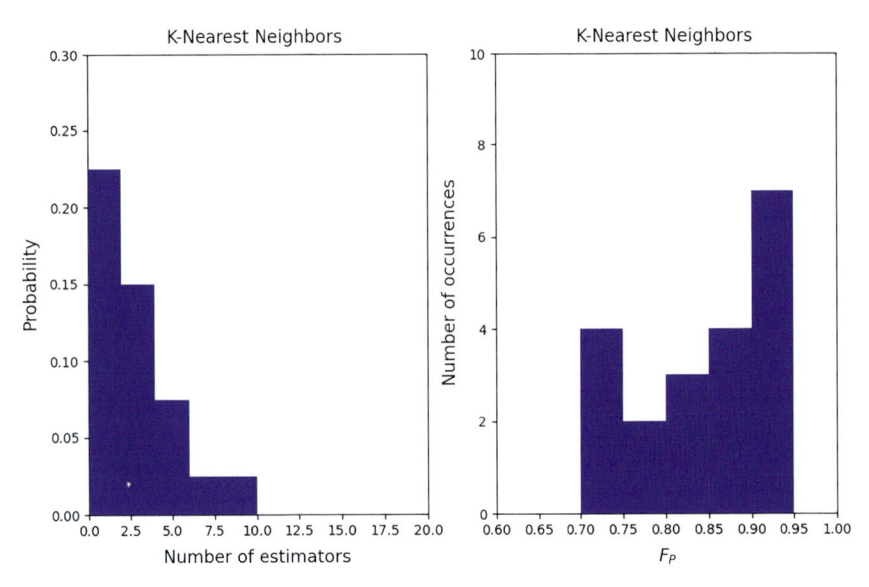

**Figure 2.1** Histograms of the hyperparameters and $F_P$ coefficient for the 21 families studied with *KNN* algorithm.

histograms of the "number of neighbors" hyperparameter and the $F_P$ coefficient. The mean value of the first parameter was $2.35 \pm 1.88$, whereas $F_P = 0.84 \pm 0.08$.

### 2.3.2.2 Decision tree algorithm

A decision tree is an essential machine learning technique representing decisions as a tree with leaf nodes denoting outcomes and branches representing decision rules (Quinlan, 1986). The dataset features to be classed are considered when making decisions. Let's say a set of data needs to be classified. The method asks a question about the data at the first level of the tree, and it proceeds to the next node by following the branch. After that, the tree is split into smaller trees until it reaches the leaf node, the last node. The hyperparameter of this approach is the number of decision nodes employed in the data assessment, or max depth, in the decision tree algorithm implementation by scikit-learn. Fig. 2.2 provides an overview of the findings. The average value of this parameter for the families under study was $7.25 \pm 2.91$, and for $F_P$, we found $F_P = 0.84 \pm 0.07$.

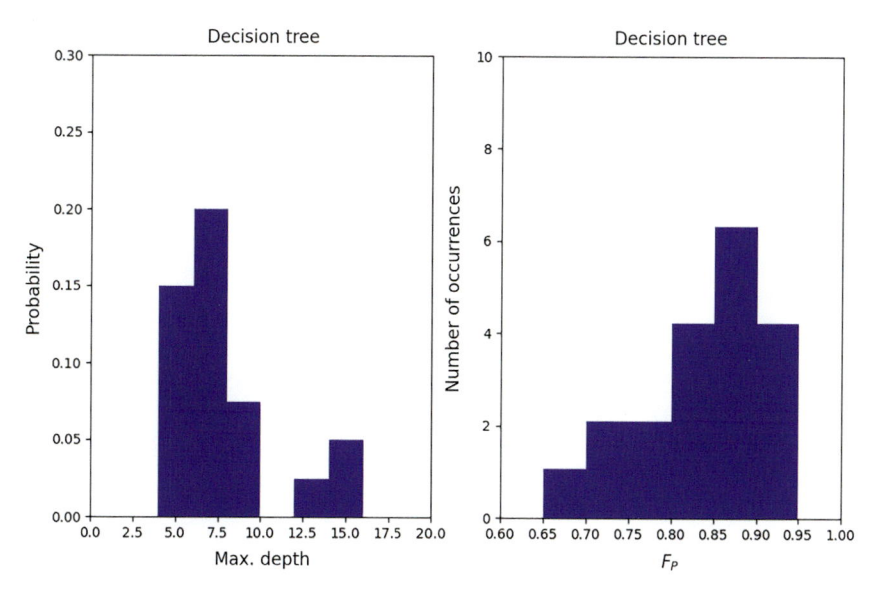

**Figure 2.2** Histograms of the hyperparameters and $F_P$ coefficient for the 21 families studied with *decision tree* algorithm.

### 2.3.3 Ensemble methods: bagging and boosting classifiers

#### 2.3.3.1 Bagging classifiers

A model aggregation method called the *Bagging* classifier seeks to lower the model variance. The bootstrap samples are replacement samples created by dividing the training data into several samples. Data repetition will occur since the bootstrap sample size is the same as the original but with 3/4 of the original values and replacements for the remaining 1/4. On each of the bootstrap samples, separate models are constructed. For the regression model, an average of the predictions is done, whereas the majority vote is the criterion for classification. In Carruba et al. (2020), the *bagging* classifier approach was employed, with the number of estimators being a hyperparameter of the model examined in the range from 0 to 100, and the optimal value of the number of neighbors for the decision trees determined in the preceding section. Fig. 2.3 displays histograms of the "number of estimators" hyperparameter and the $F_P$ coefficient. The mean value of the hyperparameter was $15.55^{+18.16}_{-15.55}$, whereas for $F_P$ we obtained $F_P = 0.81 \pm 0.09$.

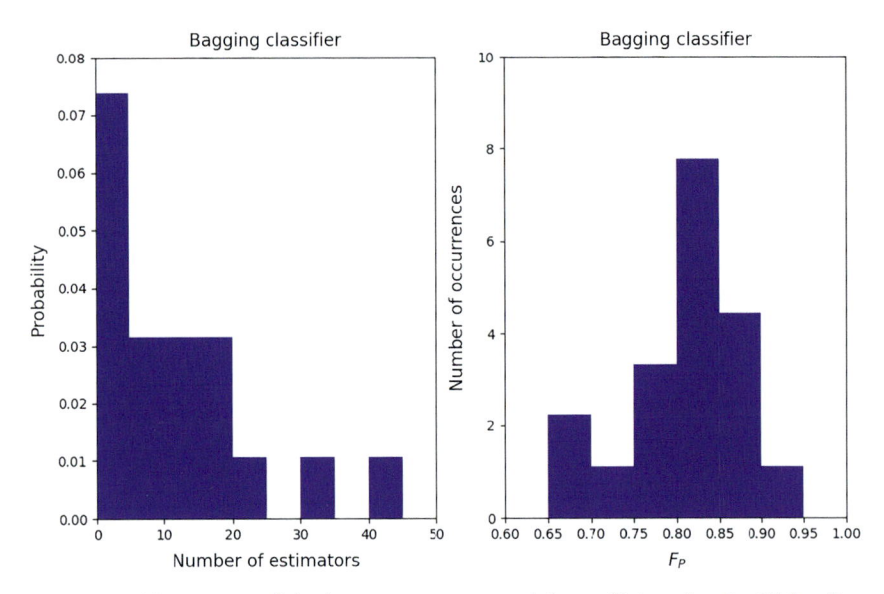

**Figure 2.3** Histograms of the hyperparameters and $F_P$ coefficient for the 21 families studied with *Bagging classifier* algorithm.

### 2.3.3.2 Random forest algorithms

*Random forest* is made up of numerous individual *decision Tree*, which functions as a classifier for *Bagging*. A distinct bootstrap sample of the same size as the training set is used to train each estimator. Every tree within the *Random forest* generates a class prediction; our model predicts the class that receives the greatest number of votes. The number of estimators utilized in the approach is a free model parameter. As in earlier sections, the *GridSearchCV* method was employed to determine the optimal values for this parameter within the range of 15 to 50. Fig. 2.4 displays histograms of the "number of estimators" hyperparameter and the $F_P$ coefficient. The hyperparameter mean value was $40.7 \pm 9.56$, and $F_P = 0.77 \pm 0.08$.

### 2.3.3.3 Extremely randomized trees (ExtraTree)

*ExtraTree* randomly creates many decision trees to make the bagging process extremely random, and then, by combining the results of each tree, finds the final answer. This allows the model's variance to be further reduced but at the expense of increased bias. The number of estimators in this approach is the hyperparameter, much like in other bagging ensemble algorithms. Once more, we employed the *GridSearchCV* method to determine this parameter's optimal value within the range of 1 to 150. Fig. 2.5

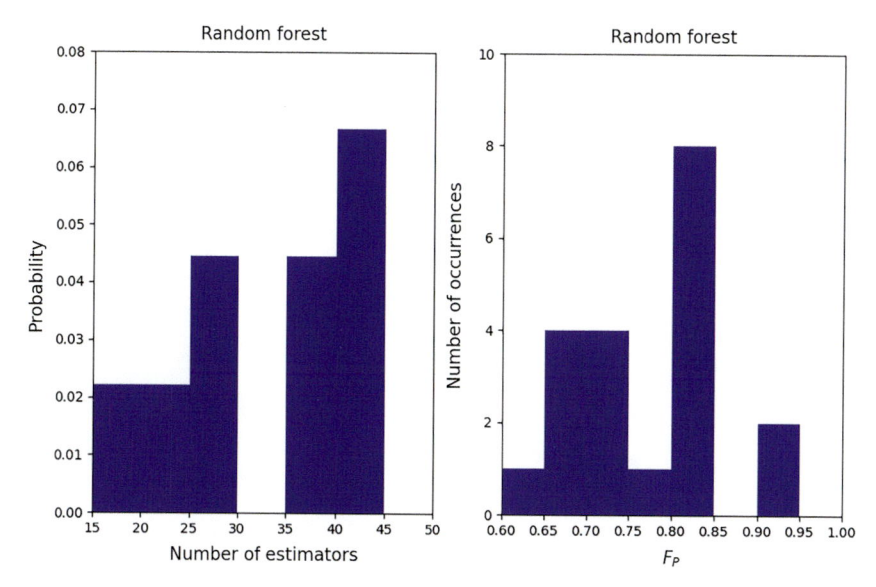

**Figure 2.4** Histograms of the hyperparameters and $F_P$ coefficient for the 21 families studied with *Random forest* algorithm.

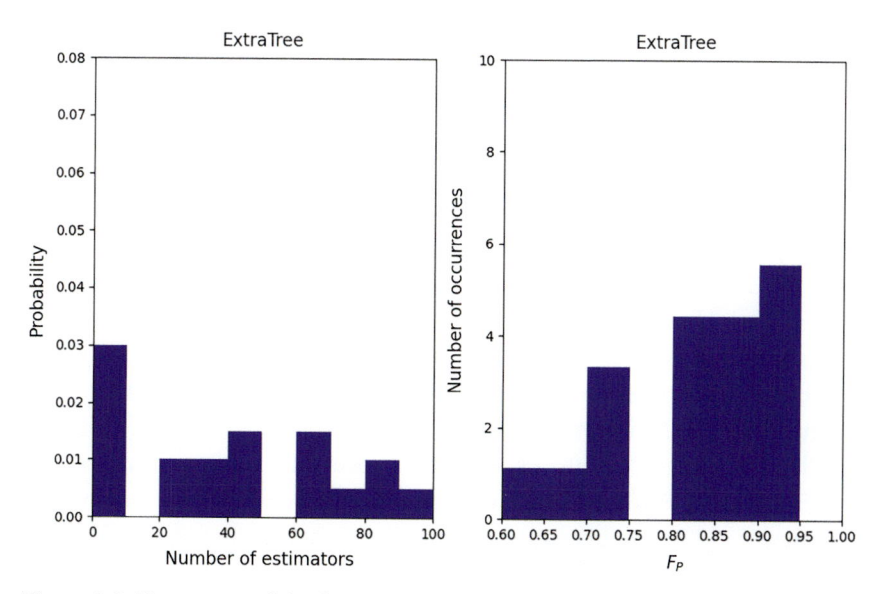

**Figure 2.5** Histograms of the hyperparameters and $F_P$ coefficient for the 21 families studied with *ExtraTree* algorithm.

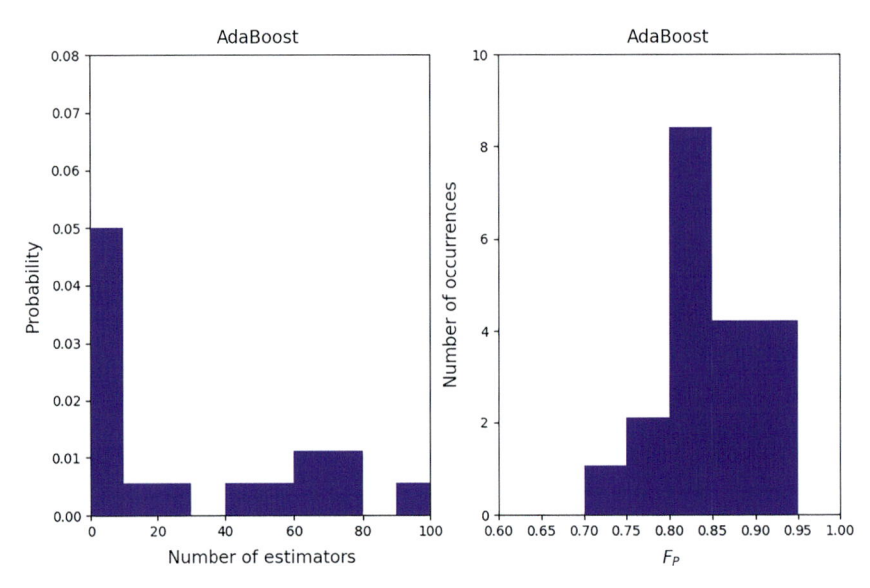

**Figure 2.6** Histograms of the hyperparameters and $F_P$ coefficient for the 21 families studied with *AdaBoost* algorithm.

displays histograms of the "number of estimators" hyperparameter and the $F_P$ coefficient. The hyperparameter has a mean value of $40.95 \pm 29.21$ and $F_P = 0.84 \pm 0.10$.

### 2.3.3.4 *Adaptive boosting (AdaBoost) algorithm*

Boosting algorithms can, therefore, convert weak learners into strong learners. They can be applied to any data classification algorithm, as their technique is based on many different classifiers and not just one; perhaps they can have higher-than-expected accuracy. *AdaBoost* Classifier combines several weak classification algorithms and transforms them into a single robust algorithm. A weak classification algorithm performs poorly, but when combined with other weak algorithms, it can perform much better. The method *AdaBoost* follows three steps. First, uniform weights are assigned to all data points. Weights are adjusted at every method iteration to the training data to minimize the weighted error function (see equations in Swamynathan, 2017), until no further corrections are needed and the final model is reached.

The number of estimators that have been improved using *GridSearchCV* to determine the ideal value between 1 and 150 is a crucial hyperparameter for this model. Fig. 2.6 shows histograms of the $F_P$ coefficient and the num-

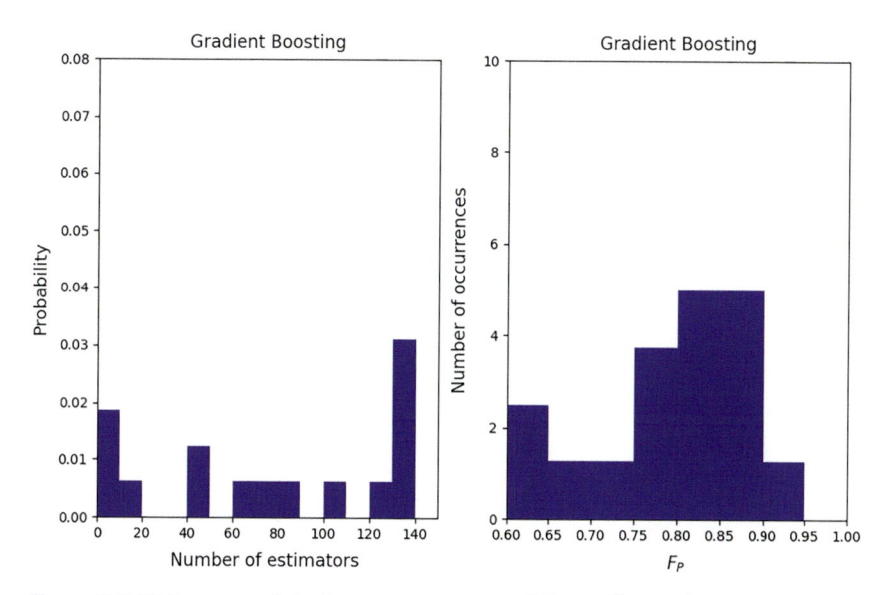

**Figure 2.7** Histograms of the hyperparameters and $F_P$ coefficient for the 21 families studied with *GBoost* algorithm.

ber of estimators hyperparameter, revealing the results of this procedure. The hyperparameter's mean value was $41.25^{+45.41}_{-41.25}$, and $F_P = 0.85 \pm 0.06$ was the value found for $F_P$.

### 2.3.3.5 Gradient boosting (GBoost) algorithm

Whereas *AdaBoost* employs high-weight data points, *GBoost* uses gradients to determine the shortcomings of a poor learner. The procedure iteratively fits a classifier to the training set of data, calculating the negative gradient, and building a new base learner using the outcome. When there is no more room for improvement, the estimator is updated, and the process terminates. The outcome of this process is displayed in Fig. 2.7, which displays histograms of the $F_P$ coefficient and the number of estimators hyperparameter. The optimal mean value of the hyperparameter number of estimators found was $91.45 \pm 52.67$. The mean value for $F_P$ was $F_P = 0.76 \pm 0.19$.

### 2.3.3.6 eXtreme gradient boosting (XGBoost) algorithm

*XGBoost* algorithm is a more regularized and extended version of a gradient boosting algorithm. The algorithm works by optimizing an objective function using the gradient descent technique. In the last few years, the algorithm has won several competitions on Kaggle, the predictive modeling and analytical competitions forum. *XGBoost* depends on several hyperparameters. Among them, there are the following:

- eta: the learning rate;
- max_depth: maximum depth of trees;
- colsample_bytree: the fraction of columns to be randomly sampled for each tree;
- Subsample: the fraction of observations to be randomly sampled for each tree algorithm;
- lambda: L2 regularization term on weights;
- alpha: L1 regularization term on weight.

The ideal values for the hyperparameters were discovered using the *GridSearchCV* to optimize the values of these parameters for each family. The method fits the data and finds the most optimal values from the studied ones for each given hyperparameter value. To get the ideal hyperparameter settings, Carruba et al. (2020) used the ranges shown in Table 2.2. They also examined values from 1 to 150 for the number of estimators hyperparameters, similar to other ensemble approaches.

Table 2.2  Range for hyperparameters (Carruba et al., 2020).

| Hyperparameters | Range for hyperparameters | Default value |
|:---:|:---:|:---:|
| eta | [0.001, 0.01, 0.1] | 0.3 |
| max_depth | [2, 5, 10, 20] | 6 |
| colsample_bytree | [0.1, 0.5, 0.8, 1] | 1 |
| Subsample | [0.1, 0.5, 1] | 1 |
| alpha | [0.1, 0.5, 1] | 1 |
| lambda | [0.1, 0.5, 1] | 1 |

The findings are displayed in Table A9 of Carruba et al. (2020). In every instance, the values of the subsample and eta parameters were equal to 0.1. The distributions' standard deviations were estimated to be the same as the mean values, and their errors for max_depth $= 5 \pm 5$, alpha$= 0.1 \pm 0.1$, colsample_bytree $= 1.0 \pm 0.1$, and lambda $= 0.5 \pm 0.4$, respectively. The approach's outcomes are displayed in Fig. 2.8, which shows histograms of the $F_P$ coefficient and the "number of estimators" hyperparameter. The hyperparameter mean value was $88.15 \pm 46.71$, and the estimators' hyperparameter for $F_P$ was found to be $F_P = 0.83 \pm 0.09$. The values were tested within a range of 1 to 150.

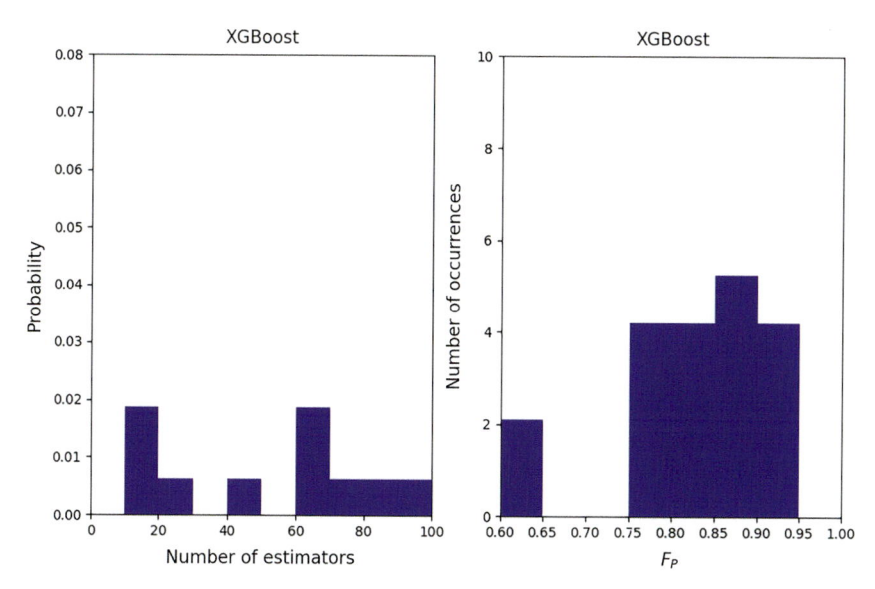

**Figure 2.8** Histograms of the hyperparameters and $F_P$ coefficient for the 21 families studied with *XGBoost* algorithm.

## 2.4. Classification of new asteroid families members

Factors including age, population, potential interactions with dwarf planets or massive asteroids, and the influence of the local dynamical environment distinguish each asteroid family. It is not unexpected that the 21 instances under investigation have been divided into different categories. One crucial factor to take into account is the size of the sample that was used to train the classification algorithm. A bigger sample size yields a more accurate projected result. The most effective approach was chosen for every family using the $F_P$ parameter. The results can be seen in Table 2.3.

*ExtraTree* algorithm presented the best results for ten families; 4 of these cases correspond to families with many members and $F_P$ values greater than 89 percent. The following best algorithm is *KNN* for four families with a medium to a small number of members (less than 150 members) with $F_P$ values greater than 74 percent. Other possible options are *XGBoost* (for three families) and *AdaBoost* (for two families), and with a single case, the *Dec. Tree* algorithm and *Bagging Class* were observed.

The $(a, e)$ projections of the best results for a family membership of the *ExtraTree* algorithm (Koronis family) and *KNN* technique (Veritas family) are shown in Fig. 2.9. Blue circles indicate the family members' orbital locations, as the machine learning system anticipated, whereas black dots

**Table 2.3** An overview of the top-performing classification algorithms under study. Family identification, the number of members with $H < 14$, the best $F_P$ coefficient value, and the name of the estimator that performed better than the other methods are the columns in that order. For more details, see table (1) of Carruba et al. (2020).

| Family Id. | Members with $H < 14$ | Best $F_P$ | Best Estimator |
|---|---|---|---|
| 20 Massalia | 4 | 0.64 | |
| 15 Eunomia | 1072 | 0.89 | |
| 668 Dora | 128 | 0.96 | |
| 847 Agnia | 98 | 0.86 | |
| 221 Eos | 2043 | 0.89 | *Extra Tree* |
| 24 Themis | 1029 | 0.95 | |
| 158 Koronis | 790 | 0.96 | |
| 375 Ursula | 285 | 0.85 | |
| 1040 Klumpkea | 195 | 0.91 | |
| 845 Naema | 18 | 0.94 | |
| 363 Padua | 39 | 0.74 | |
| 1726 Hoffmeister | 42 | 0.89 | |
| 283 Emma | 37 | 0.89 | *KNN* |
| 490 Veritas | 133 | 0.94 | |
| 410 Chloris | 45 | 0.87 | |
| 128 Nemesis | 19 | 0.81 | *XGBoost* |
| 10 Hygiea | 453 | 0.93 | |
| 163 Erigone | 16 | 0.78 | *AdaBoost* |
| 569 Misa | 11 | 0.81 | |
| 170 Maria | 392 | 0.85 | *Bagging Class* |
| 808 Merxia | 28 | 0.90 | *Dec. Tree* |

represent the family members as determined by the AFP. For both families, $F_P$ values are more significant than 94%, indicating that the two approaches perform exceptionally well.

The case of five asteroid families in the highly inclined main belt and Cybele, which have a statistically significant population of objects with $H < 14$, was studied to explore the generalizability of the two top-performing methods. The families in dispute include Sylvia (87), Euphrosyne (31), Alauda (702), Hansa (480), Barcelona (945), and Euphrosyne (31). The ideal values for the hyperparameters about the number of neighbors in *KNN* and the number of estimators in the *Extra Tree* algorithm are 2 and 41, respectively.

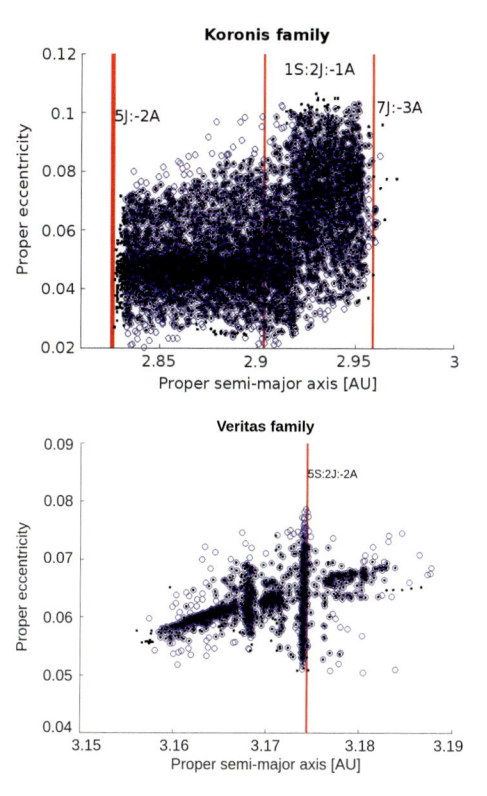

**Figure 2.9** ($a, e$) projections of the sampled (black dots) and predicted (blue circles) members of the Koronis family and Veritas family given by *KNN* and *ExtraTree* algorithm techniques, respectively. The locations of local two- and three-body mean-motion resonances are shown as vertical lines. ($a$, sin ($i$)) projections can be seen in Figure (3) of Carruba et al. (2020).

To verify the performance reported by the *ExtraTree* and *KNN* algorithms, five families of high inclination asteroids in the main belt with a dense membership and $H < 14$ were chosen to be tested: (480) Hansa, (945) Barcelona, (31) Euphrosyne, (702) Alauda, and (87) Sylvia, the latter of which is located in the Cybele region. The analysis shows that the *ExtraTree* algorithm maintains efficiency (best algorithm for locating new family members) with $F_P$ values greater than 74 percent. For more details, review Table 2 of Carruba et al. (2020). Those results are shown in Fig. 2.10. *ExtraTree* performed best in all cases, with values of $F_P$ consistently higher than 74 percent. Based on this analysis, this classification algorithm was considered the best tool for finding new family members among the studied methods.

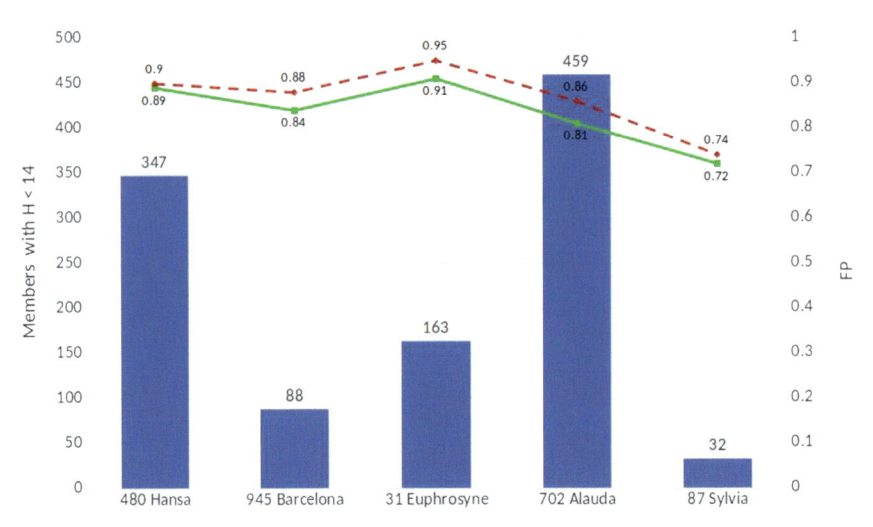

**Figure 2.10** A histogram for the following five families of steeply inclined asteroids: (480) Hansa, (945) Barcelona, (31) Euphrosyne, (702) Alauda, and (87) Sylvia. The $F_P$ value determined by the *KNN* and *ExtraTree* classifier methods are shown by green and red points on the lines, respectively.

## 2.4.1 Updating family members

New family members were identified using *Extra Tree* (optimal classification algorithm). For this goal, the region to be studied was determined in the domain of proper elements, and the procedure was modified to consider two phases. The most significant number of members in the space of proper elements was identified in the first phase. Asteroids are labeled 1 and 0 for this grouping to classify whether they are family members or the opposite. In the second phase, the machine learning algorithm is used, with this data, to train and identify asteroids that have the highest probability of belonging to the family.

Four asteroid families with a small, medium, and large number of members were selected for the quantitative analysis of this new procedure. The small family of (694) Ekard (105 members), the medium family of (480) Hansa (1484 members), the large family of (15) Eunomia (6076 members), and (832) Karin (480 objects), which is a Koronis subgroup (identified as a large family). The last family was added to determine the algorithm's efficiency in dense regions. Because there is no information about the family in the extended sampling, the efficiency of the adjustment is quantified by validating the extended family (obtained by the HCM method) with the adjusted family in terms of $d_{std}$, given by

$$d_{std} = \frac{1}{\sqrt{3}}\sqrt{\left(\frac{\sigma(a_{fitted})}{\sigma(a_{known})}\right)^2 + \left(\frac{\sigma(e_{fitted})}{\sigma(e_{known})}\right)^2 \left(\frac{\sigma(\sin(i)_{fitted})}{\sigma(\sin(i)_{known})}\right)^2}, \quad (2.4)$$

where $\sigma$ is the standard deviation in the domain of proper element $(a, e, \sin(i))$, the "known" and "fitted" subscripts refer to the families obtained for the adjustment process and with the algorithm (extended family). For distributions of family members with values of $d_{std}$ close to 1, the extended families are compatible with the fitted families.

The Ekard family had 37 new members discovered by the *Extra Tree* algorithm, whereas Hansa had 694, Eunomia had 1550, and the Karin group had 324. The orbital distribution of the latest members is in complete agreement with the old members. The standard deviation of $d_{std}$ defines the error, and the mean of $d_{std}$ overall was $0.97 \pm 0.04$. Similar outcomes apply to other asteroid families; the system can even precisely identify members of subfamilies such as Karin, scattered throughout the main belt. In summary, *Extra Tree* appears to be a dependable instrument for automatically identifying possible new members.

Fig. 2.11 shows the projections on the plane of proper elements $(a, e)$ of the families (694) Ekard, (480) Hansa, (15) Eunomia. The fourth panel presents the projection in the domain of $(a, \sin(i))$ proper elements of the (832) Karin subgroup of the Koronis family. The black dots indicate the original population (fitted family), whereas the blue circles represent the members of the families identified by the *Extra Tree* algorithm (extended family). Members recognized for the (694) Ekard family (a sparsely populated family) are scattered in regards to their eccentricity, ranging from 0.2 to 0.28. (480) Hansa, we find a median population dispersed around its eccentricity of 0.004 from approximately 2.55 to 2.75 AU; (15) Eunomia, members around their eccentricity of 0.15 with a considerable population. Whereas the young family of (832) Karin in the outer Main Belt reflects a dispersed distribution in the plane $(a, \sin(i))$ of proper elements, in other planes such as $(a, e)$, this family shows some dispersion with a slight oblique linear shape. More details about these families and their distribution in other planes are in Carruba et al. (2020). For this task, *Extra Tree* is a reliable tool for determining potential new members automatically.

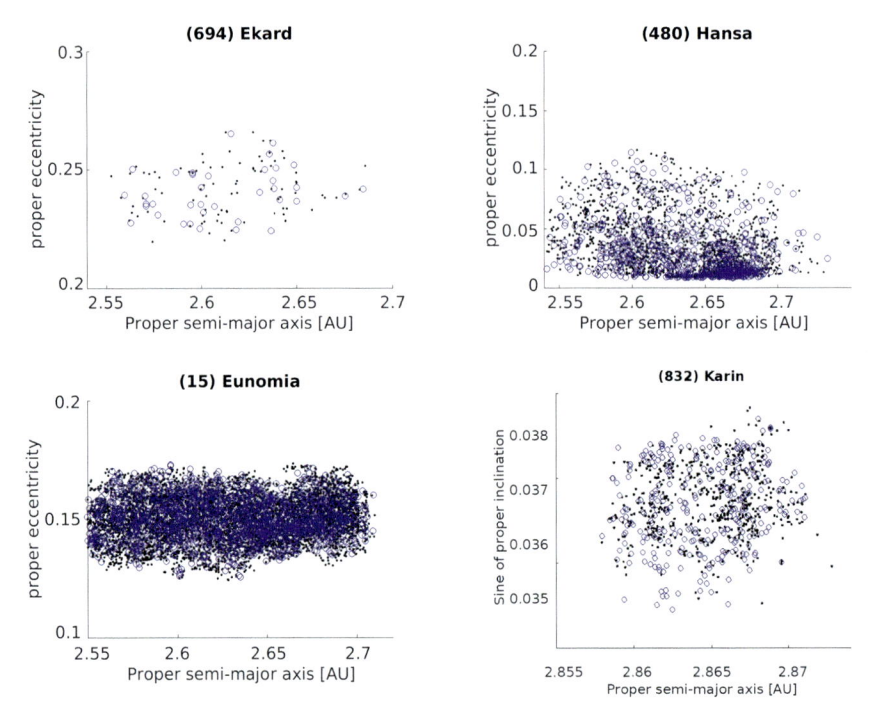

**Figure 2.11** $(a, e)$ projections of the Ekard, Hansa, and Eunomia asteroid families' known members (black dots) and expected members (blue circles). A $(a, \sin(i))$ projection of the fitted and predicted Karin subfamily members is shown in the fourth panel.

## 2.5. Genetic algorithms

As previously mentioned, Carruba et al. (2020) studied the ideal hyperparameters of each algorithm for each family, which proved to be very costly. To optimize the study for identifying the ideal classification algorithm, a tool based on a genetic algorithm was used by Lourenço and Carruba (2022). Comparing the outcomes to the metrics employed in the machine learning application test revealed that they were satisfactory. The new machine learning techniques regularly produced similar results in terms of accuracy, suggesting that this option could be quick and somewhat efficient. Thus at least concerning the new database of numbered asteroids at the Asteroid Families Portal, we will extend that previous investigation by making a direct percentage comparison with current results based on an updated asteroid database (Lourenço and Carruba, 2022; eq. (4)). The database available on the *AFP* was accessed on October 30, 2023. There were 1,052,382 registered objects in October 2023. Comparison between the $F_{P_{2020}}$ and $F_{P_{2024}}$ coefficient metrics from Carruba et al. (2020) machine learning algorithms and genetic algorithms, respectively, applied to an updated database will be given by

$$R = \frac{F_{P_{2024}} - F_{P_{2020}}}{F_{P_{2024}}}. \tag{2.5}$$

### 2.5.1 Using genetic algorithms to optimize machine learning predictions

Genetic algorithms can automatically select the best-performing method and its best parameter values, as described in subsection 1.2.3 of Chapter 1. *TPOT* (Olson et al. (2016a); Olson et al. (2016b); Le et al., 2020) is an automated machine-learning application that runs on Python and uses genetic algorithms to determine which machine-learning pipeline is best suited for a specific task. The user must supply the following essential inputs after manually cleaning their raw data: the number of generations in the genetic algorithm's training iterations, the population size (the number of individuals retained in each generation), the cross–validation CV (the number of groups into which the data sample is divided), and the random number generator seed for reproducibility (random state). The machine learning pipeline model is then automatically generated by the program.

The main goal of genetic algorithms is to find a minimum fitness function. In that function, *TPOT* might not always find a global minimum. It might not always produce the most effective machine-learning pipeline.

We reapply the machine learning algorithms from Carruba et al. (2020) and the machine learning algorithms produced by genetic algorithms to the revised database of the 21 asteroid families researched to acquire the $R$ percentual value. The updated percentual difference $R$ and the $F_P$ results for the 21 families are displayed in Table 2.4. Fig. 2.12 shows the frequency of $R$ distribution in the 21 asteroid families that were analyzed.

The histogram data and Table 2.4 indicate that the sample values are concentrated closest to the mean, indicating a consistent sample. The $R$ distribution has a moderate standard deviation. The $R$ distribution exhibits a tiny negative skew, indicating that the Carruba et al. (2020) algorithms performed marginally better. This is caused by the tail being more pronounced on the left side than on the right. Nevertheless, both machine learning outcomes are comparable, as indicated by the histogram mean ($\cong 0.024$), with little deviations on the negative side (values less than 1%) and better performances at levels above the mean.

**Table 2.4** An updated database was subjected to a table containing $F_{P_{2020}}$ and $F_{P_{2024}}$ coefficient metrics from Carruba et al. (2020) machine learning algorithms and genetic algorithms, respectively. The third column displays the percentual comparison $R$ (see Eq. (2.5)).

| Family id | $F_{P_{2020}}$ | $F_{P_{2024}}$ | $R$ |
|---|---|---|---|
| 20 Massalia | 0.869 | 0.869 | 0.000 |
| 163 Erigone | 0.671 | 0.722 | 0.071 |
| 15 Eunomia | 0.714 | 0.780 | 0.085 |
| 170 Maria | 0.765 | 0.743 | -0.030 |
| 668 Dora | 0.939 | 0.971 | 0.033 |
| 847 Agnia | 0.666 | 0.792 | 0.160 |
| 363 Padua | 0.444 | 0.469 | 0.053 |
| 1726 Hoffmeister | 0.833 | 0.880 | 0.053 |
| 410 Chloris | 0.756 | 0.662 | -0.142 |
| 808 Merxia | 0.888 | 0.922 | 0.037 |
| 128 Nemesis | 0.810 | 0.836 | 0.031 |
| 569 Misa | 0.851 | 0.852 | 0.001 |
| 221 Eos | 0.668 | 0.720 | 0.072 |
| 24 Themis | 0.825 | 0.828 | 0.004 |
| 158 Koronis | 0.744 | 0.754 | 0.013 |
| 10 Hygiea | 0.807 | 0.755 | -0.069 |
| 375 Ursula | 0.481 | 0.487 | 0.012 |
| 1040 Klumpkea | 0.562 | 0.580 | 0.031 |
| 283 Emma | 0.792 | 0.793 | 0.001 |
| 845 Naema | 0.939 | 0.962 | 0.024 |
| 490 Veritas | 0.764 | 0.809 | 0.056 |

The results show an excellent relationship between the Carruba et al. (2020) and the *TPOT* algorithms. Genetic algorithms are a secure and efficient method for discovering new members of the asteroid family, even in the face of slight declines in performance in certain families and marginal improvements in others.

## 2.6. Final remarks

The hierarchical clustering method (HCM) is typically used to observe asteroid families in the $(a, e, \sin(i))$ proper element domain. However,

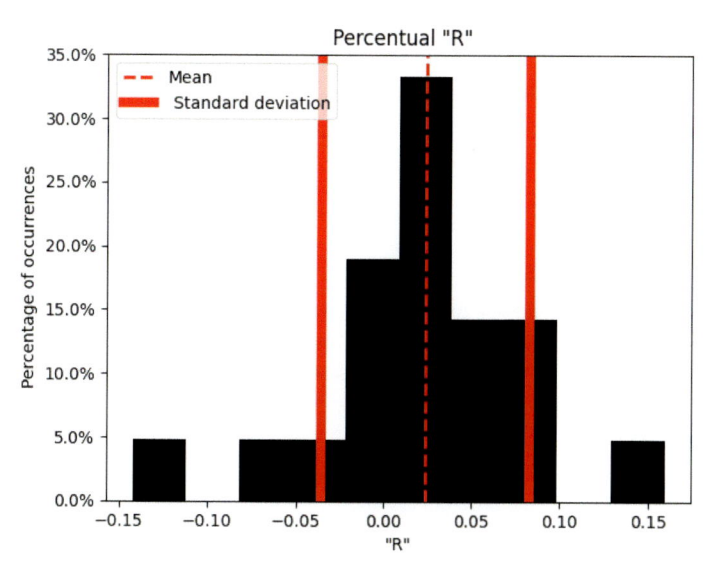

**Figure 2.12** $R$ (Eq. (2.5)) histogram. The percentual frequency of $R$ on the 21 asteroid families under study is shown graphically. The data distribution moments are as follows: skewness $\cong -0.656$, kurtosis $\cong 2.925$, mean $\cong 0.024$, standard deviation $\cong 0.060$.

in high-density locations, the approach may need help distinguishing between nearby families.

The main goal of this chapter was to provide an overview of two intriguing earlier studies (Carruba et al., 2020; Lourenço and Carruba, 2022) that used methods for potentially new asteroid family identification. These methods involved applying machine-learning classification algorithms to the orbital distribution in proper elements of 21 known family constituents. The precision values of the genetic machine learning algorithms have consistently been comparable to the same evaluative metrics used in the prior machine learning application study, indicating that this alternative technique can be adequately efficient and quick.

These strategies could be advantageous when the Vera C. Rubin surveys begin to detect millions of new asteroids following launch (Jones et al., 2015).

## 2.7. Code availability

Jupyter notebooks containing the codes used in this work are available as "classification algorithms" at https://anaconda.cloud/share/notebooks/c922e783-e91e-4bbf-8ed5-1c3fe3211512/overview and "genetic algorithms" at https://anaconda.cloud/share/notebooks/6d89fec4-a7a7-42aa-9d16-40e984e12df3/overview.

Additionally, the notebooks are in the GitHub repository available at https://solar-system-ml.github.io/book/chapter2/ML_algorithms_classification/, see Chapter 2.

## Acknowledgments

We thank Bryce T. Bolin for his careful review and constructive remarks. RD acknowledges the Financier of Studies and Projects (FINEP, grant 0527/18) and the São Paulo Research Foundation (FAPESP, grant 2016/024561-0).

## References

Carruba, V., Aljbaae, S., Domingos, R.C., Lucchini, A., Furlaneto, P., 2020. Machine learning classification of new asteroid families members. Monthly Notices of the Royal Astronomical Society 496, 540–549.

Carruba, V., Domingos, R.C., Nesvorný, D., Roig, F., Huaman, M.E., Souami, D., 2013. A multidomain approach to asteroid families' identification. Monthly Notices of the Royal Astronomical Society 433, 2075–2096.

Carruba, V., Michtchenko, T.A., 2007. A frequency approach to identifying asteroid families. Astronomy & Astrophysics 475, 1145–1158.

Carruba, V., Michtchenko, T.A., 2009. A frequency approach to identifying asteroid families - II. Families interacting with nonlinear secular resonances and low-order mean-motion resonances. Astronomy & Astrophysics 493, 267–282.

Jones, R.L., Jurić, M., Ivezić, Z., 2015. Asteroid discovery and characterization with the large synoptic survey telescope. Proceedings of the International Astronomical Union 10 (S318), 282–292. https://doi.org/10.1017/S1743921315008510.

Le, T.T., Fu, W., Moore, J.H., 2020. Scaling tree-based automated machine learning to biomedical big data with a feature set selector. Bioinformatics 36, 250–256.

Lourenço, M.V.F., Carruba, V., 2022. Genetic optimization of asteroid families' membership. Frontiers in Astronomy and Space Sciences 9. https://doi.org/10.3389/fspas.2022.988729.

Milani, A., Cellino, A., Knežević, Z., Novaković, B., Spoto, F., Paolicchi, P., 2014. Asteroid families classification: exploiting very large datasets. Icarus 239, 46–73.

Nesvorný, D., Brož, M., Carruba, V., 2015. Identification and dynamical properties of asteroid families. Asteroids IV, 297–321.

Novaković, B., Vokrouhlický, D., Spoto, F., Nesvorný, D., 2022. Asteroid families: properties, recent advances, and future opportunities. Celestial Mechanics & Dynamical Astronomy 134.

Olson, R.S., Bartley, N., Urbanowicz, R.J., Moore, J.H., 2016a. Evaluation of a tree-based pipeline optimization tool for automating data science. In: Proceedings of the Genetic and Evolutionary Computation Conference 2016, pp. 485–492. https://doi.org/10.1145/2908812.2908918.

Olson, R.S., Urbanowicz, R.J., Andrews, P.C., Lavender, N.A., Kidd, L., Moore, J.H., 2016b. Automating biomedical data science through tree-based pipeline optimization. In: Squillero, G., Burelli, P. (Eds.), Applications of Evolutionary Computation. EvoApplications 2016. In: Lecture Notes in Computer Science, vol. 9597, pp. 123–137. https://doi.org/10.1007/978-3-319-31204-0_9.

Pedregosa, F., Varoquaux, G., Gramfort, A., Michel, V., Thirion, B., Grisel, O., Blondel, M., Prettenhofer, P., Weiss, R., Dubourg, V., Vanderplas, J., Passos, A., Cournapeau,

D., Brucher, M., Perrot, M., Duchesnay, E., Louppe, G., 2011. Scikit-learn: machine learning in python. Journal of Machine Learning Research 12, 2825–2830.

Quinlan, J.R., 1986. Induction of decision trees. Machine Learning 1, 81–106.

Radović, V., Novaković, B., Carruba, V., Marčeta, D., 2017. An automatic approach to exclude interlopers from asteroid families. Monthly Notices of the Royal Astronomical Society 470, 576–591. https://doi.org/10.1093/mnras/stx1273.

Swamynathan, M., 2017. Mastering Machine Learning with Python in Six Steps. Apress. https://api.semanticscholar.org/CorpusID:1825964.

Zappalà, V., Bendjoya, P., Cellino, A., Farinella, P., Froeschlé, C., 1995. Asteroid families: search of a 12,487-asteroid sample using two different clustering techniques. Icarus 116, 291–314. https://doi.org/10.1006/icar.1995.1127.

Zappalà, V., Cellino, A., Farinella, P., Knezevic, Z., 1990. Asteroid families. I. Identification by hierarchical clustering and reliability assessment. The Astronomical Journal 100, 2030. https://doi.org/10.1086/115658.

CHAPTER THREE

# Asteroids in mean-motion resonances

**Evgeny Smirnov**
Belgrade Astronomical Observatory, Belgrade, Serbia

## 3.1. Introduction

The role of mean-motion resonances (MMRs) in the Solar System is a fundamental topic in celestial mechanics and the study of asteroid dynamics. The resonances occur when there is a simple integer ratio between the orbital periods of two or more celestial bodies, leading to gravitational interactions that can significantly influence their orbits (Murray and Dermott, 1999). Understanding these resonances is crucial for predicting the long-term evolution of asteroid orbits and their stability.

The essential role of mean-motion resonances in the dynamics of asteroids became evident when Daniel Kirkwood found gaps, which are now called *Kirkwood gaps*, in the distribution of the semimajor axes of the asteroids' orbits in the main belt (see Fig. 3.1). Further research in the field has attributed these gaps to the gravitational influence of Jupiter: the deepest minima correspond to the two-body mean-motion resonances with Jupiter with ratios 3/1 ($a \approx = 2.50$ au), 5/2 ($a \approx = 2.83$ au), 7/3 ($a \approx = 2.95$ au), and 2/1 ($a \approx = 3.28$ au) (Murray and Dermott, 1999).

However, the physical explanation of the Kirkwood gaps was unknown for the next hundred years. The analysis of resonances in general performed by Chirikov (1979) showed that the interaction of resonances can lead to *dynamical chaos*. Wisdom (1982) demonstrated this effect for the two-body mean-motion resonances with Jupiter and explained the origin of Kirkwood gaps: due to the chaotic behaviors of the orbits of the asteroids that are close to the gaps, the eccentricity could sporadically fluctuate and asteroids becomes Mars-crossers and hence, could be ejected from the corresponding gap.

The study of dynamical chaos in the Solar System has been further expanded by Milani and Nobili (1992) and Milani et al. (1997). They introduced the concept of stable chaos: the study identifies a significant number of asteroids in the main belt that exhibit chaotic behavior (and have short

*Machine Learning for Small Bodies in the Solar System*
https://doi.org/10.1016/B978-0-44-324770-5.00008-8

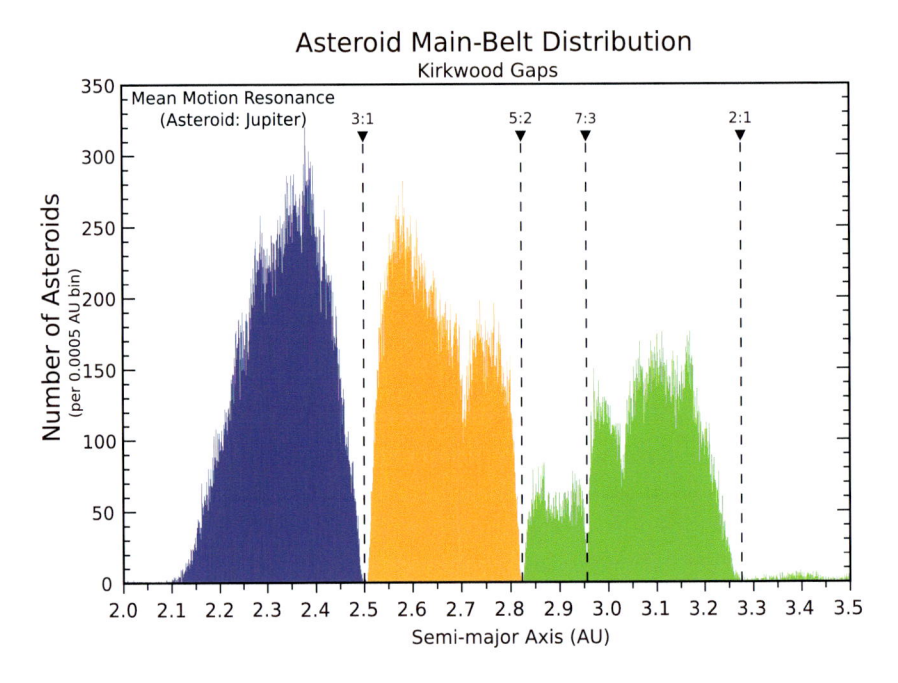

**Figure 3.1** Kirkwood gaps. The source of the image: Wikipedia (by NASA).

Lyapunov times) but maintain stable orbits over long time scales. The authors suggest that this is related to the fact that these asteroids are trapped in high-order mean-motion resonances.

The paper by Gallardo (2006) presents a comprehensive study of mean-motion resonances (MMRs) in the Solar System. It introduces a numerical method to calculate the strength of these resonances and provides an extensive atlas of MMRs constructed using this method. The author reveals several unexpectedly strong resonances and identifies asteroids trapped in these resonances.

The studies of mean-motion resonances have been further expanded to include three-body resonances, which involve two planets and an asteroid. The three-body resonances are more complex and numerous than the two-body resonances, as they can be formed by various combinations of the mean frequencies of three bodies (Nesvorný and Morbidelli, 1998; Murray et al., 1998). Nesvorný and Morbidelli (1998) even suggested that "the three-body mean motion resonances seem to be the main actors structuring the dynamics in the main asteroid belt" (Nesvorný and Morbidelli, 1998, p. 3032).

Shevchenko (2006) studied three-body MMRs and introduced an analytical method for estimating the maximum Lyapunov exponents of asteroidal motion in two scenarios: motion near ordinary or three-body mean motion resonances with planets, and motion in highly eccentric orbits with moderate planetary encounters. The methodology uses the general separatrix map theory and compares the analytical estimates of Lyapunov times with those obtained from numerical integration.

Smirnov and Shevchenko (2013) performed a massive identification of three-body mean-motion resonances with Jupiter and Saturn and two-body mean-motion resonances with Jupiter. They found that $\approx 4.4$ per cent and $\approx 1.1$ per cent of asteroids in the main belt are trapped in three-body and two-body MMRs, respectively. These results provide an argument for the significant role of three-body resonances in the dynamics of asteroids, mentioned by Nesvorný and Morbidelli (1998). This study was expanded further to three-body (Smirnov, 2017) and two-body (Smirnov and Dovgalev, 2018) resonances with all planets or their combinations.

Lastly, Tsiganis (2010) offers a comprehensive review of the dynamics of small bodies in the Solar System. This work synthesizes current knowledge on the subject, emphasizing the dynamical processes that govern the movement and interaction of minor celestial bodies, such as asteroids and comets. These studies collectively enhance our understanding of the complex dynamical systems that govern the Solar System's evolution.

For further reading on mean-motion resonances and their role in the dynamics of asteroids, the works by Murray and Dermott (1999), Morbidelli (2002), and Shevchenko (2020) are recommended. They offer a modern perspective on celestial mechanics, focusing specifically on aspects of Solar System dynamics and its mathematical underpinnings.

### 3.1.1 Machine-learning methods

Machine-learning methods are increasingly being applied to the study of asteroids and their dynamics. In this chapter, we focus on the application of machine-learning techniques to the identification of asteroids trapped in mean-motion resonances. The classical method of identifying resonant asteroids requires numerical integration, which is time-consuming. The application of machine-learning methods to this problem could significantly improve the efficiency of the identification procedure.

Smirnov and Markov (2017) focused on the identification of asteroids trapped in three-body mean-motion resonances. The authors applied different supervised machine-learning classifiers to identify resonant asteroids.

They used the synthetic proper elements of asteroids as input variables and the binary classification (resonant or nonresonant) as the output. They were able to achieve high accuracy, with the best results obtained using the k-nearest neighbors (kNN) algorithm.

Further development was performed by Carruba et al. (2021). In this study, the authors explored the application of artificial neural networks (ANNs) and random forest methods to the same problem. They focused on classifying asteroid orbits in the M1:2 mean-motion resonance with Mars. The key aspect of this work included the successful use of ANNs to classify images, which is a necessary step to identify resonant behavior. Previous studies used more traditional methods for such analysis, either manual (i.e., Nesvorný and Morbidelli, 1998) or automated (i.e., Smirnov and Markov, 2017). However, the latter had very limited accuracy, $\approx 80$ per cent. The authors improved their model later (see Carruba et al. (2022) and Section 3.6).

Another paper related to the application of machine-learning methods to MMRs is Carruba et al. (2023). The study focuses on the challenge of imbalanced datasets, which is the case for resonant asteroids, where the number of resonant asteroids is much smaller than nonresonant ones. The research provided some suggestions in terms of efficient approaches for such a problem.

To sum up: The role of mean-motion resonances in the dynamics of asteroids is well established. However, the classical method of identification of resonant asteroids is time-consuming and not cost-effective. The application of machine-learning methods to this problem could significantly improve the efficiency of the identification procedure. Let us go deeper into the classical method first and find out the details of the challenges it faces and how machine-learning methods could be applied to resolve them.

## 3.2. Mean-motion resonances

As noted earlier, a mean-motion resonance represents the commensurability between the mean frequencies of several bodies:

$$\sum_{i=1}^{N} m_i \dot{\lambda}_i \approx 0, \qquad (3.1)$$

where $N$ is the number of bodies, $\lambda_i$ are mean longitudes of all involved bodies, and $m_i$ are integers. A resonance can include multiple bodies. For asteroids, the most well-known classes are two-body mean-motion resonances involving an asteroid and a planet and three-body mean-motion resonances involving an asteroid and two planets.

To universalize, let us denote a resonance in the following way:

$$m_0 P_0 m_1 P_1 ... m_N,$$

where $m_i$ are integers with their signs, and $P_i$ are the first letters of the planets' names (R is used for Mercury, due to the ambiguity with Mars). For example, 2J−1 means the two-body MMR with Jupiter described by the commensurability $2\lambda_J - \lambda$; the notation 4J−2S+1 means the three-body MMR with Jupiter and Saturn followed by the commensurability $4\lambda_J - 2\lambda_S + \lambda$.

Note that the integers $p_i$ are omitted here. Formally, the designation $m_0 P_0 m_1 P_1 ... m_N$ represents not just one resonance, but *a multiplet* (Nesvorný and Morbidelli, 1998; Shevchenko, 2020). The separation between subresonances within a multiplet is of order $10^{-4}$ AU and is related to the different values of $p_i$ (Nesvorný and Morbidelli, 1998). Hereinafter, let us denote the multiplet by an MMR.

One might ask: how to identify whether an asteroid is trapped in a resonance? It is clear that the commensurability between the frequencies cannot be exact, due to observational and other errors. To resolve this, astronomers use a special parameter called *resonant argument* (Murray and Dermott, 1999; Morbidelli, 2002). For the planar case, it is a linear combination of angular variables of all bodies:

$$\sigma = \sum_{i=1}^{N} \lambda_i m_i + \varpi_i p_i, \tag{3.2}$$

where $N$ is the number of bodies, $\lambda_i$ and $\varpi_i$ are mean longitudes and longitudes of periapsis of all involved bodies, $m_i$ and $p_i$ are integers.

If the resonant argument (3.2) librates similarly to the libration of the pendulum, the system is in a resonance (Chirikov, 1979; Smirnov and Shevchenko, 2013; Shevchenko, 2020). If it circulates, the asteroid is not trapped in a resonance. The state bordering between libration and circulation is called *a separatrix*.

The oscillations of the resonant argument are analyzed for the given period of time, depending on the expected period of librations. For the asteroids in the main belt, it varies from tens of thousands to hundreds of thousands of years (Nesvorný and Morbidelli, 1998; Murray and Dermott, 1999; Smirnov and Shevchenko, 2013).

During the period of analysis, there are several possible cases. Sometimes, a resonant argument could librate all the time. Such a behavior is called *a pure libration* (see Subplot 1 on Fig. 3.2). If the resonant argument librates a significant time (but not all the time), the behavior is called *a transient libration* (see Subplot 2 on Fig. 3.2). In other cases, the asteroid is out of the resonance.

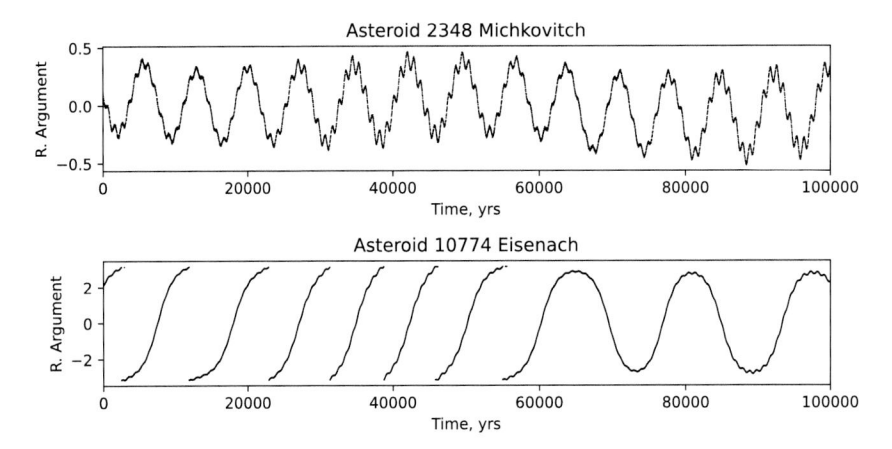

**Figure 3.2** Different types of librations: (1) the pure resonant asteroid (2348) Michkovitch in the resonance $4J-2S-1$; (2) the transient resonant asteroid (10774) Eisenach in the resonance $4J-2S-1$.

Formally, the number of resonances is infinite, due to the infinity of integers. Moreover, it is known that high-order resonances play an essential role in the dynamics of asteroids (Lemaître, 1984; Holman and Murray, 1996; Milani et al., 1997; Tsiganis, 2002; Gallardo, 2006). Therefore for numerical calculations, researchers have to add some limitations.

Firstly, without loss of generality, let us assume that the first integer $m_1$ is always positive ($m_1 > 0$) and $\gcd\{m_i\} = 1$, due to Eq. (3.1).

Secondly, the integers should satisfy the D'Alembert rule (for more details, see Morbidelli (2002)):

$$\sum_{i=1}^{N} m_i + p_i = 0. \tag{3.3}$$

Let us introduce here another important parameter—*resonant order*, denoted by $q$. It affects the strength of a resonance. Mathematically, it is an absolute value of all $m$-integers (Murray and Dermott, 1999).

$$q = \left| \sum_{i=1}^{N} m_i \right|. \tag{3.4}$$

Thirdly, the order of a resonance is limited by a positive integer:

$$q = \left| \sum_{i=1}^{N} m_i \right| < x, x \in \mathbb{N}. \tag{3.5}$$

Fourthly, the absolute values of integers in Eq. (3.2) should be limited. There are several approaches used (Nesvorný and Morbidelli, 1998; Smirnov and Shevchenko, 2013; Smirnov and Dovgalev, 2018). The first approach suggests limiting each of the integers either by some values:

$$|m_i| < y_i, y_i \in \mathbb{N}. \tag{3.6}$$

Another option is to limit the sum of the absolute values of integers:

$$\sum_{i=1}^{N} |m_i| < y, y \in \mathbb{N}. \tag{3.7}$$

Based on the four groups of limitations above, it is possible to perform an identification whether an asteroid is trapped in a resonance.

## 3.3. Identification of resonant asteroids

To identify whether an asteroid is trapped in a mean-motion resonance, one should perform several steps. Firstly, it is necessary to determine what are the resonances in which the asteroid could be trapped.

From the definition, it follows that a resonance represents a commensurability between the orbital periods, and hence semimajor axes.

For the given two-body resonance with the planet P, the formula for the resonant value of the semimajor axis $a_r$ follows from the third Kepler's law and Eq. (3.1):

$$a_r \approx (1 + m_P)^{-1/3} \left( \frac{m_1}{-m_2} \right)^{2/3}, \tag{3.8}$$

where $m_P$ is the mass of the planet in solar units.

The case with three-body mean-motion resonances is more complicated. To calculate the resonant value of the semimajor axis, one can use a simple formula provided by Gallardo (2014) (under the assumption of unperturbed Keplerian motions) or a more accurate iterative process developed by Smirnov and Shevchenko (2013).

However, as mentioned above, the commensurabilities are never exact, due to the observational errors. Thus even if a real asteroid's value of semimajor axis is close to the resonant one, one cannot conclude that the asteroid is trapped in the resonance. To claim that, it is necessary to analyze the behavior of the resonant angle.

This procedure is called *dynamical identification* (Smirnov and Shevchenko, 2013). It has the following steps:

1. Integrate the equation of motion for the given period.
2. Calculate the values of the resonant argument.
3. Identify whether or not the resonant argument librates.

Note that there might be a false-positive identification when the libration of the resonant argument is not related to a mean-motion resonance. To check this, there should be similar librations of the semimajor axis: the frequencies of oscillations for these variables should match (Smirnov, 2017; Smirnov and Dovgalev, 2018).

The most difficult part of the dynamical identification is related to the identification of librations of the resonant argument. A human can easily state whether the resonant argument librates. Such an approach was used earlier, i.e., by Murray and Holman (1997) and Nesvorný and Morbidelli (1998). However, when there are multiple objects and multiple resonances, it becomes a time-consuming task.

Smirnov and Shevchenko (2013) suggested a simple automatic algorithm that analyzed how often the resonant argument crossed the limits $(-\pi$ and $\pi)$. This approach resolves the problem. However, its disadvantage is in its accuracy, which is about 80 per cent.

To improve the accuracy of the identification procedure, Smirnov et al. (2017) suggested the analysis of periodograms of a resonant argument and semimajor axis and a cross-periodogram between a resonant argument and semimajor axis. This method improved the accuracy by up to $\approx 90$ per cent. It is available in a developed Python package called `resonances` (Smirnov, 2023).

The outcomes of the dynamical identification performed by the package `resonances` are shown in Fig. 3.3, where vertical gray lines in Panels 4 and 5 illustrate the range of peak intervals. The green and red lines denote standard critical thresholds set at 0.10 and 0.05, respectively. Frequencies surpassing the green line threshold are deemed to be accurately identified, whereas those falling beneath the red line are considered likely to be false positives.

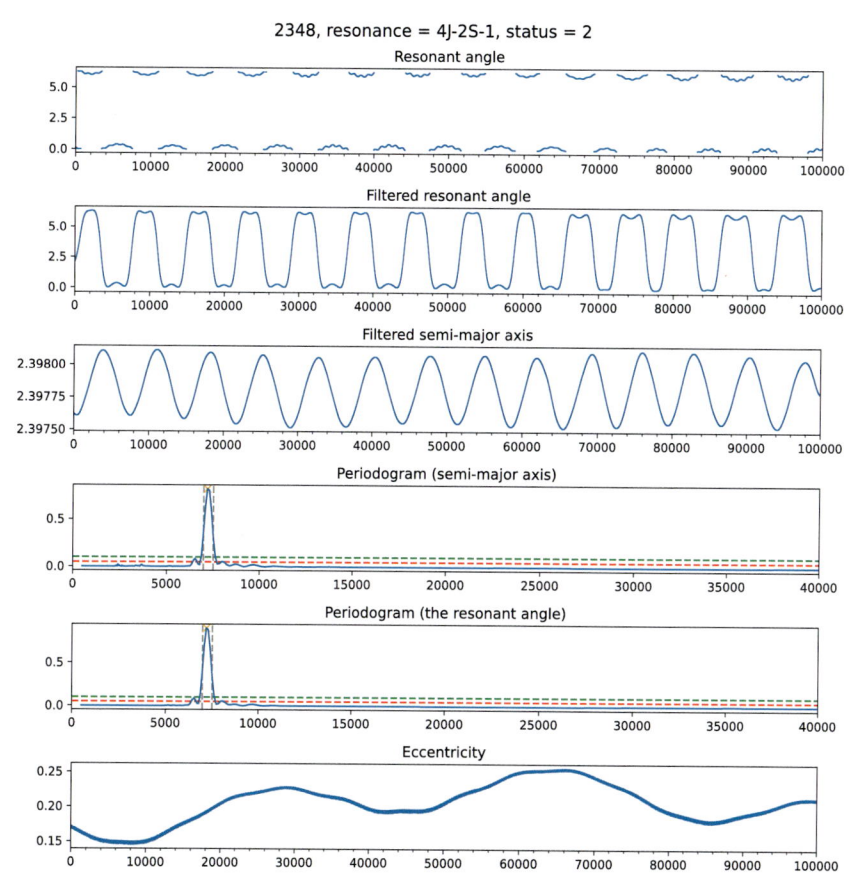

**Figure 3.3** The plot is composed of several panels showcasing various aspects of the dynamics of the asteroid 2348 Michkovitch trapped in the resonance 4J−2S−1: (1) displays the resonant angle, (2) shows the resonant angle after filtering, (3) illustrates the filtered semimajor axis, (4–5) present periodograms for both the semimajor axis and the resonant angle, and (6) depicts the eccentricity.

### 3.3.1 Results

Smirnov et al. (2017) and Smirnov and Dovgalev (2018) performed the dynamical identification of the resonant asteroids in the Solar System. They found that $\approx 5$ per cent of them are trapped in the two-body mean-motion resonances and $\approx 14$ per cent in the three-body mean-motion resonances. A lot of the resonant asteroids are trapped in the MMRs with Jupiter ($\approx$ 54 per cent of the total amount of the two-body resonant asteroids) and Jupiter and Saturn ($\approx 29$ per cent of the total amount of the three-body resonant asteroids). The top twelve most populated resonances are shown in Table 3.1.

**Table 3.1** Top twelve most populated mean-motion resonances in the Solar System. T+P means the number of transient + pure resonant asteroids, P represents the number of pure resonant asteroids, $a_{res}$ stands for the resonant value of the semi-major axis.

| Resonance | $a_{res}$ | T+P | P |
|---:|---|---|---|
| 1J−1 | 5.2008 | 4275 | 4273 |
| 1M−2 | 2.4187 | 2539 | 523 |
| 4J−6U−1 | 2.4189 | 2163 | 11 |
| 3J−2 | 3.9690 | 1954 | 1856 |
| 6S−3N−1 | 3.0748 | 1488 | 21 |
| 5J−2S−2 | 3.1747 | 1396 | 307 |
| 8J−3 | 2.7045 | 1268 | 79 |
| 3J−2S−1 | 3.0801 | 1267 | 251 |
| 4J−2S−1 | 2.3981 | 1233 | 1050 |
| 3J−1S−1 | 2.7530 | 1059 | 397 |
| 4J−3S−1 | 2.6235 | 1006 | 195 |
| 11J−5 | 3.0746 | 956 | 42 |

Though the identification process for asteroids trapped in MMRs is accurate, it is particularly time-consuming for three-body MMRs, due to the large number of such resonances. The dynamical identification of one asteroid takes about $1 - 10$ minutes on a modern computer, depending on the integrator used and the number of MMRs considered. It is not a problem for a single asteroid, but it becomes a challenge when one needs to analyze hundreds of thousands of asteroids. Thus another approach could be helpful.

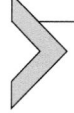

## 3.4. Supervised learning for resonance identification

### 3.4.1 Challenges of the classical method

With the growing number of known asteroids exceeding 1,000,000 objects, the classical approach (Nesvorný and Morbidelli, 1998; Smirnov and Shevchenko, 2013; Smirnov and Markov, 2017) of identifying mean-motion resonances in the main asteroid belt faces significant computational challenges. A possible solution to these challenges could be machine-learning techniques.

The regular method for the dynamical identification of MMRs requires numerical integration, often spanning over 100,000 years. Also, for each of the asteroids, there is a possibility of being trapped in multiple resonances, each with several subresonances. This necessitates the computation of various resonant arguments. Such an approach is resource-intensive and impractical for statistical tasks or simulations.

The identification of MMRs is further complicated when some Keplerian elements of a celestial body are unknown or poorly defined, i.e., in the case of exoplanetary systems. The absence of complete and accurate orbital data presents a significant challenge to the classical approach, necessitating more innovative methods for resonance identification.

The determination of whether a resonant argument librates requires also either human analysis or automated procedures. Manual analysis is time-consuming and not cost-effective, and automated methods are prone to errors (Smirnov, 2023). Combining a machine-learning approach with the traditional one has the potential to enhance the accuracy and efficiency of resonance identification, reducing the need for extensive human involvement and mitigating the shortcomings of purely automated systems.

### 3.4.2 Supervised learning

The natural way of applying machine-learning techniques to the resonance identification problem is to use supervised learning. The general form of the task is the following: the input is a set of features $\mathbf{X}$ representing an asteroid (i.e, proper or osculating elements or physical parameters); the output is the result of the classification. In the simplest form, it could be boolean: 1 if the asteroid is trapped in a resonance and 0 otherwise. One might envision a more complex output, where there are several statuses: 2: pure libration, 1: transient libration, 0: no libration, $-1$: requires human attention. Smirnov and Markov (2017) have shown that the best results could be achieved when there are less inputs and the output is boolean.

Note that a Boolean output does not imply that it is impossible to have a full classification of the type of libration: instead of using one classifier, one can create a group of classifiers. The first one will determine whether the asteroid is resonant. The second one will identify what is the type of libration. Several models in this case provide better accuracy (Smirnov and Markov, 2017).

Also, it is essential to determine what "better" means in the context of this problem. Classical machine learning typically uses several metrics: accuracy, precision, recall, or F-score (Ivezić et al., 2020; Pedregosa et al., 2011; Olson and Delen, 2008). These metrics have been discussed in Chapter 1. In our case, the most important metric is recall. It is preferable to have some false positives, which can be eliminated by the classical method (since there are not too many, it is not time-consuming), rather than risk failing to identify some resonant asteroids.

Now let us outline the details of the methods of classical machine learning that could be used to resolve the problem. A brief introduction to these classifiers is performed in Chapter 1, Section 1.2. However, for this chapter, we will examine several distinct supervised machine learning (ML) techniques in more detail.

### 3.4.3 Designations

The input variables are symbolized by $\mathbf{X}$, with individual components indicated as $X_j$ for vector $\mathbf{X}$. In our analysis, these inputs primarily include the synthetic proper elements of an asteroid ($a$, $e$, $\sin i$) and their derivatives ($n$: proper mean motion). The output $G$ is binary, with $G \in \{-1, 1\}$. Here, $-1$ signifies nonresonant motion, whereas 1 indicates resonant motion of an asteroid. The training dataset consists of pairs $\mathbf{X^i} = (\mathbf{x_i}, g_i)$, $i = 1 \dots N$. Predicted outcomes are represented by $\hat{G}$.

The primary goal is to accurately predict the output $G$, symbolized as $\hat{G}$, for a given set of input vectors $\mathbf{X^i}$ and a predefined training dataset $(\mathbf{x_i}, g_i)$, $i = 1 \dots N$ (Hastie et al., 2009).

Now, let us discuss what are the ML classifiers employed in this chapter.

### 3.4.4 k-nearest neighbors

The *k-nearest neighbors (kNN)* method assesses the proximity of an unknown asteroid to known resonant and nonresonant asteroids in the feature space. It then classifies the object based on the majority class among its k-nearest neighbors. The kNN classification of $\hat{G}$ is given by

$$\hat{G}(x) = \frac{1}{k} \sum_{\mathbf{x_i} \in N_k(x)} w_i(x)g_i, \qquad (3.9)$$

where $N_k(x)$ signifies the neighborhood of $x$, determined by the k-nearest objects $x_i$ in the training set, and $w_i(x)$ is the weighting function (Hastie et al., 2009). To find the *nearest* objects, distance calculation is necessary, represented by the metric

$$\rho : \mathbf{X} \times \mathbf{X} \to [0, +\infty). \qquad (3.10)$$

The commonly used metric is Euclidean:

$$\rho(\mathbf{x_i}, \mathbf{x_j}) = \sqrt{\sum_k (x_i^{(k)} - x_j^{(k)})^2}. \qquad (3.11)$$

Another commonly used metric is the cosine distance, which is particularly useful in high-dimensional spaces. It is computed as

$$\rho(\mathbf{x_i}, \mathbf{x_j}) = 1 - \frac{\mathbf{x_i} \cdot \mathbf{x_j}}{||\mathbf{x_i}||_2 ||\mathbf{x_j}||_2}. \qquad (3.12)$$

Though the weighting function could be very simple like a geometric sequence, it could also be distance-based, known as the *Parzen window* method (Parzen, 1962), defined as

$$w_i(\mathbf{x}) = K\left(\frac{\rho(\mathbf{x}, \mathbf{x_i})}{h}\right), \qquad (3.13)$$

where $K(z)$ is the kernel function, and $h$ is the bandwidth. Smirnov and Markov (2017) found that the kNN method with a Parzen window, where the width equals the distance to the farthest neighbor, provided the best overall result.

## 3.4.5 Decision tree

*Decision trees* classify asteroids by creating a tree-like model of decisions. The decision at each node of the tree is based on the value of a certain input feature, leading to a classification at the leaves (Bowles, 2015).

The decision tree method segments the feature space into rectangles. Initially, the space is divided into two regions, choosing a split point for the best fit. This process is recursively applied to the resulting subregions. For a node $m$ and a region $R_m$ containing $N_m$ objects, the class proportion $\hat{p}_{mk}$ is

$$\hat{p}_{mk} = \frac{1}{N_m} \sum_{\mathbf{x_i} \in R_m} I(g_i = k). \tag{3.14}$$

Classification of the node $m$ is then determined as the majority class $\hat{G}(m) = \arg\max_j \hat{p}_{mj}$.

The overall error or *impurity* could be estimated as the following:

- Gini index: $\sum_{i \neq j} \hat{p}_{mi} \hat{p}_{mj}$, simplifying to $2p(1-p)$ for two classes.
- Entropy: $-\sum_{i=1}^{G} \hat{p}_{mi} \log \hat{p}_{mi}$, reducing to $-p \log p - (1-p) \log(1-p)$ for binary classification.

Building decision trees involves minimizing these impurity measures (Hastie et al., 2009).

### 3.4.6 Logistic regression

Logistic regression is a statistical model that estimates the probability of a binary outcome based on one or more predictor variables. In our case, the outcome is whether or not an asteroid is trapped in a resonance. The probability that a given asteroid $G$ is resonant is modeled as

$$P(\hat{G} = 1 | \mathbf{X}) = \frac{1}{1 + e^{-(\beta_0 + \sum_{i=1}^{n} \beta_i \mathbf{x_i})}},$$

where $P(\hat{G} = 1 | \mathbf{X})$ is the probability that asteroid $G$ is resonant, given its features, and $\beta_i$ are the model coefficients computed from the test data. The decision rule for classification is given by

$$\hat{G} = \begin{cases} +1, & \text{if } P(\hat{G} = 1 | \mathbf{X}) > 0.5 \\ -1, & \text{otherwise.} \end{cases}$$

This rule assigns an asteroid to the resonant class (+1) if the predicted probability is greater than 0.5, and to the nonresonant class (–1) otherwise.

The logistic regression model is trained using maximum likelihood estimation, which seeks to find the coefficients $\beta_i$, which maximize the likelihood of the observed training data $\mathbf{X}^i$.

### 3.4.7 Ensemble learning methods

In this section, we will explore other types of machine-learning methods, which are called *ensemble learning methods*. They combine multiple classifiers to improve the overall performance of the model. The two ensemble methods used in this chapter are *gradient boosting* and *random forest* (Hastie et al., 2009; Breiman, 2001).

*Gradient boosting* classifies data through a combination of multiple weighted classifiers. Decision trees are often used (Hastie et al., 2009). The method relies on the "voting" mechanism of classifiers, with the final classification determined by a linear combination of these votes, refined through gradient descent (Bowles, 2015; Hastie et al., 2009). This approach significantly enhances the effectiveness of even weak classifiers.

An example of gradient boosting is the "AdaBoost.M1" algorithm, which involves multiple weak classifiers $\hat{G}_m(x)$, with $m = 1 \dots M$ and assumes that an output could be one of two values: $\{-1, 1\}$. The result of gradient boosting is then

$$\hat{G}(\mathbf{x}) = \text{sign}\left(\sum_{m=1}^{M} w_m \hat{G}_m(\mathbf{x})\right), \tag{3.15}$$

where $w_m$ are the weights calculated and assigned by the boosting algorithm. Training data $(\mathbf{x_i}, g_i)$ are dynamically weighted, with misclassified objects receiving increased weights in subsequent iterations, focusing new classifiers on previously misclassified objects (Freund and Schapire, 1997; Hastie et al., 2009).

**Random forest** is a machine-learning model, which uses a multitude of decision trees during training and outputs the class that is the mode of the classes (classification) or mean prediction (regression) of the individual trees. In our case, the outcome is whether or not an asteroid is trapped in a resonance.

Let $T$ be the total number of trees in the forest, and $h(\mathbf{x}, \Theta_k)$ be the decision tree, where $\Theta_k$ are the parameters of the tree learned in the training process. The output of the random forest model for an input vector $\mathbf{x}$ is given by

$$\hat{G}(\mathbf{x}) = \frac{1}{T} \sum_{k=1}^{T} h(\mathbf{x}, \Theta_k). \tag{3.16}$$

In the case of classification, the final class is typically taken as the mode of the classes output by individual trees (Breiman, 2001).

### 3.4.8 Naïve Bayes

Naïve Bayes is a probabilistic machine-learning model that is used for classification tasks. The idea of the model is the assumption of independence between every pair of features (Rish et al., 2001). Given a class variable $y$ and a dependent feature vector $x_1$ through $x_n$, Bayes' theorem states the following relationship:

$$P(y|x_1, ..., x_n) = \frac{P(y)P(x_1, ..., x_n|y)}{P(x_1, ..., x_n)}. \tag{3.17}$$

Using the naive independence assumption that

$$P(x_i|y, x_1, ..., x_{i-1}, x_{i+1}, ..., x_n) = P(x_i|y), \tag{3.18}$$

for all $i$, this relationship is simplified to

$$P(y|x_1, ..., x_n) = \frac{P(y) \prod_{i=1}^{n} P(x_i|y)}{P(x_1, ..., x_n)}. \tag{3.19}$$

Despite its simplicity, Naïve Bayes can be accurate in many scenarios, including asteroids classification.

## 3.5. Experiments with supervised learning

When one designs a study that is aimed to identify resonant asteroids by using machine-learning techniques, there are the following questions:

1. What method (k-nearest neighbors, decision tree, logistic regression, gradient boosting, random forest, or naïve Bayes) has the best performance?

2. What are the features (parameters) of an asteroid or its orbit that should be included in a model to achieve the highest values of recall and $F_1$ score?

3. How many asteroids should be included in a test set?

4. What are the parameters of the best method?

Let us talk about the last question. There are multiple ways of how to fine-tune the hyperparameters of a model. The most common approaches are the following: (1) the grid search, (2) manual brute-force search, (3) genetic algorithms (Carruba et al., 2021; Smirnov, 2024). The first approach is the simplest and fastest one. Furthermore, the `scikit-learn` library has already a built-in module GridSearchCV that supports almost all the classifiers discussed earlier. However, it has some limitations in flexibility and does not allow varying easily the size of the training set or the features used.

The manual brute-force search is the most flexible one. It allows the user to vary any parameter of the model and the experiment itself. Thus the results obtained by this method should be the best. However, it is time-consuming and requires a lot of computational resources.

Genetic algorithms are the most advanced approach. They mimic the process of natural selection applied to the classifiers. More details are available in Chapter 1. Within the problem of the identification of resonant asteroids, the genetic algorithms can find the optimal values of hyperparameters quicker than the manual brute-force search. However, they also require some fine-tuning of their hyperparameters and, sometimes, could be stuck in a local minimum. Lourenço and Carruba (2022) implemented this approach for asteroid families and demonstrated that the results were comparable with those that were obtained by the manual selection. The authors used the *Tpot Python* library (Le et al., 2020).

In this chapter, we will use mostly the manual brute-force search.

### 3.5.1 Comparing the performance of different methods

Firstly, let us eliminate the methods that have the worst performance. To do this, we will use the following experimental setup:

1. The training set will contain the first 50 resonant asteroids and the first 17, 000 nonresonant asteroids.
2. Following Smirnov and Markov (2017), we will use the k-nearest neighbors method with $k = 13$ and $p = 3$ and gradient boosting with $n = 10$.
3. For decision tree, let us set the value of depth automatically and use the Gini index.
4. For random forest, we will use $n = 10$.
5. As features, we will use the proper elements of an asteroid: semimajor axis ($a$), eccentricity ($e$), inclination ($i$), and mean motion ($n$). To resolve the problem of underfitting or overfitting for the first iteration, we will use two options: (1) all features as inputs and (2) mean motion as the only input.
6. The test set will contain the next 50, 000 asteroids.

The results of the experiment are presented in Table 3.2. They indicate that the best performance ($F_1$ score) is achieved by the naïve Bayes method, gradient boosting, and k-nearest neighbors. The highest value of recall is achieved by the k-nearest neighbors and naïve Bayes and the best precision by gradient boosting and random forest.

**Table 3.2** Explanation of the columns: Method: method used; Feat.: features used (all means that $a$, $e$, $n$, and $sinI$ are used); TP: true positives; FP: false positives; TN: true negatives; FN: false negatives; Acc.: accuracy; Prec.: precision; Rec.: recall; F1: $F_1$ score. The three best results for precision, recall, and $F_1$ score are highlighted in bold.

| Method | Feat. | TP | FP | TN | FN | Acc. | Prec. | Rec. | F1 |
|---|---|---|---|---|---|---|---|---|---|
| kNN | all | 57 | 30 | 47979 | 61 | 0.998 | 0.655 | 0.483 | 0.556 |
| kNN | n | 109 | 37 | 47972 | 9 | 0.999 | 0.747 | **0.924** | 0.826 |
| DT | all | 96 | 27 | 47982 | 22 | 0.999 | 0.78 | 0.814 | 0.797 |
| DT | n | 100 | 28 | 47981 | 18 | 0.999 | 0.781 | 0.847 | 0.813 |
| LR | all | 86 | 17358 | 30651 | 32 | 0.639 | 0.005 | 0.729 | 0.010 |
| LR | n | 118 | 19127 | 28882 | 0 | 0.603 | 0.006 | 1.000 | 0.012 |
| GBoost | all | 107 | 25 | 47984 | 11 | 0.999 | **0.811** | 0.907 | **0.856** |
| GBoost | n | 2 | 8 | 48001 | 116 | 0.997 | 0.2 | 0.017 | 0.031 |
| RF | all | 99 | 23 | 47986 | 19 | 0.999 | **0.811** | 0.839 | 0.825 |
| RF | n | 100 | 27 | 47982 | 18 | 0.999 | **0.787** | 0.847 | 0.816 |
| NB | all | 117 | 45 | 47964 | 1 | 0.999 | 0.722 | **0.992** | **0.836** |
| NB | n | 116 | 40 | 47969 | 2 | 0.999 | 0.744 | **0.983** | **0.847** |

Based on Table 3.2, we can eliminate logistic regression: although it has amazing recall, the precision is very low ($< 0.1$), which makes this method useless. Also, though the decision tree method has acceptable metrics, other classifiers have better results. For the sake of this study, we will eliminate this method. However, for a real-life experiment, it should be studied further.

## 3.5.2 The role of features

Even from Table 3.2, it is clear that the combination of features has a significant impact on the performance of the classifiers. For example, when there is only mean motion used, k-nearest neighbors has the value of recall equals 92 per cent, which is pretty good. However, when all features are used, it goes significantly down to 48 per cent.

The selection of initial data elements is essential in shaping the accuracy of classification models. The principle of "Garbage in, garbage out" is a well-known concept in the field of computer science and machine learning. It refers to the fact that the quality of the output is determined by the quality of the input. If the input data is poor or irrelevant, the output, even from the most sophisticated models, will also be poor or irrelevant. In our case, this principle emphasizes the importance of feature selection.

Let us perform the following experiment:

1. Choose four best methods: k-nearest neighbors, gradient boosting, naïve Bayes, and random forest.
2. Vary all possible combinations of features: $a$, $e$, $i$, and $n$.
3. Assess recall and $F_1$ score for each combination.

The results are in Table 3.3. The bold values indicate the best results for recall and $F_1$ score for each method.

**Table 3.3** Recall and $F_1$ scores for k-nearest neighbors, gradient boosting, and naïve Bayes for different combinations of features. Three (or more if there are equal numbers) best results for recall and $F_1$ score per method are highlighted in bold.

| Features | kNN | | GBoost | | NB | | RF | |
|---|---|---|---|---|---|---|---|---|
| | **Rec.** | $F_1$ | **Rec.** | $F_1$ | **Rec.** | $F_1$ | **Rec.** | $F_1$ |
| a | 0.924 | **0.848** | 0.288 | 0.41 | **0.992** | 0.836 | **0.873** | 0.831 |
| n | 0.924 | **0.826** | 0.017 | 0.031 | 0.983 | **0.847** | 0.856 | 0.815 |
| e | 0.008 | 0.013 | 0.0 | 0.0 | 0.0 | 0.0 | 0.025 | 0.028 |
| sin $i$ | 0.0 | 0.0 | 0.0 | 0.0 | 0.0 | 0.0 | 0.0 | 0.0 |
| a, n | **0.949** | **0.855** | 0.288 | 0.415 | **0.992** | 0.836 | **0.881** | **0.846** |
| a, e | 0.415 | 0.508 | 0.288 | 0.41 | 0.983 | 0.838 | **0.873** | **0.841** |
| a, sin $i$ | 0.924 | 0.79 | 0.847 | 0.816 | 0.975 | 0.83 | 0.856 | 0.821 |
| e, n | 0.415 | 0.508 | 0.017 | 0.031 | 0.975 | **0.852** | 0.856 | **0.852** |
| e, sin $i$ | 0.0 | 0.0 | 0.0 | 0.0 | 0.0 | 0.0 | 0.0 | 0.0 |
| sin $i$, n | 0.924 | 0.787 | **0.932** | **0.846** | 0.966 | **0.844** | 0.839 | 0.802 |
| a, e, n | 0.542 | 0.59 | 0.288 | 0.41 | **0.992** | 0.836 | **0.873** | 0.841 |
| a, e, sin $i$ | 0.254 | 0.349 | 0.847 | 0.816 | 0.966 | 0.832 | 0.678 | 0.744 |
| a, sin $i$, n | **0.949** | 0.803 | **0.907** | **0.853** | **0.992** | 0.836 | 0.831 | 0.813 |
| e, sin $i$, n | 0.297 | 0.393 | **0.932** | **0.846** | 0.958 | 0.846 | 0.627 | 0.688 |
| a, e, sin $i$, n | 0.483 | 0.556 | **0.907** | **0.853** | **0.992** | 0.836 | **0.881** | 0.832 |

Based on Table 3.3, one can conclude the following:

- For k-nearest neighbors, the best results are achieved when there are two features: semimajor axis ($a$) and mean motion ($n$). Acceptable results could be achieved even if one of these features is used. The inclusion of eccentricity ($e$) or inclination ($i$) negatively impacts the performance of the model.
- For gradient boosting, the best results are achieved when there are two features: mean motion ($n$) and inclination (sin $i$). Adding more features does not improve the performance of the model at all.
- For naïve Bayes, there is no combination that has the best recall and $F_1$ score at the same time. Thus one has to choose either a better recall or

a better $F_1$ score. Considering that in our case, recall is more important, the best combination is when either semimajor axis ($a$; the best recall) or mean motion ($n$; good $F_1$ score) is used. Adding more features either does not improve the performance or makes it even worse.

- For random forest, the best result is achieved when there are two features: semimajor axis ($a$) and mean motion ($n$). The second result is for the combination of semimajor axis ($a$) and eccentricity $e$. Acceptable results could be achieved even if only the semimajor axis is used.

The results for k–nearest neighbors are consistent with the results of Smirnov and Markov (2017): while the best combination in this paper was $e$ and $n$ or $a$, $e$, and $n$, the second-best was $a$ and $n$. The values of metrics are close. The difference could be related to the usage of different training sets and different parameters of the classifiers.

Overall, this experiment shows that all the methods are good enough to be tested further.

### 3.5.3 The role of training data volume

The volume of training data is a critical factor in the performance of machine-learning models, to avoid overfitting or underfitting (Everitt and Skrondal, 2010). In our case, it means that it is necessary to determine how many asteroids should be in the test dataset to achieve the best performance (recall or $F_1$ score).

Smirnov and Markov (2017) explored this by employing datasets of various sizes—containing 50, 100, and 200 resonant asteroids—and applying the k-nearest neighbors method with eccentricity ($e$) and mean motion ($n$) as inputs. Key metrics, such as true positives (TP), false positives (FP), true negatives (TN), false negatives (FN), and their derivatives were analyzed. The main results are shown in Table 3.4.

**Table 3.4** Influence of training data volume (first column) on the performance of the k-nearest neighbors method with eccentricity ($e$) and mean motion ($n$) as inputs.

| Training set | Prec. | Rec. | $F_1$ |
|---:|---|---|---|
| 50 | 0.907 | 0.980 | 0.942 |
| 100 | 0.899 | 0.980 | 0.938 |
| 200 | 0.929 | 0.985 | 0.956 |

The analysis of the data in Table 3.4 surprisingly reveals minimal variance across different training set sizes. Increasing the number of objects in the test dataset increases precision, but it has virtually no effect on accuracy and recall. Moreover, it seems that there is a plateau in the learning curve, where increasing the number of training examples does not significantly enhance the model's performance.

To address this question, let us conduct the following experiment:

1. Choose four classifiers studied earlier: k-nearest neighbors, gradient boosting, naïve Bayes, and random forest.
2. Take first $n$ resonant asteroids, $n = [\![1, 100]\!]$, and all corresponding non-resonant asteroids with numbers lower than the maximum number of any resonant asteroid taken.
3. Run a simulation with the best combination of features for each method and given $n$.

The results are presented in Figs. 3.4, 3.5, 3.6, and 3.7 for k-nearest neighbors, gradient boosting, naïve Bayes, and random forest, respectively. The horizontal axis represents the number of resonant asteroids in the training set, whereas the vertical axis represents the values of precision, recall, and $F_1$ score.

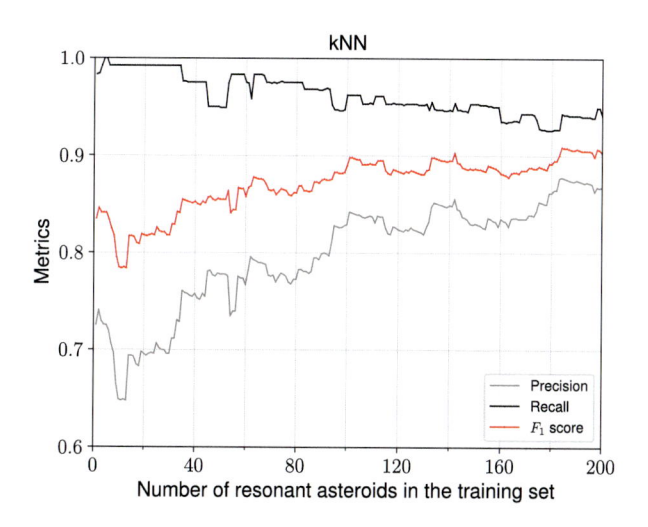

**Figure 3.4** Performance of k-nearest neighbors vs the volume of the training set.

For k-nearest neighbors, the good volume of the training data representing the tradeoff between recall and precision is around $n = 60$: for $n = 63$, the values of recall and $F_1$ score equal 0.983 and 0.878, respectively. As expected, precision and $F_1$ score increase as the volume of the training set, but recall decreases: the more neighbors are used, the better classification is

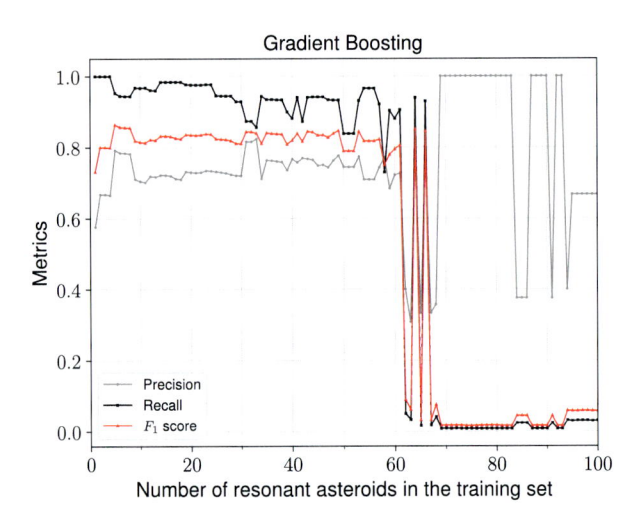

**Figure 3.5** Performance of the gradient boosting method vs the volume of the training set.

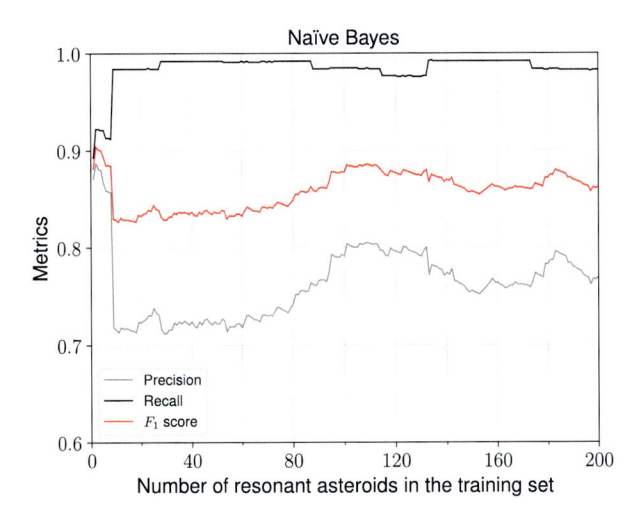

**Figure 3.6** Performance of the naïve Bayes method vs the volume of the training set.

achieved. Given the fact that we want to get high values of recall, the best volume of the training set is 63 resonant asteroids.

For gradient boosting, there is a clear threshold for $n > 61$, due to overfitting. Thus one should choose smaller values of $n$. Surprisingly, the best results are for $n = 5$: recall and $F_1$ score equal $0.953$ and $0.846$, respectively.

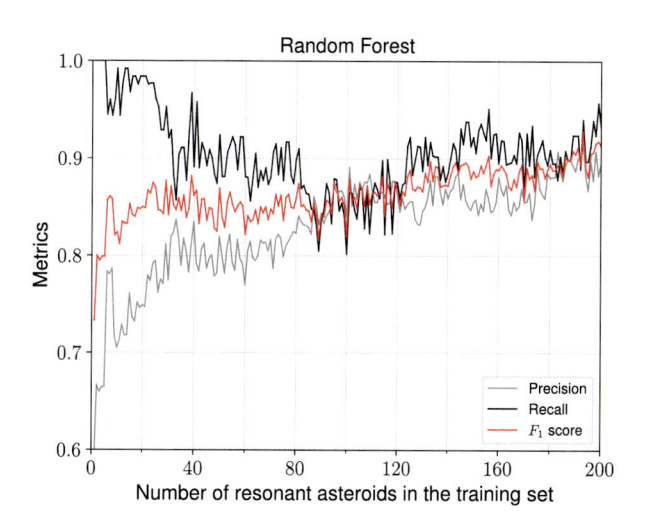

**Figure 3.7** Performance of the random forest method vs the volume of the training set.

The same is valid for naïve Bayes: small volumes of the training set outperforms the larger ones. The best overall results are for $n = 2$: recall and $F_1$ score equal 0.922 and 0.904, respectively. Comparable results are only for $n > 100$. However, if one wants to achieve the highest value of recall, the best volume of the training set is 87 resonant asteroids: the values are 0.992 and 0.863 for recall and $F_1$ score, respectively.

For random forest, there is a different picture: one could achieve amazing recall ($\approx 1.0$) with a small training set. However, precision is moderate—$\approx 0.7$. Overall, the $F_1$ score increases as the volume of the training set. Therefore, this is almost the only method that gains benefits from the increase of the training set. For example, for $n = 195$, the values of recall and $F_1$ score equal 0.948 and 0.924, respectively. This seems to be the best overall result.

A brief summary of the best results for each method is presented in Table 3.5.

### 3.5.4 Parameters of the best methods

The final question is: What are the parameters of the best methods? To demonstrate how one could answer this question, let us conduct the following experiment:

**Table 3.5** The best numbers of resonant asteroids in the training set and the corresponding values of precision, recall, and $F_1$ score for kNN, gradient boosting, naïve Bayes, and random forest.

| Method | $n$ | Precision | Recall | $F_1$ Score |
|---|---|---|---|---|
| k-Nearest Neighbors | 63 | 0.793 | 0.983 | 0.878 |
| Gradient Boosting | 5 | 0.791 | 0.953 | 0.846 |
| Naïve Bayes | 2 | 0.887 | 0.922 | 0.904 |
| Naïve Bayes | 87 | 0.764 | 0.992 | 0.863 |
| Random Forest | 2 | 0.667 | 1.000 | 0.800 |
| Random Forest | 195 | 0.902 | 0.948 | 0.924 |

1. Given the results of the previous experiment (see Table 3.5), we will focus on k-nearest neighbors only because (a) gradient boosting has lower recall and $F_1$ score and (b) naïve Bayes has no parameters to vary.
2. For kNN, let us vary the number of neighbors $k = [\![1, 100]\!]$ and the distance metric $p$. For the latter, let us use three options: Minkowski, Euclidean, and cosine distances.

The results are presented in Table 3.6: the first five rows demonstrate the best values of recall, whereas the last five rows are sorted by $F_1$ score. The best recall is for $k = 26$ and Euclidean distance: recall and $F_1$ score equal 0.991 and 0.856, respectively. The best $F_1$ score is for $k = 16$ and $p = 1$: recall and $F_1$ score equal 0.983 and 0.878, respectively. However,

**Table 3.6** The best parameters for the k-nearest neighbors classifier. $k$ is the number of neighbors, $p$ is the order of distance metric, TP is true positives, FP is false positives, TN is true negatives, FN is false negatives, Prec. is precision, Rec. is recall, and $F_1$ is $F_1$ score. The horizontal line separates the best five results for recall and $F_1$ score, respectively.

| k | p | TP | FP | TN | FN | Prec. | Rec. | F1 |
|---|---|---|---|---|---|---|---|---|
| 26 | 2 | 116 | 38 | 48051 | 1 | 0.753 | 0.991 | 0.856 |
| 27 | 2 | 116 | 39 | 48050 | 1 | 0.748 | 0.991 | 0.853 |
| 28 | 2 | 116 | 41 | 48048 | 1 | 0.739 | 0.991 | 0.847 |
| 29 | 2 | 116 | 41 | 48048 | 1 | 0.739 | 0.991 | 0.847 |
| 30 | 2 | 116 | 42 | 48047 | 1 | 0.734 | 0.991 | 0.844 |
| 16 | 1 | 115 | 30 | 48059 | 2 | 0.793 | 0.983 | 0.878 |
| 9 | 2 | 113 | 28 | 48061 | 4 | 0.801 | 0.966 | 0.876 |
| 11 | 2 | 112 | 28 | 48061 | 5 | 0.800 | 0.957 | 0.872 |
| 12 | 1 | 113 | 30 | 48059 | 4 | 0.790 | 0.966 | 0.869 |
| 13 | 1 | 113 | 30 | 48059 | 4 | 0.790 | 0.966 | 0.869 |

one might conclude that the number of neighbors could be even lower for the latter: even for $k = 9$ or $k = 11$, the results are close to the best.

### 3.5.5 Summary

In this section, we have explored the performance of different machine-learning methods for the classification of resonant and nonresonant asteroids. We varied the volume of the training set, the number of features, and the parameters of the best methods to achieve the highest value of recall, $F_1$ score, and precision.

We have found that the best methods are k-nearest neighbors, naïve Bayes, gradient boosting, and random forest. Each method has its own set of parameters of the variables, leading to the best performance. The highest achieved value of recall is 0.992 for naïve Bayes, whereas the highest value of $F_1$ score is 0.878 for k-nearest neighbors. The highest value of precision is 0.887 naïve Bayes.

The decision tree method has comparable results. Thus it could be used as well. However, it should be fine-tuned for a specific task to achieve the comparable results. Logistic regression and balanced random forest have the unacceptable performance, and hence should not be used for similar tasks.

Note that the same approach could be used for other problems as well when there is a group of asteroids or other objects that should be classified. In general, we did not use any information about the resonances, except for the fact that the asteroids are trapped in them. Thus the same approach could be used for the classification of asteroid families or groups, which is discussed in Chapter 2.

Time resources required for the training of the models are not significant: it takes less than a second to train one model. The same is valid for the prediction: it takes less than a second to predict the class of all asteroids in the main belt. Therefore such an approach could be used for any simulation that has multiple virtual asteroids and where the traditional approach requires high computational resources.

It is worth mentioning that though in this chapter we wanted to demonstrate how one could design and run a machine-learning experiment with the aim of classifying resonant and nonresonant asteroids, there are other approaches. For example, one could use a grid search to find the best parameters for the classifiers. The Python package "scikit-learn" has a built-in function for this purpose. However, the approach used in this chapter is more general and works even for the cases when grid search is impossible, due to the high number of variables to vary.

Note that each resonance (or other asteroid's group) has its own set of parameters. Thus it is necessary to train a separate model for each resonance. However, the training of the model is fast, so it is not a problem.

## 3.6. Other applications

Besides the main task of the classification of resonant and nonresonant asteroids, machine-learning methods could be used for other applications as well. For example, one of the tasks required for the classical approach is the identification of whether or not a resonant angle librates.

The automation of this task is not trivial. The very first attempt was made by Smirnov and Shevchenko (2013). The authors created a simple algorithm that uses the data about intersecting borders by resonant angle. Its precision was 80 per cent. Though it was enough for that statistical study, it should be improved for more sophisticated simulations.

The next attempt was made by Smirnov et al. (2017) and Smirnov and Dovgalev (2018). The authors used periodograms and cross-periodograms to enhance the precision of the algorithm. The results were better: the precision was 90 per cent.

A completely different approach was chosen by Carruba et al. (2021). The authors used artificial neural networks (ANN) to classify resonant angles. The methodology involved converting the time-varying resonant angle of asteroids into images. These images were then used as inputs for the ANN, which was trained to identify librations.

The results demonstrate strong dependence on the volume of the training size and type of libration (pure, transient, or circulating). For example, the value of recall could vary from 0.65 to 1.0, and $F_1$ score varies from $\approx 0.68$ to 1.0 (Carruba et al., 2021).

Thus the study by Carruba et al. (2021) demonstrates a totally different approach: instead of using the raw data about the resonant angle, the authors try to simulate human perception of the resonant angle. However, they did not provide a comparison with the "raw data approach," which could be interesting.

It seems that the performance achieved is lower than in the classical method.[1] However, it does not mean that such an approach is useless. On the contrary, its main advantage is that it uses a totally different method.

---

[1] To be accurate, the MMRs used in Carruba et al. (2021) and Smirnov (2023) are different. Furthermore, the method has been improved in Carruba et al. (2022). More details are available in Chapter 4.

Thus it could be used in combination with the classical approach to enhance the precision of the algorithm. The simplest option is to perform a classification task using both methods. If the results of both methods are the same, then the classification is correct. If the results are different, then one could perform classification manually. This approach perfectly fits the idea of using machine-learning methods as a tool for reducing time spent by human beings.

## 3.7. Conclusions

Let us summarize the main outcomes of this chapter.

Firstly, we have explored the performance of different machine-learning methods for the classification of resonant and nonresonant asteroids. The results are promising and demonstrate that machine-learning methods could be used for such a task. We have evaluated the following classifiers: k–nearest neighbors, decision tree, logistic regression, gradient boosting, random forest, and naïve Bayes. For each method, we varied the volume of the training set, the number of features, and the parameters of the method to achieve the highest value of recall and $F_1$ score.

Secondly, we have found that the best methods are k–nearest neighbors, naïve Bayes, gradient boosting, and random forest. Decision tree has comparable results. However, it requires extra fine-tuning, which is out of the scope of this chapter. Each method has its own set of parameters of the variables, leading to the best performance. The highest achieved value of recall is 0.992 for naïve Bayes, whereas the highest value of the $F_1$ score is 0.904. The highest value of precision is 0.887 naïve Bayes.

Thirdly, we have found that the performance of the methods depends on the volume of the training set. Often, the best results are achieved when the volume of the training set is small. For gradient boosting and naïve Bayes, the best performance is achieved when the training set contains $n \in [2, 5]$ resonant asteroids. For k–nearest neighbors, it is higher and contains $n = 63$ objects or more.

Fourthly, we have found that the performance of the methods depends on the number of features. For k–nearest neighbors and random forest, the best results are achieved when there are two features: semimajor axis ($a$) and mean motion ($n$). For the latter, the combination containing all features has acceptable performance too. Naïve Bayes has a different story: there is no single combination that has the best recall and $F_1$ score at the same time.

The best recall could be achieved with one feature—semimajor axis $a$; the good $F_1$ score is with mean motion ($n$).

Fifthly, we have found that the performance of the methods depends on the parameters of the methods. For kNN, we found that the good value of $k$ is around $k = 60$ or small values ($k \in [3, 16]$).

Sixthly, the results depend on the size of the training set. For the k-nearest neighbors classifier, more is better. However, acceptable results can be obtained with approximately 60 resonant asteroids in the training set. For gradient boosting, the best value is $n = 5$. For naïve Bayes, the best overall results are for $n < 5$. Note that the concrete values of parameters depend on the task and the training set.

Finally, we have found that machine-learning methods could be used for other applications as well. For example, one could use them for the identification of whether or not a resonant angle librates. The results are promising, but the performance is lower than in the classical method. However, if this approach is combined with the classical one, it could benefit both methods.

## 3.8. Code availability

The code and supplementary materials are available at

- https://github.com/solar-system-ml/book (docs/chapter3 folder),
- https://solar-system-ml.github.io/book/chapter3/index/.

## References

Bowles, M., 2015. Machine Learning in Python: Essential Techniques for Predictive Analysis. John Wiley & Sons, Inc, Indianapolis, IN. OCLC: ocn889736073.

Breiman, L., 2001. Random forests. Machine Learning 45, 5–32. https://doi.org/10.1023/A:1010933404324. http://link.springer.com/10.1023/A:1010933404324.

Carruba, V., Aljbaae, S., Caritá, G., Lourenço, M.V.F., Martins, B.S., Alves, A.A., 2023. Imbalanced classification applied to asteroid resonant dynamics. Frontiers in Astronomy and Space Sciences 10, 1196223. https://doi.org/10.3389/fspas.2023.1196223. https://www.frontiersin.org/articles/10.3389/fspas.2023.1196223/full.

Carruba, V., Aljbaae, S., Domingos, R.C., Barletta, W., 2021. Artificial neural network classification of asteroids in the M1:2 mean-motion resonance with Mars. Monthly Notices of the Royal Astronomical Society 504, 692–700. https://doi.org/10.1093/mnras/stab914.

Carruba, V., Aljbaae, S., Domingos, R.d.C., Huaman, M., Martins, B., 2022. Identifying the population of stable ${\nu}\_6$ resonant asteroids using large databases. Monthly Notices of the Royal Astronomical Society 514, 4803–4815. https://doi.org/10.1093/mnras/stac1699. arXiv:2203.15763.

Chirikov, B.V., 1979. A universal instability of many-dimensional oscillator systems. Physics Reports 52, 263–379.

Everitt, B., Skrondal, A., 2010. The Cambridge Dictionary of Statistics, 4th edition ed. Cambridge University Press, Cambridge, UK. OCLC: 668966755.

Freund, Y., Schapire, R.E., 1997. A decision-theoretic generalization of on-line learning and an application to boosting. Journal of Computer and System Sciences 55, 119–139. https://doi.org/10.1006/jcss.1997.1504. https://linkinghub.elsevier.com/retrieve/pii/S002200009791504X.

Gallardo, T., 2006. The occurrence of high-order resonances and Kozai mechanism in the scattered disk. Icarus 181, 205–217. https://doi.org/10.1016/j.icarus.2005.11.011.

Gallardo, T., 2014. Atlas of three body mean motion resonances in the solar system. Icarus 231, 273–286. https://doi.org/10.1016/j.icarus.2013.12.020. arXiv:0708.2080.

Hastie, T., Tibshirani, R., Friedman, J., 2009. The Elements of Statistical Learning. Springer Series in Statistics. Springer New York, New York, NY. https://doi.org/10.1007/978-0-387-84858-7. http://link.springer.com/10.1007/978-0-387-84858-7.

Holman, M.J., Murray, N.W., 1996. Chaos in high-order mean resonances in the outer asteroid belt. The Astronomical Journal 112, 1278–1293. 1996AJ....112.1278H.

Ivezić, Z., Connolly, A.J., VanderPlas, J.T., Gray, A., 2020. Statistics, Data Mining, and Machine Learning in Astronomy: a Practical Python Guide for the Analysis of Survey Data. Princeton University Press.

Le, T.T., Fu, W., Moore, J.H., 2020. Scaling tree-based automated machine learning to biomedical big data with a feature set selector. Bioinformatics 36, 250–256.

Lemaître, A., 1984. High-order resonances in the restricted three-body problem. Celestial Mechanics 32, 109–126. https://doi.org/10.1007/BF01231119. http://link.springer.com/10.1007/BF01231119.

Lourenço, M.V.F., Carruba, V., 2022. Genetic optimization of asteroid families' membership. Frontiers in Astronomy and Space Sciences 9, 988729. https://doi.org/10.3389/fspas.2022.988729.

Milani, A., Nobili, A., Knezevic, Z., 1997. Stable chaos in the asteroid belt. Icarus 125, 13–31. https://doi.org/10.1006/icar.1996.5582.

Milani, A., Nobili, A.M., 1992. An example of stable chaos in the Solar System. Nature 357, 569–571. https://doi.org/10.1038/357569a0. https://www.nature.com/articles/357569a0.

Morbidelli, A., 2002. Modern Celestial Mechanics: Aspects of Solar System Dynamics. Advances in Astronomy and Astrophysics, vol. 5. Taylor & Francis, London. OCLC: 248084916.

Murray, C.D., Dermott, S.F., 1999. Solar System Dynamics. Cambridge Univ. Press, Cambridge, UK. OCLC: 908618105.

Murray, N., Holman, M., 1997. Diffusive chaos in the outer asteroid belt. The Astronomical Journal 114, 1246. https://doi.org/10.1086/118558.

Murray, N., Holman, M., Potter, M., 1998. On the origin of chaos in the asteroid belt. The Astronomical Journal 116, 2583–2589. https://doi.org/10.1086/300586.

Nesvorný, D., Morbidelli, A., 1998. Three-body mean motion resonances and the chaotic structure of the asteroid belt. The Astronomical Journal 116, 3029–3037. https://doi.org/10.1086/300632.

Olson, D.L., Delen, D., 2008. Advanced Data Mining Techniques. Springer, Berlin Heidelberg.

Parzen, E., 1962. On estimation of a probability density function and mode. The Annals of Mathematical Statistics 33, 1065–1076. Publisher: JSTOR.

Pedregosa, F., Varoquaux, G., Gramfort, A., Michel, V., Thirion, B., Grisel, O., Blondel, M., Prettenhofer, P., Weiss, R., Dubourg, V., Vanderplas, J., Passos, A., Cournapeau, D., Brucher, M., Perrot, M., Duchesnay, E., 2011. Scikit-learn: machine learning in Python. Journal of Machine Learning Research 12, 2825–2830.

Rish, I., et al., 2001. An empirical study of the naive Bayes classifier. In: IJCAI 2001 Workshop on Empirical Methods in Artificial Intelligence, pp. 41–46. Issue: 22.

Shevchenko, I.I., 2006. On the Lyapunov exponents of the asteroidal motion subject to resonances and encounters. Proceedings of the International Astronomical Union 2, 15–30. https://doi.org/10.1017/S174392130700302X.

Shevchenko, I.I., 2020. Dynamical Chaos in Planetary Systems. Astrophysics and Space Science Library, vol. 463. Springer International Publishing, Cham. https://doi.org/10.1007/978-3-030-52144-8. http://link.springer.com/10.1007/978-3-030-52144-8.

Smirnov, E., 2017. Asteroids in three-body mean-motion resonances with Jupiter and Mars. Solar System Research 51. https://doi.org/10.1134/S003809461702006X.

Smirnov, E., 2023. A new python package for identifying celestial bodies trapped in mean-motion resonances. Astronomy and Computing. https://doi.org/10.1016/j.ascom.2023.100707. https://linkinghub.elsevier.com/retrieve/pii/S2213133723000227.

Smirnov, E., 2024. A comparative analysis of machine learning classifiers in the classification of resonant asteroids. Icarus, 116058. https://doi.org/10.1016/j.icarus.2024.116058. https://www.sciencedirect.com/science/article/pii/S0019103524001180.

Smirnov, E., Dovgalev, I., Popova, E., 2017. Asteroids in three-body mean motion resonances with planets. Icarus. https://doi.org/10.1016/j.icarus.2017.09.032.

Smirnov, E., Markov, A., 2017. Identification of asteroids trapped inside three-body mean motion resonances: a machine-learning approach. Monthly Notices of the Royal Astronomical Society 469. https://doi.org/10.1093/mnras/stx999.

Smirnov, E.A., Dovgalev, I.S., 2018. Identification of asteroids in two-body resonances. Solar System Research 52, 347–354. https://doi.org/10.1134/S0038094618040056.

Smirnov, E.A., Shevchenko, I.I., 2013. Massive identification of asteroids in three-body resonances. Icarus 222, 220–228. https://doi.org/10.1016/j.icarus.2012.10.034. arXiv:1206.1451.

Tsiganis, K., 2002. Stable chaos in high-order Jovian resonances. Icarus 155, 454–474. https://doi.org/10.1006/icar.2001.6737. https://linkinghub.elsevier.com/retrieve/pii/S0019103501967375.

Tsiganis, K., 2010. Dynamics of small bodies in the solar system. The European Physical Journal Special Topics 186, 67–89. https://doi.org/10.1140/epjst/e2010-01260-9.

Wisdom, J., 1982. The origin of the Kirkwood gaps: a mapping for asteroidal motion near the 3/1 commensurability. The Astronomical Journal 87, 577–593.

# CHAPTER FOUR

# Asteroid families interacting with secular resonances

**Valerio Carruba**

São Paulo State University (UNESP), Department of Mathematics, Guaratinguetá, SP, Brazil

## 4.1. Introduction

A secular resonance occurs when the precession frequencies of an asteroid's pericenter $g$ and node $s$ longitudes and the basic frequencies of planetary theory coincide. Mean-motion resonances have a characteristic V-shape in proper $(a, e)$ domains, with $a$ being the proper semimajor axis, and $e$ the proper eccentricity, and this makes identifying asteroids interacting with such resonances visually easier. Secular resonances, however, have a more complicated tridimensional structure in proper $(a, e, \sin(i))$ domains, which makes the identification of asteroids interacting with such resonances a more complicated problem.

Yet, mapping the secular structure of the asteroid's main belt is of paramount importance. Some asteroids belong to asteroid families, which are clusters of objects that are identifiable in domains of proper elements (Bendjoya and Zappalà, 2002), and which are thought to have mostly formed from the collisions of two asteroids. After the collision, the fragments forming the families are ejected at characteristic velocities, typically matching the escape velocities of the parent body. Since the family-forming event, the family members' orbits then dynamically evolve because of gravitational forces, i.e., for example, through the interaction with mean-motion and secular resonance, close encounters with main belt massive bodies (Carruba et al., 2003), and nongravitational forces, such as the Yarkovsky and YORP (Vokrouhlický et al., 2013) forces, which makes retrieving the original ejection velocity field of asteroid families a rather complex task.

However, for asteroid families that interact with secular resonances, constraints from secular dynamics can be used to obtain reliable information on the original ejection velocity field, since for these families, conserved quantities associated with the secular dynamics have not changed since the family-forming event (Carruba et al., 2018). Numerical methods based on backward numerical simulations, such as the backward integration method

*Machine Learning for Small Bodies in the Solar System*
https://doi.org/10.1016/B978-0-44-324770-5.00009-X

(BIM (Nesvorný et al., 2002)), and the close encounters method (CEM (Pravec et al., 2010)), can be used to date the family age with a precision not available for more evolved groups, if the family is younger than 7 Myr. For families that are both young and interacting with secular resonances, both the age and the ejection velocity field can be estimated with high precision, making these groups of particular interest among all the known asteroid families.

Yet, as discussed, identifying asteroid families that interact with secular resonance is not a trivial endeavor. Though methods based on range intervals for asteroids' frequencies can be used to select asteroids most likely to interact with secular resonances, ultimately, the identification of the resonant status of an object is performed by analyzing the resonant angle of the asteroid. This angle, which is a combination of the angles of the asteroid and planets involved in the resonance (see Section 4.2 for more details on the definition of this angle for various resonances), will oscillate around an equilibrium point if the asteroid is in a resonant state, and will circulate from $0°$ to $360°$ degrees if the asteroid is not in resonance. Since the number of asteroids that may interact with a secular resonance is of the order of several thousands, performing a classification of resonant asteroids involves studying several thousand plots, which, for a human observer, might be a daunting task.

Recently, machine learning (ML) and deep learning methods based on either the proper element distribution or computer vision approaches have been introduced for the purpose of automatically identifying the resonant status of large populations of asteroids. The main goal of this chapter is to revise these new, modern approaches to problems of secular dynamics, and to provide the reader with useful tools to address the problems of classifications of asteroids near secular resonances. We will start by reviewing what secular resonances are and where they are located in the asteroid's main belt.

## 4.2. Secular resonances: an overview

Secular resonances occur when there is a commensurability between the precession frequencies of the longitudes of pericenter $g$ and node $s$ of an asteroid and the fundamental frequencies of planetary theory $g_i = \dot{\varpi}_i$ and $s_i = \dot{\Omega}_i$, where $i$ is a suffix identifying the planet, going from 2 for Venus up to 8 for Neptune, and whose values are reported in Table 4.1 (Celletti and Perozzi, 2020). The frequencies associated with a given resonance must

**Table 4.1** The fundamental planetary frequencies in the Solar System.

| $g$ Frequencies | Value [" $yr^{-1}$] | $s$ Frequencies | Value [" $yr^{-1}$] |
|---|---|---|---|
| $g_2$ | 7.456 | $s_2$ | −7.080 |
| $g_3$ | 17.365 | $s_3$ | −18.852 |
| $g_4$ | 18.002 | $s_4$ | −17.633 |
| $g_5$ | 4.257 | – | – |
| $g_6$ | 28.243 | $s_6$ | −26.345 |
| $g_7$ | 3.093 | $s_7$ | −2.996 |
| $g_8$ | 0.669 | $s_8$ | −0.692 |

satisfy the relationship

$$pg + qs + \sum_i (p_i \dot{g}_i + q_i \dot{s}_i) = 0, \tag{4.1}$$

where the numbers $p, q, p_i, q_i$ must adhere to the D'Alembert requirements for valid arguments, which state that the sum of the coefficients must be zero, and that the sum of the nodal longitude frequencies must be even. The combinations from Eq. (4.1), which solely involve the frequency of the asteroid perihelion and node, are referred to as "pericenter resonances" or "$g$–type resonances" and "node resonances" or "$s$–type resonances," respectively. When resonances involve a direct commensurability between the frequency of an asteroid and a planet, just as in the case of the $g - g_6$ resonance, they are called linear secular resonances. They can be identified by the arguments $v_i$, with $i$ going from 2 to 8 for the planets, for $g$–type resonances, or $v_{1i}$ for $s$–type resonances. Higher-order secular resonances are often called nonlinear resonances, and can be identified in terms of combinations of arguments for linear resonances. For example, the $g - g_6 + s - s_6$ can also be identified as $v_6 + v_{16}$.

When secular resonances involve combinations of both the $g$ and $s$ frequencies, they can be referred to as mixed. A special type of mixed resonances is for those that belong to the $z_k$ series, which is defined as $z_k = k(g - g_6) + s - s_6$, where $k = 1, 2, 3$, etc. Their arguments can also be written as $z_k = kv_6 + v_{16}$. The orbital location of the most dynamically diffusive secular resonances has been recently identified in works such as Knežević (2021). A list of $g$–type, $s$–type, mixed and $z_k$ resonances is provided in Tables 4.2, 4.3, 4.4, and 4.5, respectively,

**Table 4.2** The three most diffusive $g$−type secular resonances in the main belt, as listed by Huaman et al. (2017). We present the resonant argument in terms of frequencies, in terms of combinations of linear secular resonances, the central value of the asteroidal $g$ frequency associated with each resonance, and the number of asteroids, numbered and multiopposition, likely to be affected by the resonances.

| Res. argument frequencies | Res. argument linear resonances | Frequency value ["/yr] | Numbered ast. | Multiopp. ast. |
|---|---|---|---|---|
| $g - 2g_6 + g_5$ | $2v_6 - v_5$ | 52.229 | 15267 | 6422 |
| $g - 3g_6 + 2g_5$ | $3v_6 - 2v_5$ | 76.215 | 1780 | 928 |
| $2g - 3g_5 + g_6$ | $3v_5 - v_6$ | -7.736 | 15 | 7 |

**Table 4.3** The most diffusive $s$−type secular resonances in the main belt, according to Knežević (2021). The format is the same as in Table 4.2.

| Res. id. | Res. argument frequencies | Res. argument linear resonances | Frequency ["  yr$^{-1}$] | Numb. ast. | Multiopp. ast. |
|---|---|---|---|---|---|
| 1 | $s - s_4$ | $v_{14}$ | -17.633 | 1 | 6 |
| 2 | $s - 2 \cdot s_8 + s_7 - g_5 + g_6$ | $2 \cdot v_{18} - v_{17} + v_5 - v_6$ | -21.683 | 1578 | 1950 |
| 3 | $2 \cdot s - s_4 - s_6$ | $v_{16} + v_{14}$ | -21.989 | 1833 | 2135 |
| 4 | $s - 2 \cdot s_4 + s_7 - g_6 + g_4$ | $2 \cdot v_{14} - v_{17} + v + 6 - v_4$ | -22.029 | 1853 | 2129 |
| 5 | $s - 2 \cdot s_6 + s_7 - g_6 + g_8$ | $2 \cdot v_{16} - v_{17} + v_6 - v_8$ | -22.120 | 1925 | 2157 |
| 6 | $s - s_3 - g_5 - g_6 + 2 \cdot g_4$ | $v_3 + v_5 + v_6 - 2 \cdot v_4$ | -22.357 | 1730 | 1857 |
| 7 | $s - s_3 - g_8 + g_5$ | $v_{13} + v_8 - v_5$ | -22.439 | 1707 | 1841 |
| 8 | $s - s_6 - g5 + g_8$ | $v_{16} + v_5 - v_8$ | -22.758 | 1476 | 1530 |
| 9 | $s - s_6$ | $v_{16}$ | -26.345 | 20 | 24 |
| 10 | $s - s_6 - g_5 + g_6$ | $v_{16} + v_5 - v_6$ | -50.332 | 4899 | 3498 |
| 11 | $s - s_6 - 2 \cdot g_5 + 2 \cdot g_6$ | $v_{16} + 2 \cdot v_5 - 2 \cdot v_6$ | -74.319 | 2588 | 2170 |

**Table 4.4** The most diffusive secular resonances that are combinations of both $g$ and $s$ in the main belt, according to Knežević (2021). See the caption of Table 4.2 for the meaning of the discussed quantities.

| Res. id. | Res. argument frequencies | Res. argument linear resonances | Frequency ["  yr$^{-1}$] | Numb. ast. | Multiopp. ast. |
|---|---|---|---|---|---|
| 1 | $g - g_5 - 2s + 2s_6$ | $v_5 - 2 \cdot v_{16}$ | 56.946 | 16 | 16 |
| 2 | $g - g_5 + 2s - 2s_6$ | $v_5 + 2 \cdot v_{16}$ | -48.434 | 6207 | 4150 |
| 3 | $g - g_5 - s + s_6$ | $v_5 - v_{16}$ | 30.601 | 44 | 22 |
| 4 | $g - g_6 - s + s_6$ | $v_6 - v_{16}$ | 54.588 | 28 | 15 |
| 5 | $2g - 2g_5 + s - s_7$ | $2 \cdot v_5 + v_{17}$ | 5.516 | 47 | 51 |
| 6 | $2g - g_5 - g_6 + s - s_6$ | $v_5 + v_6 + v_{16}$ | 6.154 | 63 | 0 |

While identifying asteroids in mean-motion resonance is usually a straightforward process in proper elements domains, such as the $(a, e)$ plane, since in such a domain mean-motion resonances have a V-shape geometry, identifying asteroids in secular resonances is a more complicated issue.

**Table 4.5** The most diffusive $zk-$type secular resonances in the main belt, as listed in Knežević (2021).

| Res. id. | Res. argument frequencies | Res. argument linear resonances | Frequency ["yr$^{-1}$] | Numb. ast. | Multiopp. ast. |
|---|---|---|---|---|---|
| 1 | $g - g_6 + s - s_6$ | $\nu_6 + \nu_{16}$ | 1.898 | 17795 | 11640 |
| 2 | $2 \cdot (g - g_6) + s - s_6$ | $2 \cdot \nu_6 + \nu_{16}$ | 30.141 | 343 | 326 |
| 3 | $3 \cdot (g - g_6) + s - s_6$ | $3 \cdot \nu_6 + \nu_{16}$ | 58.384 | 114 | 113 |

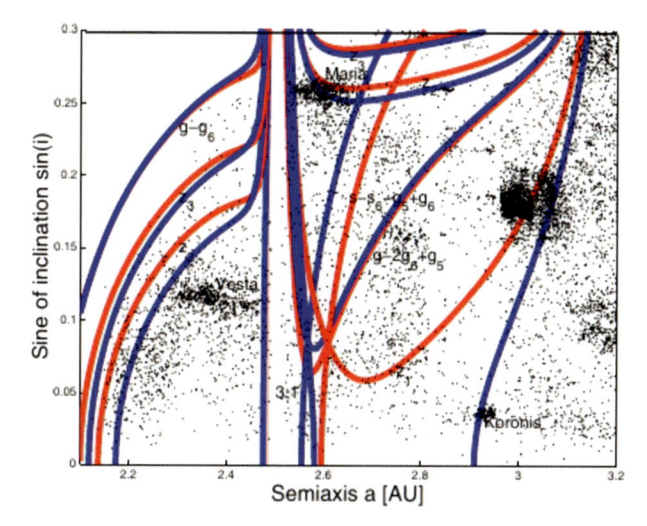

**Figure 4.1** The primary secular resonances in the asteroid belt. The proper elements a and sin(i) of the asteroids with proper e falling within the range of ±0.025 of the central value are superimposed with the secular resonances derived for the fixed value of 0.10 of the proper eccentricity. The primary asteroid families in the area are indicated by their names. The resonance locations are shown by the red lines at the lower edge and by the blue curves at the upper edge. For further information, refer to Figs. 7, 8, and 9 in Milani and Knežević (1994). Adapted from figure (1) of Carruba and Michtchenko (2007) and reproduced with permission from the authors and A&A (©A&A).

Fig. 4.1 displays the location of the primary secular resonances in the main belt in the proper $(a, \sin(i))$ domain. The reader may notice that the shape of a single secular resonance may change dramatically, depending on the value of the eccentricity. Overall, secular resonances have a complicated tri-dimensional structure in the proper $(a, e, \sin(i))$ domain.

For this reason, Carruba and Michtchenko (2007, 2009) suggested identifying asteroids interacting with secular resonances in domain of proper frequencies, where such resonances appear as straight lines. Fig. 4.2 displays the distribution of asteroids with absolute magnitudes $H < 13$ in the $(n, g)$

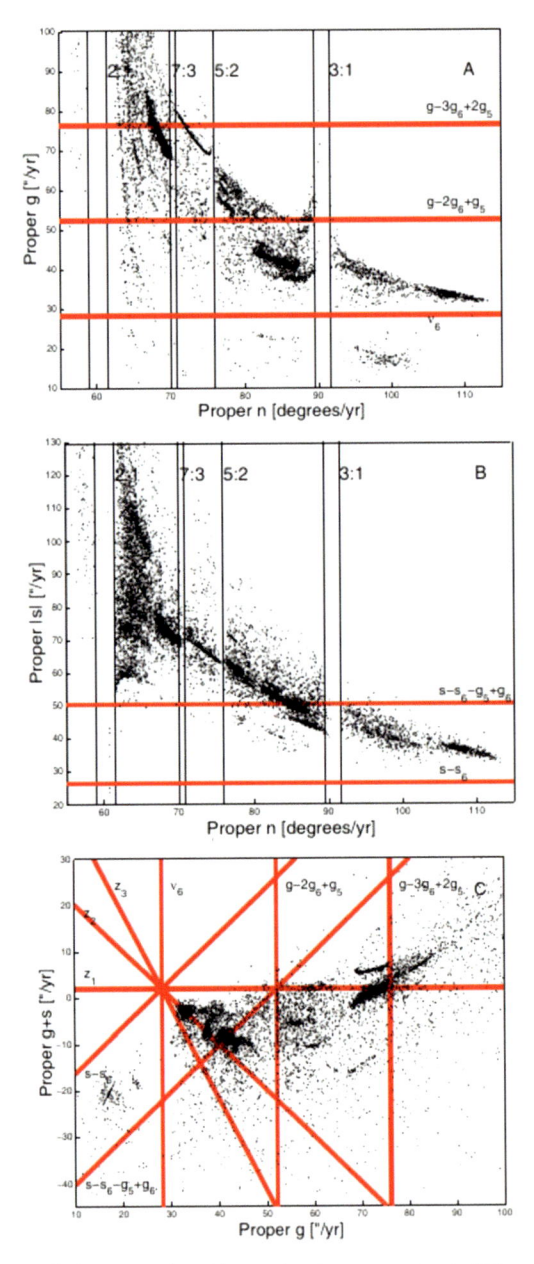

**Figure 4.2** All main belt asteroids with an absolute magnitude $H$ less than 13 are projected in the $(n, g)$ (panel A), $(n, |s|)$ (panel B), and $(g, g + s)$ (panel C) planes. Panels A and B indicate the sites of secular resonances of the pericenter and node, respectively, whereas horizontal lines represent the locations of mixed pericenter and node resonances in panel C. Adapted from Figure (2) of Carruba and Michtchenko (2007) and reproduced with permission from the authors and A&A (©A&A).

(panel A), $(n, |s|)$ (panel B), and $(g, g + s)$ (panel C) planes. Resonances of pericenter, node, and mixed secular resonances of pericenter and node appear as straight lines in such domains. Other plots, such as the proper (a,g), (a,|s|), etc., have been later suggested in the literature, having the advantage of allowing for an easier representation of both mean-motion and secular resonances. See Carruba et al. (2018) for a more in-depth discussion of these kinds of representations.

## 4.3. Asteroid families interacting with secular resonances

Asteroid families can be identified as clusters in domains of proper elements or frequencies. In the original application of the hierarchical clustering method (HCM, (Zappalà et al., 1990)), the distance between a given asteroid and a neighbor is first computed. The second asteroid is considered to be a member of the family if its distance from the first is less than a critical threshold parameter, known as $d_{cutoff}$. The process is iterated until no new family members are discovered, and it is then repeated with the second asteroid on the list. The selection of an acceptable metric in the three-dimensional element space is a crucial step in this process. Zappalà et al. (1990) define the distance as

$$d_1 = na\sqrt{k_1(\frac{\Delta a}{a})^2 + k_2(\Delta e)^2 + k_3(\Delta \sin (i))^2}, \qquad (4.2)$$

where $n$ is the mean-motion of the first asteroid in the family, and $\Delta x$, with $x = a, e$, or $\sin(i)$ are the differences in the asteroids proper elements. The most commonly used choices for the weighting factors are $k_1 = \frac{5}{4}, k_2 = 2, k_3 = 2$, but other choices that, for instance, give higher weights to the asteroids inclinations are possible. Interested readers could find more details on the subject in Bendjoya and Zappalà (2002).

Asteroid families can also be identified in domains of proper frequencies, rather than proper elements. Carruba and Michtchenko (2007) searched families interacting with $(g + s)$-type secular resonances in a proper $(n, g, g + s)$ domain, using a distance metric:

$$d_2 = \sqrt{h_1(\frac{\Delta n}{h_0})^2 + h_2(\Delta g)^2 + h3(\Delta (g + s))^2}, \qquad (4.3)$$

where the simplest choice for the weights $h_i(i = 1, 2, 3)$ is to take them equal to 1, and $h_0$ is a normalization factor of dimension $1°/arcsec$. Distances in frequency domains have units of arcsec yr$^{-1}$. Families within this domain could, for suitable $d_2$ choices, be connected to the objects that drifted in secular resonances of the $(g + s)$-type, and that the traditional HCM in the appropriate element domain did not identify as family members. An example of this would be the Eos family and its interaction with the $z_1$ resonance. Carruba and Michtchenko (2009) also looked into additional families interacting with different kinds of secular resonances, such as $s, g, g + s, g - s, 2g + s$, and $3g + s$, as well as the proper distance metrics to analyze each scenario. Interested readers can find more information in those two papers (Carruba and Michtchenko, 2007, 2009).

More recently, ML methods have been proposed to identify asteroid families with unsupervised and supervised approaches, which are the subject of Chapter 2 in this book. Here, we will briefly summarize these approaches. ML-HCM was introduced in Carruba et al. (2019) to identify asteroid families in low-number density regions of the main belt, among the highly inclined population. With an accuracy, defined as the fraction of accurate prediction with respect to the total number of data, of more than 89.5%, ML-HCM can reliably identify family members, and it can retrieve every asteroid that was previously classified as a family member using the traditional HCM approach.

Supervised ML methods were later on applied in (Carruba et al., 2020a). In this work, the orbital distribution in proper $(a, e, \sin(i))$ of previously identified family constituents was utilized as the basis for ML classification techniques to find additional members of the family. After comparing the results of nine classification algorithms from both standalone and ensemble methods, the authors discovered that the very randomized trees (ExtraTree) method produced the best results, retrieving up to 97% of the family members that were detected using traditional HCM.

### 4.3.1 Constraints on the initial ejection velocity field from conserved quantities of secular dynamics

The resonant nature of asteroid families interacting with secular resonances permits us to obtain constraints on the original ejection velocity field of these groups, not available for other asteroid families. At the simplest level of perturbation theory, the $\nu_6$ resonance is characterized by the conservation of the quantities $K_1 = \sqrt{a}$ and $K_2 = \sqrt{a(1 - e^2)}(1 - \cos(i))$. Therefore the quantity

$$K_2' = \frac{K_2}{K_1} = \sqrt{(1 - e^2)}(1 - \cos(i)), \tag{4.4}$$

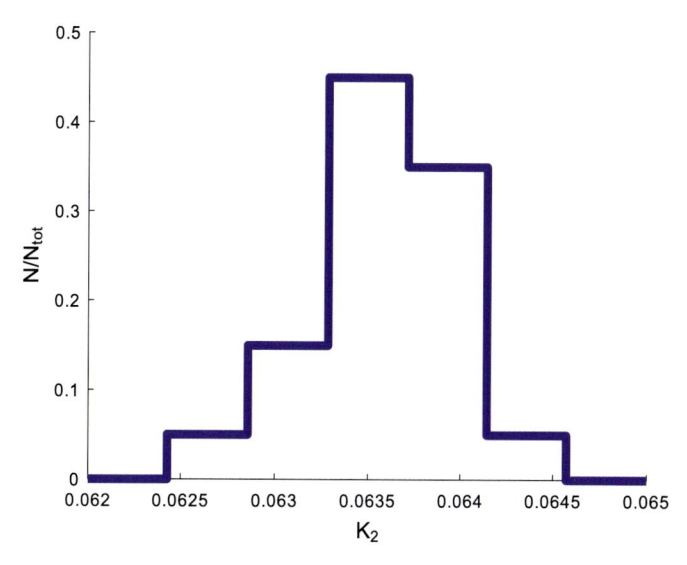

**Figure 4.3** The distribution of the $K'_2$ quantity for the Tina family members, as computed in Carruba and Morbidelli (2011). Adapted from Figure (10) of the same work and reproduced with permission from the authors and MNRAS (©MNRAS).

is conserved, even when nonconservative forces, such as the Yarkovsky effect are considered (Carruba and Morbidelli, 2011). Thus information about the initial $K'_2$ amount is preserved in the present distribution, which is depicted in Fig. 4.3 for Tina asteroids. Considering the ejection velocity field that gave rise to the Tina family as isotropic, Gaussian, centered at zero, and with a standard deviation that varied with size, we get the following relationship:

$$V_{SD} = V_{EJ} \left( \frac{5 \ km}{D} \right), \qquad (4.5)$$

where $D$ is the asteroid diameter. $V_{EJ}$ is a free parameter, which is usually of the order of the escape velocity from the parent body. The $K'_2$ values of simulated asteroid families with various $V_{EJ}$ parameters can then be used to best-fit the currently observed $K'_2$ values. Next, a $\chi^2$-like variable for each family can be introduced, which is defined as follows:

$$\chi^2 = \sum_{i=1}^{N_{int}} \frac{(q_i - p_i)^2}{q_i}, \qquad (4.6)$$

where $p_i$ is the number of synthetic family members in the same i-th interval, $q_i$ is the number of real objects in the i-th interval in $K'_2$, and $N_{int}$

is the number of interval utilized for the values of $K_2'$. The minimum of the $\chi^2$ variable and standard properties of its distribution can then be used to estimate the value of the $V_{EJ}$ parameters that best-fits the current $K_2'$ distribution, with its error. Interested readers can find more information in (Vokrouhlický et al., 2006b), where this method was first introduced.

Such an approach can also be used for asteroid families interacting with $g-$type nonlinear secular resonances. For families interacting with $z_1$ and $z_2$ resonances, the $K_2'$ quantity has to be changed to $K_2' = \sqrt{(1 - e^2)}(2 - \cos(i))$, and $K_2' = \sqrt{(1 - e^2)}(3 - \cos(i))$, respectively. More information on these topics can be found in Carruba et al. (2018).

## 4.3.2 The role of young families, time-reversal numerical methods for dating asteroid families

There are several ways to date asteroid families. Among the most popular is one that fits the V-shape of family members' slope in the (a, 1/D) domain (Spoto et al., 2015). Other approaches make use of Monte Carlo simulations of the dynamical diffusion of family members, or Yarko–Yorp models, where both the diffusion caused by Yarkovsky and YORP effects are accounted for (Vokrouhlický et al., 2006a). For these methods to function well, a sample of asteroids with at least 100 members is needed. Techniques based on time-reversal numerical simulations, such as the backward integration method (BIM (Nesvorný et al., 2002)) and the close encounters method (CEM (Pravec et al., 2010)), can yield precise age estimates for smaller, younger clusters.

Time-reversal numerical simulations are used in the BIM technique to establish past differences between the longitudes of the claimed parent body and the pericenter $\varpi$ and node $\Omega$ of family members. These differences ought to decrease to nearly nothing during the family formation phase. BIM can identify families with members up to 18 years old, according to Radović et al. (2017). The way the CEM technique works is by creating several clones of the original body as well as the other family members. The main body's escape velocity and the Hill's radius are utilized to determine the cutoff values for the relative encounter speeds and distances. The method registers the times when two clones interact closely, at low relative speeds and distances. The median values of these intervals are utilized to determine the age of the pair, whereas the error is estimated using the range between the 25th and 75th quantiles of the distribution.[1] For further details on these two approaches, see Carruba et al. (2020b) techniques section. CEM can identify asteroid families younger than around 7 million years old.

Lastly, fission pairs may arise inside very young asteroid families, as described in detail in Carruba et al. (2020b). Fission pair candidates are asteroids with a mass ratio (see Equation (5) in Carruba et al. (2020b)) of less than 0.3 and relative distances in proper elements, ascertained using Equation (4) in Carruba et al. (2020b), of less than 5 $m \times s^{-1}$.

## 4.4. Machine learning methods for identifying asteroids interacting with secular resonances: an overview

Until the year 2020, the standard approach for identifying asteroids in secular resonance was the following: First, one selected asteroids with frequency values close to those of the studied resonance. For the $v_6 = g - g_6$ resonance, that would mean identifying asteroids with $g = g_6 \pm \Delta g$, with $\Delta g$ being a threshold value, usually of the order of 0.2 arcsec $yr^{-1}$. Then, numerical simulations were performed on these asteroids under the gravitational influence of all planets, and the time series for the relevant resonant argument was obtained and plotted. Again, for the $v_6$ secular resonance, that would be $\varpi - \varpi_6$. Depending on the visual inspection of the time behavior of this angle, an asteroid could have been classified as being in a circulating orbit, if its resonant argument covered all values from $0°$ to $360°$, on a librating orbit if it oscillated around an equilibrium point, which could be $0°$ or $180°$, or on a switching orbit if the angles alternated phases of circulation and libration. A sample of such resonant argument's plots is shown in Fig. 4.4. Again, interested readers can find more details on these procedures in Carruba et al. (2018).

A problem with these methods is that the number of asteroids interacting with secular resonances is significantly increasing with new results from astronomical surveys. As a consequence, already a method based on a visual analysis of resonant arguments plots involves the analysis of several thousands of images, which is a vexing tax for a human observer. This problem,

---

[1] Quantiles are values that split sorted data or a probability distribution into equal parts. For instance, the 75th quantile is the value at which 25% of the answers lie above that value and 75% of the answers lie below that value.

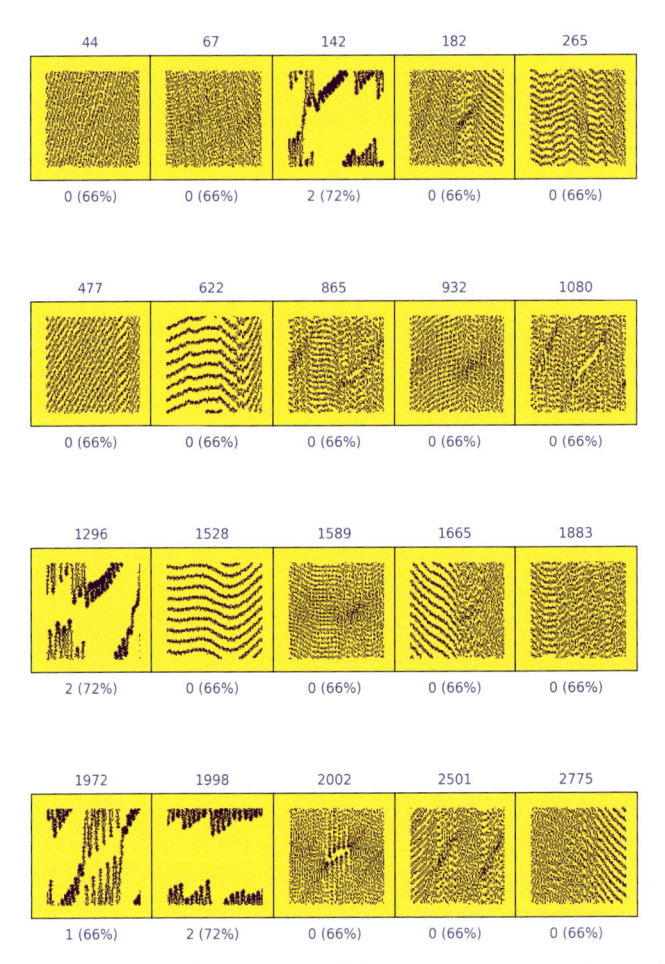

**Figure 4.4** Twenty pictures showing asteroids' resonance arguments that are impacted by Mars's M1:2 mean motion resonance. The asteroid identity is indicated by the number above each resonant argument, and the class is shown by the number at the bottom (0 for circulation, 1 for switching orbits, and 2 for librating ones), along with the corresponding probability that the asteroid belongs to that class, as determined by the Carruba et al. (2021b) ANN model. Reproduced with permission from the authors and MNRAS (©MNRAS).

which is already quite severe, will become much worse when millions of new asteroids will be found by the Vera C. Rubin scans once they begin to operate (Jones et al., 2015).

Machine learning methods for the identification of populations of asteroids in librating states of secular resonances have been developed in the last years to obviate to this issue.[2] The next sections will discuss ML and computer vision approaches that have been introduced in the last few years.

## 4.5. ML models based on the proper elements distributions

Populations of asteroids in librating states of secular resonances can be predicted by using standard ML supervised models trained on a detected population of librators. Before computer vision methods, which can build easily updated databases of images of asteroids affected by secular resonance, an ML method based on a dataset obtained purely by visual confirmation of images of resonant arguments for the $z_1$ and $z_2$ secular resonances was explored in Carruba et al. (2021a).

The proper element distribution $(a, e, \sin(i))$ of asteroids affected by these two resonances was used to train ML models selected using genetic algorithms approaches (Chen et al., 2004), as implemented in the *Tpot Python* library (Trang et al., 2019; Olson et al., 2016). Genetic algorithms imitate the process of natural evolution. There are several creatures in the actual world. The strongest ones endure and procreate. When they produce kids, genetic mixing and competition occur for some of the mutations. Then, it is done again for the next generation. Hyperparameter tuning follows the same principle in that some models with hyperparameter settings, such as the number of decision trees in a random forest algorithm, or the learning rate in a perceptron neural network, are made. A scoring mechanism, for instance, accuracy, can be used to select the best model, and new models that closely resemble the top models can then be developed. One can prevent attaining a local minimum by adding some randomness, and the procedure is repeated until a predetermined set of requirements are satisfied. Interested readers can find more information on genetic algorithms in Chapter 1.

The behavior of selected algorithms was tested using a pooling sample approach (Ishida et al., 2019; Settles, 2012). The datasets of asteroids near the $z_1$ and $z_2$ resonances were divided into a training, pool, and a test set, approximately divided in proportions of 40%, 40%, and 20%, respectively.

---

[2] Alternative methods based on periodograms of resonant arguments have been recently developed for identification of asteroids interacting with mean-motion resonances (Smirnov and Shevchenko, 2013; Smirnov, 2023). At present, no such method has yet been applied to asteroids interacting with secular resonances.

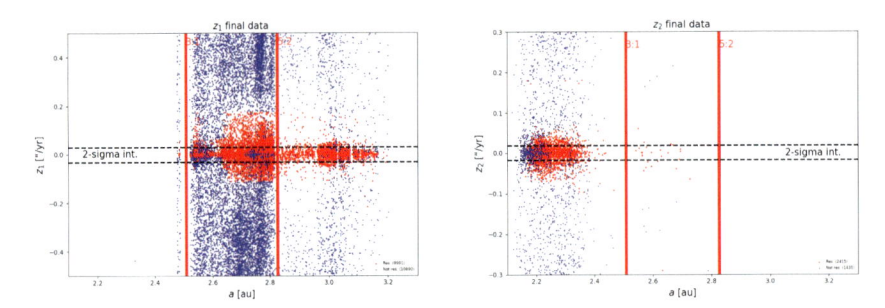

**Figure 4.5** A projection of the unlabeled asteroids in the $(a, z_1)$ (left panel) and $(a, z_2)$ (right panel). Full blue dots indicate asteroids likely to be in circulating orbits, and full red dots indicate asteroids forecast to be resonant. The 2-sigma interval calculated for the training set is displayed by the black dashed lines. Important mean-motion resonance locations are indicated by vertical lines. Adapted from Figure (8) of Carruba et al. (2021a) and reproduced with permission from the authors and CMDA (©CMDA).

The pooling sample is used to randomly choose asteroids. After being removed from the pooling sample and added to the training sample, the new asteroid is then used to predict the labels of the test sample bodies using the updated training data. After that, the procedure is repeated until the pooled sample runs out. This method was applied to confirm for which size of the training set the best results were obtained.

The results of these methods are displayed in Fig. 4.5. The dataset of resonant asteroids in the $z_1$ and $z_2$ secular resonances was later used to identify resonant subgroups using standard hierarchical clustering methods (HCM), three of which were deemed good candidates to be possible asteroid families. HCM is a clustering method, which allows identifying groups of objects in proper element domains. More details on this method were provided in Section 4.3.

Some of the databases currently available for asteroids interacting with secular resonances can be affected by a high imbalance between asteroids in different orbital classes. This is the case, for instance, of asteroids in aligned states of the $\nu_6$ secular resonance, which are in a 1:270 imbalance ratio with asteroids in other classes. This high imbalance ratio can influence the performance of traditional ML algorithms, which were not built to deal with such extreme imbalances. To solve this issue, several ways have recently been developed, including cost-sensitive tactics, methods that oversample the minority class, undersample the majority class, or combinations of both. For instance, one could use techniques such as random oversampling, which involves randomly duplicating examples from the minority class until its size matches the size of the majority class, SMOTE (synthetic minority over-sampling technique) oversampling, which is an extension of random oversampling that generates synthetic examples for the minority class, rather than simply duplicating existing examples, random undersampling, which reduces the number of examples in the majority class to achieve a balanced dataset, or Tomek undersampling, which works by removing examples from the majority class that are close to the minority class in the feature space.

Carruba et al. (2023) recently investigated the most effective strategy for increasing the performance of ML algorithms for known resonant asteroidal databases. Cost-sensitive approaches were discovered to either improve or have no effect on the outcome of ML methods and should be employed wherever possible. SMOTE oversampling plus Tomek undersampling, SMOTE oversampling, random oversampling, and undersampling performed best for the tested databases. Testing these strategies first among all the available approaches may save time and effort in future investigations with imbalanced asteroidal databases. Interested readers can find more detail on the oversampling and undersampling methods in Carruba et al. (2023). The main results of this work were that such methods should only be applied for cases of severe imbalance ratios, i.e., ratios higher than 1:100.

Later works on the identification of asteroids affected by secular resonances all used computer vision approaches. Since the proper element distribution of librating asteroids may change if there are larger samples of new bodies are found, computer vision methods could provide more robust and exportable results then those described in this section. Such methods will be discussed in the next section.

## 4.6. Computer vision models

Since the work of Carruba et al. (2021b), identification of asteroids in secular resonances is now more commonly performed using artificial neural networks (ANN) models and, more recently, convolutional neural networks (CNN) approaches (Carruba et al., 2022b,a). In this section, we will review these computer vision approaches, provide some background on these methods, and discuss recent applications.

## 4.6.1 ANN models

Neural networks' building blocks, or perceptron models, are essential to deep learning. They are considered to be the most basic type of artificial neural network, having been first introduced by Rosenblatt (1958).

A perceptron is a mathematical model that generates an output according to an activation function after applying weights to a set of input values. The activation function decides whether or not the model should operate based on the weighted sum of the inputs, and the weights establish the significance of each input. The weighted sum is calculated by multiplying each input by its corresponding weight and summing them up. The activation function then applies a nonlinear transformation to the weighted sum to produce the output. A step function, which yields 1 if the weighted sum is above a threshold and 0 otherwise, is the most widely used activation function for these architectures. The perceptron's weights are changed throughout the training phase based on a learning algorithm known as the learning rule. For a given set of training examples, the learning rule seeks to minimize the error between the desired output and the anticipated result. The weights are updated iteratively until the perceptron achieves the desired level of accuracy.

By creating multilayer perceptrons (MLPs) or artificial neural networks (ANNs) with numerous layers of interconnected perceptrons, deep learning expands the concept of perceptrons. The network can learn increasingly complicated representations of the input data since the outputs of one layer are used as inputs for the subsequent layer. Deep learning models can learn and extract high-level features from unstructured data, such as text or images, thanks to this hierarchical structure.

In conclusion, perceptron models are the fundamental building blocks of neural networks and are the basis of deep learning. Deep learning models may learn intricate patterns and representations from data by merging numerous perceptrons in layered architectures. This results in state-of-the-art performance in a variety of fields, such as computer vision, natural language processing, and speech recognition.

### 4.6.1.1 Applications to secular dynamics

The first application of an ANN model was carried out in Carruba et al. (2021b), where an ANN model was applied to identify images of resonant arguments of asteroids interacting with the M1:2 mean-motion resonance with Mars. To identify resonant argument images, the authors developed a four-layer model with a flatten, an inner, a hidden, and an output layers,

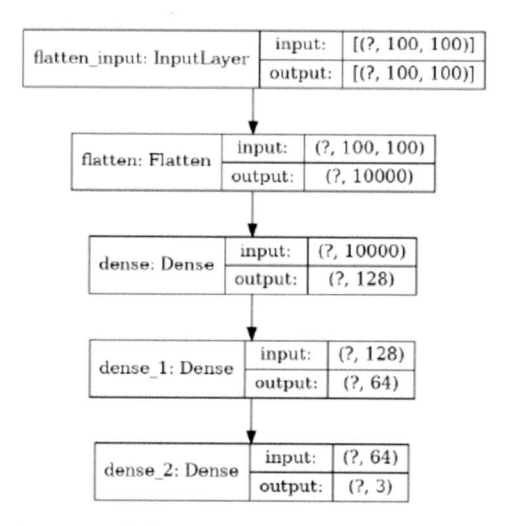

**Figure 4.6** The architecture of the ANN model introduced in Carruba et al. (2021b). Reproduced from Figure (5) of the same paper with permission from the authors and MNRAS (©MNRAS).

whose architecture is shown in Fig. 4.6. The image matrices were converted into arrays via the flatten layer. With half the neurons of the inner layer, the hidden layer searched for more complicated features in the argument images, while the inner layer looked for simpler patterns. The final categorization for the three potential classes—circulation, switching, and librating orbits—was carried out by the output layer, which had three nodes.

The approach employed in that work was a five-step procedure in which

- The asteroid orbits are numerically integrated under the gravitational influences of the planets.
- The resonant arguments are computed.
- Images of the time dependency of resonant arguments are created.
- The ANN learns from the training label picture data.
- Predictions on test images are collected, and classifications of test data images are generated.

The efficiency of the model was then measured using standard ML metrics, such as *accuracy, precision,* and *recall. Accuracy* was above 99.6%, and *precision* and *recall* were both above 90% if the training sample contained more than 4000 images. This basic framework for the analysis of images of resonant arguments has been maintained in more advanced works; it will be discussed in the next subsection.

## 4.6.2 CNN models

CNNs, or convolutional neural networks, are a sort of neural network model that were created to cope with two-dimensional image input. Their name is derived from the model's convolutional layer, which is a crucial layer. Convolution is a linear technique that involves multiplying a two-dimensional array of weights, also known as a filter, by an input array. The input and filter's filter-sized patch elements are multiplied element by element, and the results are then summed. It is customary to choose a smaller filter, then the input, to avoid repetitions in the outcome of this process. With the same filter size, the filter is applied successively from left to right and top to bottom to each overlapping piece of input data.

Methodically applying the same filter on an image is a powerful idea. If a filter is designed to detect a certain type of input feature, meticulously applying the filter throughout the image allows the feature to be discovered wherever in the data. The general interest in whether a feature exists, rather than its position, is defined as "translation invariance."

To generate a single value, the filter is multiplied once with a filter-sized chunk of the input array. When the filter is applied uniformly throughout the input array, it produces a two-dimensional array of output values representing the input filtering. As a result, this procedure's two-dimensional output array is known as a feature map. After constructing the feature map, each value can be processed through a nonlinearity, such as a rectified linear unit (ReLU). The ReLU activation function is supposed to be 0 if the input value is negative. It is intended to represent the input value itself for positive or zero values: $y = x$.

The disadvantage of convolutional layer feature map output is that it captures the precise position of features in the input. Even little changes in the feature's position in the input picture will result in a different feature map. This difficulty is frequently solved by inserting a pooling layer after the convolutional layer. In certain models, this sequence may be repeated once or more. The pooling layer generates new pooled feature maps with the same amount of features by acting individually on each feature map. Pooling is analogous to applying a filter to feature maps in that a pooling technique must be chosen. Pooling operations or filters, for example, are usually utilized with a stride of 2 pixels and are typically smaller then feature maps.

This means that the pooling layer will always compress each feature map by a factor of two, i.e., each dimension will be halved, yielding a feature map with a quarter the number of pixels or values. When a pooling layer is applied on a 6×6 (36 pixel) feature map, it produces a 3×3 (9 pixel) map. The operation of pooling is specified rather than learnt, by which we mean that the average pooling function returns the average value for each patch on the feature map under consideration, whereas the maximum pooling function returns the maximum value. A summary of the features found in the input is created using a pooling layer and creating downsampled or pooled feature maps. Their benefit is that even little changes in the position of the feature in the input picked up by the convolutional layer result in the feature remaining in the same location on the pooled feature map. The pooling layer helps the model achieve local translation invariance.

### 4.6.2.1 CNN models: history

LeNet-5 was the first effective image classification model, as detailed in LeCun et al. (1998). The system was developed to solve a handwritten character recognition issue and was evaluated using the MNIST standard dataset, reaching a classification accuracy of roughly 99.2% (or an error rate of 0.8%). The MNIST database is a widely used dataset in the field of machine learning and computer vision. It consists of a collection of hand-written digits from 0 to 9, each represented as a grayscale image of size 28×28 pixels. The dataset is commonly used as a benchmark for developing and evaluating machine learning algorithms, particularly for image classification tasks. The idea behind LeNet-5 was eventually recognized as the core of a larger system known as graph transformer networks.

Advances in image recognition came later, courtesy to the ImageNet project and its sponsored computer vision competition, the ImageNet Large Scale Visual Recognition Challenge (ILSVRC). The ImageNet project is a vast visual database intended for use in the development of visual object identification software. Several millions of photographs have been hand-annotated to indicate which images are featured in the project. At least one million of the pictures contain bounding boxes. Since 2010, the ILSVRC has been an annual software contest, in which software applications compete to accurately identify and recognize objects and sceneries. Krizhevsky et al. (2012) created AlexNet to compete in the ILSVRC-2010 competition for categorizing images of things into 1000 distinct categories. Before the advent of AlexNet, this problem was thought to be too difficult for the then-current capabilities of computer vision systems. AlexNet's success sparked a competition, which created a flood of alternative models, many of which were applied to the same ILSVRC challenge as the original AlexNet problem.

The purpose of the Simonyan and Zisserman (2014) article was to standardize convolutional network architectural designs, while simultaneously producing higher-performing models. VGG, after their lab's name, the Visual Geometry Group at Oxford, was built and tested for the ILSVRC-2014 competition. Their employment of numerous small filters has raised the bar. Most, but not all, convolutional layers are followed by maximum pooling layers. Another unique characteristic is the large number of filters, which grows with model depth, starting with 64 and rising to 128, 256, and 512 filters at the feature extraction end of the model. The VGG model includes several architecture variants, the most common of which are the VGG-16 and VGG-19, called after the number of learnt layers, 16 and 19, respectively. These architectures provided the highest performance and depth among the several VGG variants produced and evaluated. Finally, the VGG study was among the first to make model weights openly available, setting a precedent for computer vision researchers. As a result, pretrained models, such as VGG are commonly employed as a starting point for transfer learning. The architecture of the VGG model as specified by Brownlee (2020) is depicted in Fig. 4.7.

In their 2015 study titled "Going Deeper with Convolutions," Szegedy et al. (2015) made significant improvements in the use of convolutional layers. The study's authors suggested an architecture called inception, as well as a network named GoogLeNet, which earned the first place in the ILSVRC competition's 2014 round. The most essential invention behind the inception models, and the basis for their name, is the inception module. This is a stack of parallel convolutional layers with variable filter sizes (for example, $1\times1$, $3\times3$, and $5\times5$), as well as a pooling layer with a maximum of $3\times3$, the output of which is then concatenated. The number of filters (depth or channels) rapidly rises in a naive implementation of the inception model, especially when inception modules are stacked, which is a problem with this model's implementations. Convolutions on several filters using larger filter sizes (such as 3 and 5) can be computationally costly. In the inception model, $1\times1$ convolutional layers are utilized before the $3\times3$ and $5\times5$ convolutional layers, as well as after the pooling layer, to reduce the number of filters. Fig. 4.8 displays a typical Inception model architecture.

The 2015 ILSVRC competition was won by a residual network, or ResNet, introduced in He et al. (2015) as an unusually deep model. Their

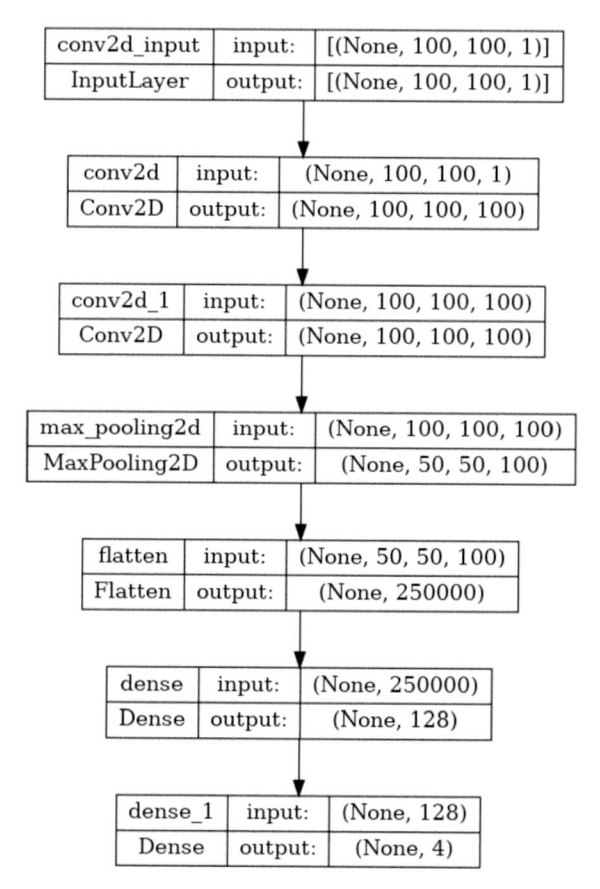

**Figure 4.7** The illustration shows a common VGG model design. The [(None, 100, 100, 1)] initial input dimension relates to the size of pictures with 100×100 pixels and just one color channel (1), coupled with a black and white image. This input is sent into the first convolutional layer, which processes it and feeds its output to the next layer until it reaches the final layer, which has a dimension of 4, corresponding to the four potential classes investigated in this case. Reproduced with permission from the authors and CMDA (©CMDA).

model is extremely remarkable, with 152 layers. The model's development relies heavily on the notion of leftover blocks that employ shortcut connections. These are simple connections in which the input is transmitted to a deeper layer, while staying unchanged (i.e., the layer below is skipped). A residual block is a pattern that combines the block's output and input by activating two convolutional layers. This is known as the shortcut connection. If the block's input shape differs from the form of the block's output,

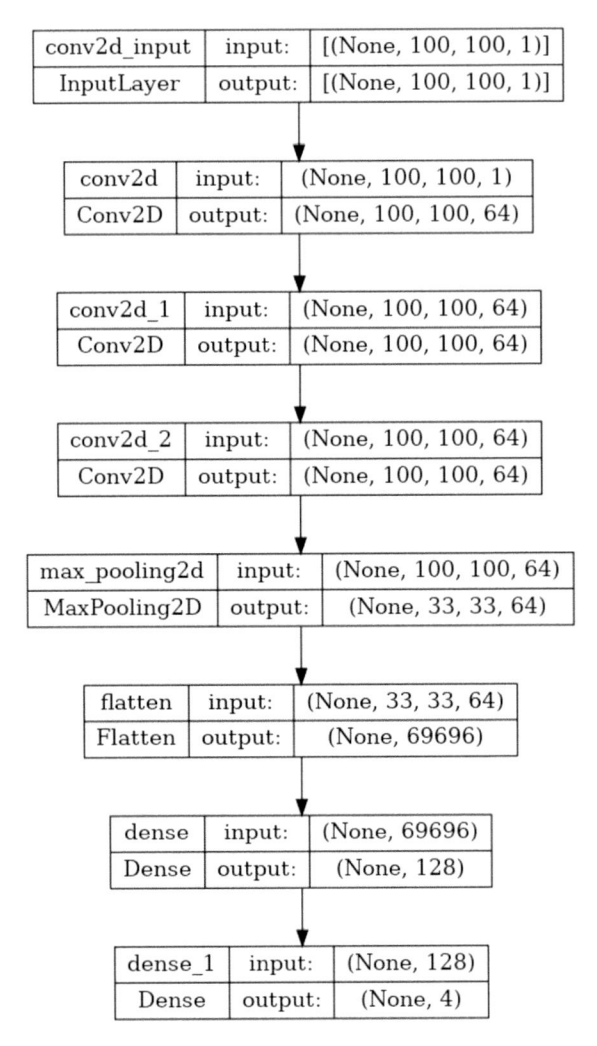

**Figure 4.8** An example of an architecture of the Inception model. The model uses a stack of convolutional layers at the start of the design. Reproduced with permission from the authors and CMDA (©CMDA).

a projected version of the input is employed. A residual network defining residual blocks is then built by introducing shortcut links to the ordinary network. The shortcut connection's input typically has the same form as the residual block's output. Fig. 4.9 displays a typical ResNet model design. More information can be found in the Keras API reference (https://keras.io/api/, (Chollet et al., 2018)).

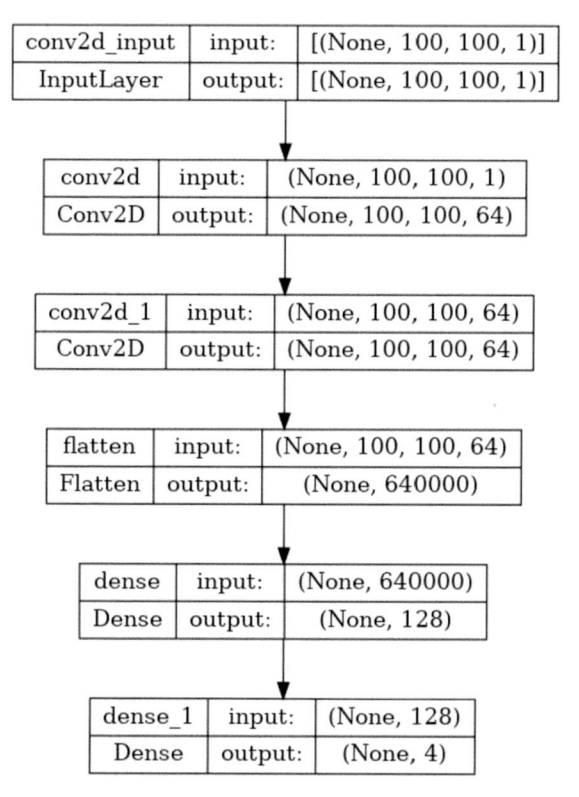

**Figure 4.9** A typical architecture of the ResNet model. The model employs residual blocks, which are patterns made up of two convolutional layers activated by ReLU functions. Reproduced with permission from the authors and CMDA (©CMDA).

### 4.6.3 Model's metrics

Two metrics are frequently used to assess the effectiveness of deep learning models: accuracy and cross-entropy loss. *Accuracy* predicts the fraction of correct predictions:

$$Accuracy = \frac{Number\ of\ correct\ predictions}{Total\ number\ of\ predictions}. \tag{4.7}$$

For binary classification, this can be defined in terms of positive and negative classifications:

$$Accuracy = \frac{TP + TN}{TP + TN + FP + FN}, \tag{4.8}$$

where $TP$ = True Positives, $TN$ = True Negatives, $FP$ = False Positives, and $FN$ = False Negatives. Again, this concept can be extended to cases of multiple class classification, in the form of a $Sparse Categorical Accuracy$ (Chollet et al., 2018). Loss/cost functions are utilized in machine learning or deep learning tasks to maximize the model during training. Reducing the loss function is the aim in most cases. If the loss is smaller, the model works better. One of the most commonly used loss functions is the cross–entropy loss. To understand how it works, we can consider the example of a CNN model that aims at classifying an image as either a dog, cat, horse, or cheetah. Let us assume that the model classifies the animal's picture with probabilities $p_i$ with values from 0 to 1. We can calculate the loss by penalizing probabilities that are more different from the expected value of 1. Differences close to 1 receive a large score, whereas smaller differences are less penalized. Mathematically, this can be expressed as

$$L_{CE} = -\sum_{i=1}^{n} t_i log(p_i), \tag{4.9}$$

where $p_i$ is the probability for the $i - th$ class and $t_i$ is the truth label. For cases with multiple classes, the *sparse categorical cross-entropy* and *categorical cross-entropy* can be defined. More information is available in the Keras API reference (https://keras.io/api/, (Chollet et al., 2018)).

### 4.6.4 The role of regularization

One of the most prominent shortcoming of deep learning models is their tendency to overfit the training data. Overfitting happens when the neural network model learns every detail of the training test. Though the model application may yield ostensibly great numbers for cross–entropy loss and accuracy, its real ability to generate credible predictions on test sets other than the training set is rather constrained. Overfitting can arise in models with a high number of trainable parameters or in models that have been trained for an extended period. Underfitting, on the other hand, may occur when the training data may still be improved. Overfitting is more prevalent than underfitting in the most recent generation of $ANN$ models.

A validation set can be used to check for overfitting in the model. This is a subset of the database that may be used to test the model that was learned on the training set. It is normally made up of around 20% of the whole

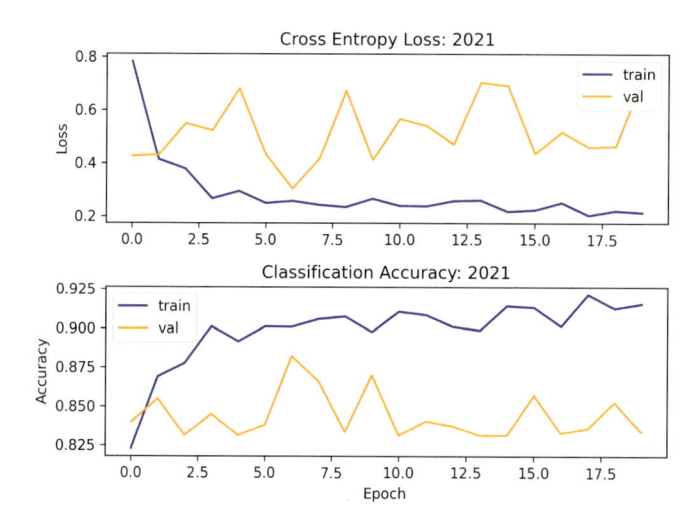

**Figure 4.10** Cross entropy loss (top panel) and classification accuracy (bottom panel) for a Carruba et al. (2022b) *ANN* model application as a function of epoch. The mismatch between the training and validation measurements, as well as the fact that the training metric's accuracy is always higher, suggest that this model may be impacted by overfitting. Adapted from Figure (5) of Carruba et al. (2022a) and reproduced with permission from the authors and CMDA (©CMDA).

database. Though the values of the training cross–entropy loss and accuracy will fall and approach one if the model overfits the training set, the same will not be observed when same metrics are applied to the validation set.

Fig. 4.10 depicts a typical example of overfitting. Methods such as data augmentation, dropout, and batch normalization can be used to improve the model's real performance. In general, the smaller the gap between the training and validation sets' cross-entropy loss and accuracy scores, the better the model's actual performance.

*Data augmentation* involves making copies of the samples in the training dataset but with small, random modifications. This increases the training dataset and has a regularizing impact by enabling the model to learn the same broad characteristics in a more general manner. Among the random augmentations that might be useful are horizontal image flipping, small image shifts, and perhaps little cropping or zooming alterations. An example of data augmentation using the *ImageDataGenerator* class in Keras for a picture from the $\nu_6$ database is shown in Fig. 4.11. Basic augmentations, such as horizontal picture flipping and 10% height and width shifts, are included in our sample.

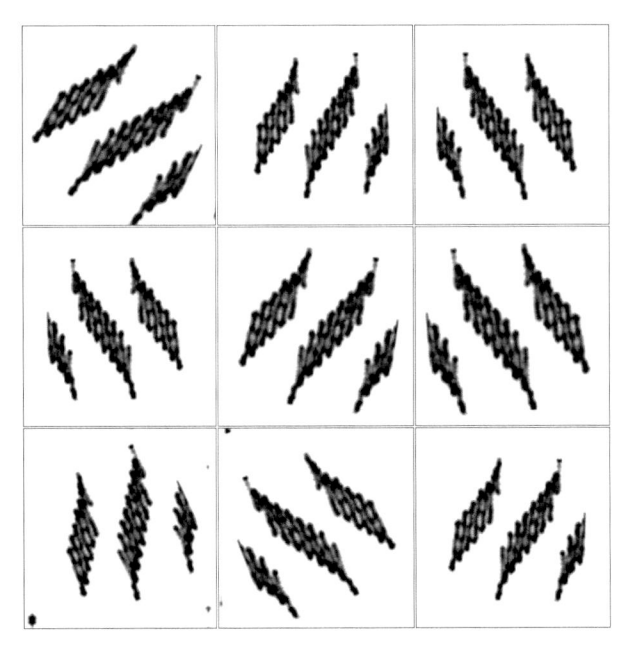

**Figure 4.11** An example of data augmentation methods being used on a picture from the $nu_6$ database. Three random flips, three random rotations, and three random zooms were used. Adapted from Figure (6) of Carruba et al. (2022a) and reproduced with permission from the authors and CMDA (©CMDA).

A simple technique that will randomly remove nodes from the network is *dropout* (Srivastava et al., 2014). This means that for each training example, a random subset of the neurons is ignored or "dropped out" with a certain probability. By dropping out neurons, the network becomes less reliant on the output of any single neuron and instead learns more robust and generalized representations. Dropout effectively creates an ensemble of multiple networks, as different subsets of neurons are dropped out during each training iteration. It has a regularizing impact since the surviving nodes have to adjust to fill up the gaps left by the removed nodes. *Dropout* may be implemented by adding new layers to the model, where a parameter controls the number of nodes eliminated.

Finally, *batch normalization* is an additional technique to lessen overfitting. The transformation used in *batch normalization* maintains the output mean and standard deviation near 0 and 1, respectively. Further details on the technique are available to interested readers in Brownlee (2020).

### 4.6.5 Computer vision applications

CNN models based on the VGG, Inception, and ResNet architectures have been recently used by Carruba et al. (2022a) to model large databases of asteroids interacting with the M1:2 mean-motion resonance with Mars and with the $\nu_6$ secular resonance (more than 6200 images of resonant arguments for both databases). The VGG model, with and without regularizations, was the most efficient method in terms of cross entropy loss function and accuracy to predict labels of these large datasets, and the less affected by overfitting issues.

More recently, Carruba et al. (2024) investigated the three most diffusive $g$-type nonlinear secular resonances using $ANN$ and $CNN$ models. $CNN$ models were obtained for the $g - 2g_6 + g_5$ secular resonance for more than 2100 images of resonant asteroids, and, for this resonance, the Inception architecture was shown to be the most efficient model. This study resulted in the identification of 12 additional likely dynamical groupings among the resonant population, including the 5507 and 170776 families, both of which have estimated ages of less than 7 Myr. These are the first two young families discovered in the resonant configurations of the investigated resonances, allowing constraints to be put on their original ejection velocity field.

## 4.7. Conclusions and future developments

In this work, we first revised the nomenclature and location of the dynamically most important resonances in the main belt (see Section 4.2). We then revised methods used to identify asteroid families in domains of proper elements or frequencies (Section 4.3). We saw what constraints on the initial ejection velocity field can be obtained from secular dynamics (subsection 4.3.1), and revised methods based on backward numerical integrations used to date young asteroid families (subsection 4.3.2).

We then introduced ML methods for identifying asteroids that interact with secular resonances (Section 4.4), either based on the asteroids proper elements distribution (Section 4.5), or on computer vision approaches to classify images of asteroids resonant arguments (Section 4.6). Computer vision approaches could be based on $ANN$ models (subsections 4.6.1, 4.6.1.1), or on $CNN$ models (subsection 4.6.2). For the latter, we revised some recent history in the introduction of architectures such as the VGG, Inception, and ResNet models (subsection 4.6.2.1), we defined metrics to assess these models' performances (subsection 4.6.3), and we discussed some approaches to avoid the effects of overfitting (subsection 4.6.4). Recent applications of *CNN* models were discussed in subsection 4.6.5, while in this section we presented our conclusions.

*ML* and *AI* approaches permitted significant advances in the field of asteroid secular dynamics, allowing for fast and reliable identification of asteroidal populations in such orbital configurations. Tasks that once required weeks of tedious visual analysis of thousands of images of resonant arguments can now be performed in seconds, with high accuracies.

Based on the analysis of the current state-of-the-art, some foreseeable trends for this field of research would include i) investigating secular resonances that have not yet been studied with an ML approach; and ii) the use of other architectures for computer vision approach to the identification of asteroids in such configurations.

Concerning the first line of research, $s-$type resonances, $(g - s)$-type, and other higher-order resonances have not yet been studied with any ML or deep learning approaches, and they are obvious candidates for future research.

The second approach would involve the use of new architecture, such as the vision transformer model (ViT). In 2021, Google researchers introduced the ViT architecture for image categorization problems (Dosovitskiy et al., 2020). Its foundation is the Transformer, a potent architecture that was first created for applications involving natural language processing. ViT's primary concept is to use the Transformer to transform images by treating them like a series of patches. Each patch in the grid-like representation of the image is regarded as a token, much like a word in natural language processing. After that, a conventional Transformer encoder is supplied with the patches in a sequence that has been flattened. The ultimate output is then generated by passing the output of the last Transformer layer through a classification head.

ViT's ability to perform at the cutting edge of image classification tasks with a far smaller computing footprint then conventional convolutional neural network (CNN) designs is one of its main advantages. This is because the Transformer's self-attention mechanism makes it possible to analyze long-range relationships in images more effectively, which is crucial for handling large-scale picture datasets. ViT has been used to classify images for a variety of purposes, such as medical imaging, object recognition, and scene classification, and it has frequently shown outstanding results.

In astronomy, most recent applications have been in the field of galaxy morphology, in works like Lin et al. (2022) or Carrasco-Davis and Carrasco (2021). There are no currently known applications for the study of asteroids interacting with secular resonances, which remains an open and interesting area for possible future research.

## 4.8. Code availability

The Carruba et al. (2021b) *ANN* model is available at this GitHub repository:

https://github.com/valeriocarruba/ANN_Classification_of_M12_resonant_argument_images/

The *CNN* models introduced in Carruba et al. (2022a) can be obtained from

https://github.com/valeriocarruba/CNN-Optimization

A Jupyter notebook for the same *CNN* models is available at these links:

Docs: https://solar-system-ml.github.io/book/chapter4/
Repository: https://github.com/solar-system-ml/book/tree/main/docs/chapter4

## Acknowledgments

We acknowledge support from the Brazilian National Research Council (CNPq, grant 304168/2021-1).

## References

Bendjoya, P., Zappalà, V., 2002. Asteroid family identification. In: Asteroids III, pp. 613–618.

Brownlee, J., 2020. Deep Learning for Computer Vision: Image Classification, Object Detection, and Face Recognition in Python. Ed. Machine Learning Mastery, San Juan, PR, USA.

Carrasco-Davis, B., Carrasco, E.R., 2021. Using transformers to predict physical parameters of galaxies. Monthly Notices of the Royal Astronomical Society. Letters 506, L1–L5.

Carruba, V., Aljbaae, S., Caritá, G., Domingos, R.C., Martins, B., 2022a. Optimization of artificial neural networks models applied to the identification of images of asteroids' resonant arguments. Celestial Mechanics & Dynamical Astronomy 134, 59. https://doi.org/10.1007/s10569-022-10110-7. arXiv:2207.14181.

Carruba, V., Aljbaae, S., Caritá, G., Lourenço, M.V.F., Martins, B.S., Alves, A.A., 2023. Imbalanced classification applied to asteroid resonant dynamics. Frontiers in Astronomy and Space Sciences 10, 1196223. https://doi.org/10.3389/fspas.2023.1196223.

Carruba, V., Aljbaae, S., Domingos, R.C., 2021a. Identification of asteroid groups in the $z_1$ and $z_2$ nonlinear secular resonances through genetic algorithms. Celestial Mechanics & Dynamical Astronomy 133, 24. https://doi.org/10.1007/s10569-021-10021-z.

Carruba, V., Aljbaae, S., Domingos, R.C., Barletta, W., 2021b. Artificial neural network classification of asteroids in the M1:2 mean-motion resonance with Mars. Monthly Notices of the Royal Astronomical Society 504, 692–700. https://doi.org/10.1093/mnras/stab914. arXiv:2103.15586.

Carruba, V., Aljbaae, S., Domingos, R.C., Huaman, M., Martins, B., 2022b. Identifying the population of stable $\nu_6$ resonant asteroids using large data bases. Monthly Notices of the Royal Astronomical Society 514, 4803–4815. https://doi.org/10.1093/mnras/stac1699. arXiv:2203.15763.

Carruba, V., Aljbaae, S., Domingos, R.C., Lucchini, A., Furlaneto, P., 2020a. Machine learning classification of new asteroid families members. Monthly Notices of the Royal Astronomical Society 496, 540–549. https://doi.org/10.1093/mnras/staa1463.

Carruba, V., Aljbaae, S., Knežević, Z., Mahlke, M., Masiero, J.R., Roig, F., Domingos, R.C., Huaman, M., Alves, A., Martins, B.S., Caritá, G., Lourenço, M., Destouni, S.C., 2024. On the identification of the first two young asteroid families in g-type non-linear secular resonances. Monthly Notices of the Royal Astronomical Society 528, 796–814. https://doi.org/10.1093/mnras/stad3968.

Carruba, V., Aljbaae, S., Lucchini, A., 2019. Machine-learning identification of asteroid groups. Monthly Notices of the Royal Astronomical Society 488, 1377–1386. https://doi.org/10.1093/mnras/stz1795.

Carruba, V., Burns, J.A., Bottke, W., Nesvorný, D., 2003. Orbital evolution of the gefion and adeona asteroid families: close encounters with massive asteroids and the Yarkovsky effect. Icarus 162, 308–327.

Carruba, V., Michtchenko, T., 2007. A frequency approach to identifying asteroid families. Astronomy & Astrophysics 475, 1145–1158. https://doi.org/10.1051/0004-6361:20077689.

Carruba, V., Michtchenko, T.A., 2009. A frequency approach to identifying asteroid families. II. Families interacting with nonlinear secular resonances and low-order mean-motion resonances. Astronomy & Astrophysics 493, 267–282. https://doi.org/10.1051/0004-6361:200809852.

Carruba, V., Morbidelli, A., 2011. On the first $\nu_6$ anti-aligned librating asteroid family of tina. Monthly Notices of the Royal Astronomical Society 412, 2040–2051.

Carruba, V., Spoto, F., Barletta, W., Aljbaae, S., Fazenda, Á.L., Martins, B., 2020b. The population of rotational fission clusters inside asteroid collisional families. Nature Astronomy 4, 83–88. https://doi.org/10.1038/s41550-019-0887-8.

Carruba, V., Vokrouhlický, D., Novaković, B., 2018. Asteroid families interacting with secular resonances. Planetary and Space Sciences 157, 72–81. https://doi.org/10.1016/j.pss.2018.03.009. arXiv:1804.00505.

Celletti, A., Perozzi, E., 2020. Celestial Mechanics: The Waltz of the Planets. Springer.

Chen, P.W., Wang, J.Y., Lee, H., 2004. Model selection of svms using ga approach. In: 2004 IEEE International Joint Conference on Neural Networks (IEEE Cat. No.04CH37541), vol. 3, pp. 2035–2040.

Chollet F., et al., 2018. Keras: the Python Deep Learning library.

Dosovitskiy, A., Beyer, L., Kolesnikov, A., Weissenborn, D., Zhai, X., Unterthiner, T., Dehghani, M., Minderer, M., Heigold, G., Gelly, S., Uszkoreit, J., Houlsby, N., 2020. An image is worth 16x16 words: transformers for image recognition at scale. arXiv preprint arXiv:2010.11929.

He, K., Zhang, X., Ren, S., Sun, J., 2015. Deep residual learning for image recognition. https://doi.org/10.48550/ARXIV.1512.03385. https://arxiv.org/abs/1512.03385, 2015.

Huaman, M., Carruba, V., Domingos, R.C., Aljbaae, S., 2017. The asteroid population in g-type non-linear secular resonances. Monthly Notices of the Royal Astronomical Society 468, 4982–4991. https://doi.org/10.1093/mnras/stx843.

Ishida, E.E.O., Beck, R., González-Gaitán, S., de Souza, R.S., Krone-Martins, A., Barrett, J.W., Kennamer, N., Vilalta, R., Burgess, J.M., Quint, B., Vitorelli, A.Z., Mahabal, A., Gangler, E., COIN Collaboration, 2019. Optimizing spectroscopic follow-up strategies for supernova photometric classification with active learning. Monthly Notices of the Royal Astronomical Society 483, 2–18. https://doi.org/10.1093/mnras/sty3015. arXiv:1804.03765.

Jones, R.L., Jurić, M., Ivezić, Z., 2015. Asteroid discovery and characterization with the large synoptic survey telescope. In: Proceedings of the International Astronomical Union, vol. 10, pp. 282–292. https://doi.org/10.1017/s1743921315008510.

Knežević, Z., 2021. Survey of secular resonances in the asteroid belt. Serbian Academy of Sciences and Arts.

Krizhevsky, A., Sutskever, I., Hinton, G.E., 2012. Imagenet classification with deep convolutional neural networks. In: Pereira, F., Burges, C., Bottou, L., Weinberger, K. (Eds.), Advances in Neural Information Processing Systems. Curran Associates, Inc., p. 1.

LeCun, Y., Bottou, L., Bengio, Y., Haffner, P., 1998. Gradient-based learning applied to document recognition. In: Proceedings of the IEEE, pp. 2278–2324. http://citeseerx.ist.psu.edu/viewdoc/summary?doi=10.1.1.42.7665.

Lin, J.Y.Y., Liao, S.M., Huang, H.J., Kuo, W.T., Ou, O.H.M., 2022. Galaxy morphological classification with efficient vision transformer. arXiv:2110.01024.

Milani, A., Knežević, Z., 1994. Asteroid proper elements and the dynamical structure of the asteroid main belt. Icarus 107, 219.

Nesvorný, D., Bottke, William F. J., Dones, L., Levison, H.F., 2002. The recent breakup of an asteroid in the main-belt region. Nature 417, 720–771. https://doi.org/10.1038/nature00789.

Olson, R.S., Urbanowicz, R.J., Andrews, P.C., Lavender, N.A., Creis Kidd, L., Moore, J.H., 2016. Automating biomedical data science through tree-based pipeline optimization. Applications of Evolutionary Computation, 123–137. arXiv:1601.07925.

Pravec, P., Vokrouhlický, D., Polishook, D., Scheeres, D.J., Harris, A.W., Galád, A., Vaduvescu, O., Pozo, F., Barr, A., Longa, P., Vachier, F., Colas, F., Pray, D.P., Pollock, J., Reichart, D., Ivarsen, K., Haislip, J., Lacluyze, A., Kušnirák, P., Henych, T., Marchis, F., Macomber, B., Jacobson, S.A., Krugly, Y.N., Sergeev, A.V., Leroy, A., 2010. Formation of asteroid pairs by rotational fission. Nature 466, 1085–1088. https://doi.org/10.1038/nature09315. arXiv:1009.2770.

Radović, V., Novaković, B., Carruba, V., Marčeta, D., 2017. An automatic approach to exclude interlopers from asteroid families. Monthly Notices of the Royal Astronomical Society 470, 576–591. https://doi.org/10.1093/mnras/stx1273. arXiv:1705.09226.

Rosenblatt, F., 1958. The perceptron: a probabilistic model for information storage and organization in the brain. Psychological Review 65, 386–408.

Settles, B., 2012. Active Learning. Synthesis Lectures on Artificial Intelligence and Machine Learning. Morgan & Claypool Publishers.

Simonyan, K., Zisserman, A., 2014. Very deep convolutional networks for large-scale image recognition. arXiv:e-prints arXiv:1409.1556.

Smirnov, E.A., 2023. A new python package for identifying celestial bodies trapped in mean-motion resonances. Astronomy and Computing 43, 100707. https://doi.org/10.1016/j.ascom.2023.100707.

Smirnov, E.A., Shevchenko, I.I., 2013. Massive identification of asteroids in three-body resonances. Icarus 222, 220–228. https://doi.org/10.1016/j.icarus.2012.10.034. arXiv: 1206.1451.

Spoto, F., Milani, A., Knežević, Z., 2015. Asteroid family ages. Icarus 257, 275–289. https://doi.org/10.1016/j.icarus.2015.04.041. arXiv:1504.05461.

Srivastava, N., Hinton, G.E., Krizhevsky, A., Sutskever, I., Salakhutdinov, R., 2014. Dropout: a simple way to prevent neural networks from overfitting. Journal of Machine Learning Research 15, 1929–1958.

Szegedy, C., Liu, W., Jia, Y., Sermanet, P., Reed, S., Anguelov, D., Erhan, D., Vanhoucke, V., Rabinovich, A., 2015. Going deeper with convolutions. In: Proceedings of the IEEE Conference on Computer Vision and Pattern Recognition, pp. 1–9.

Trang, T.L., Weixuan, F., Jason, H.M., 2019. Scaling tree-based automated machine learning to biomedical big data with a feature set selector. Bioinformatics 36, 250–256. https://doi.org/10.1093/bioinformatics/btz470. arXiv:https://academic.oup.com/bioinformatics/article-pdf/36/1/250/31813758/btz470.pdf.

Vokrouhlický, D., Brož, M., Bottke, W.F., Nesvorný, D., Morbidelli, A., 2006a. Yarkovsky/YORP chronology of asteroid families. Icarus 182, 118–142. https://doi.org/10.1016/j.icarus.2005.12.010.

Vokrouhlický, D., Brož, M., Bottke, W.F., Nesvorný, D., Morbidelli, A., 2006b. The peculiar case of the agnia asteroid family. Icarus 183, 349–361.

Vokrouhlický, D., Bottke Jr., W.F., Nesvorný, D., 2013. The Yarkovsky and YORP effects: implications for asteroid dynamics. Annual Review of Earth and Planetary Sciences 41, 39–63.

Zappalà, V., Cellino, A., Farinella, P., Knežević, Z., 1990. Asteroid families. I - Identification by hierarchical clustering and reliability assessment. Astronomical Journal 100, 2030–2046.

# CHAPTER FIVE

# Neural networks in celestial dynamics: capabilities, advantages, and challenges in orbital dynamics around asteroids

**Safwan Aljbaae**[a,b]

[a]National Institute for Space and Research (INPE), Division of Graduate Studies, São José dos Campos, SP, Brazil
[b]Make The Way, São Paulo, SP, Brazil

## 5.1. Introduction

Astronomy and space exploration depend heavily on the ability to comprehend the dynamics of celestial bodies and how they interact. Accurate knowledge of orbital dynamics around small celestial bodies is essential for planning spacecraft trajectories, ensuring their safety during missions, and conducting precise observations of these objects. This knowledge is pivotal in the examination of asteroids and comets, evaluating potential impact hazards, and probing the intricacies of planets and moons.

Predicting spacecraft behavior near small celestial bodies is quite challenging, primarily due to the irregular shapes of these bodies, which differ from the typical spherical ones. In irregularly shaped bodies, the gravitational forces acting on a spacecraft are influenced by variations in mass distribution across the nonuniform surface. This is in contrast to regularly shaped bodies, where gravitational forces are more uniformly distributed due to symmetrical mass structures. The irregular shapes introduce gravitational anomalies, necessitating advanced calculations and modeling for accurate predictions of spacecraft behavior in these challenging environments. Additionally, we must consider factors related to solar activity, particularly the impact of solar flux. Solar flux, in this context, refers to the energy and radiation emitted by the Sun. It plays a significant role in influencing how spacecraft behave in the vicinity of these celestial objects, making it an essential factor to consider in our predictions.

In this chapter, we will provide a structured exploration of the dynamic realms of space exploration and astronomy, where celestial bodies undergo

*Machine Learning for Small Bodies in the Solar System*
https://doi.org/10.1016/B978-0-44-324770-5.00010-6

constant and intricate motion. The upcoming sections will delve into the applications of time series forecasting in understanding the dynamics of spacecraft. Specifically, we will explore neural networks and their applicability, discuss convolutional neural networks (CNNs) and recurrent neural networks (RNNs) for time series, elaborate on time series forecasting with neural networks, and examine their applicability to celestial dynamics. The conclusion will provide insights into the motion relative to (99942) Apophis and predicting solar flux, followed by a summary of key findings, reflections on broader implications, and identification of future research directions.

Machine learning has proven its effectiveness in the realm of orbital dynamics through diverse and successful applications. Notably, but not exclusively, machine learning has found success in addressing intricate challenges within aerospace sciences. Dachwald (2004); Hennes et al. (2016); Mereta et al. (2017) utilized artificial neural networks (ANNs) to calculate low-thrust trajectories to asteroids. Additional applications include precise pinpoint landing computation (Sánchez-Sánchez and Izzo, 2018) and orbit prediction (Peng and Bai, 2018). Moreover, ANNs were effectively employed in the preliminary design phase of numerous asteroid rendezvous missions (Viavattene and Ceriotti, 2020; Viavattene et al., 2022).

Given that orbital dynamics and celestial body behavior are inherently time-dependent, with changes in orbital parameters and celestial body motion unfolding over time, time series forecasting becomes a pertinent approach to accurately model and predict these phenomena. Consequently, this chapter aims to explore the potential of neural networks in characterizing spacecraft motion around asteroids and predicting solar flux using historical data. Specifically, the focus involves delving into convolutional neural networks (CNNs) and recurrent neural networks (RNNs) as tools for analyzing time series data in the context of celestial dynamics.

## 5.2. Neural networks and their components

In this section, we initiate a foundational exploration of neural networks, unveiling their fundamental components, architecture, and the pivotal role they play in the analysis of celestial dynamics.

Artificial neural networks belong to a category of machine learning models engineered to discern patterns, glean insights from data, and formulate predictions or decisions. Inspired by rudimentary mathematical models of the brain, these networks adeptly represent intricate nonlinear relationships between input and output variables. Conceptually, envision a neural network as an interconnected network of nodes, or "neurons," organized into layers, comprising the following (see also Fig. 5.1):

1. Input layer: Serving as the point of entry, this layer receives the initial data for processing. Each node in the input layer represents a separate aspect of the incoming data.
2. Hidden layers: Nestled between the input and output layers, one or more hidden layers operate discreetly from direct observation. Nodes within these layers meticulously process input data, refining their calculations through adjusted weights during the learning process.
3. Output layer: Concluding the network, this layer generates the final output or prediction. The quantity of nodes in this layer varies according to the specific task the neural network addresses (e.g., classification, regression).

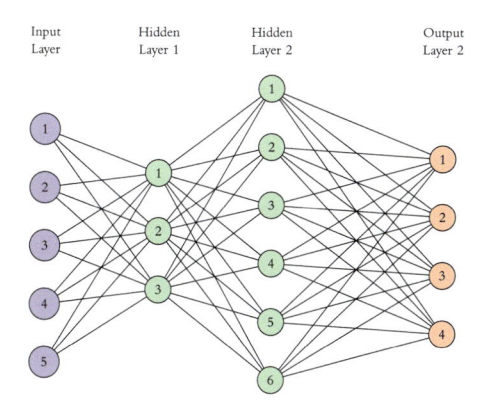

**Figure 5.1** A two-layered neural network with three hidden layers and hidden neurons for sex.

The connections between nodes have associated weights, and the network learns by adjusting these weights based on the input data and the desired output. Initially, weights are assigned random values. During the training process, the neural network learns from a labeled dataset, where both input features and their corresponding correct outputs are known. The network's prediction is compared to the actual output, and an error (the difference between the predicted and actual values) is calculated. The error is then used to update the weights through a process called backpropagation. Backpropagation involves propagating the error backward through the network adjusting the weights in a way that minimizes the error. This is usually accomplished iteratively by utilizing optimization algorithms, such as gradient descent, to update the model parameters and advance towards the minimum. This allows the neural network to generalize well to new, unseen data. Each neuron in a layer gets inputs from the layer before it, which are then multiplied by the appropriate weights. The sum of these weighted inputs is then passed through an activation function to determine whether a neuron should be activated, i.e., whether the information it receives is relevant and should be passed on to the next layer in the network. An activation function's main objective is to add nonlinearity to the network. Without a nonlinear activation function, a neural network, no matter how deep, would be equivalent to a linear model, as the composition of linear functions remains linear. Introducing nonlinearity allows neural networks to model complex relationships and make them more expressive, enabling them to learn from and adapt to intricate patterns in the data.

Based on their design, neural networks may be divided into several groups. Recurrent neural networks (RNN) and feedforward neural networks (FNN) are two common varieties.

## 5.2.1 FNN and RNN overview

In this chapter, we build upon the introductory concepts of machine learning (ML) and Artificial Intelligence (AI), discussed in Chapter 1. Interested readers could find more details in that chapter and references therein.

### 5.2.1.1 Feedforward neural networks (FNN)

Feedforward neural networks (FNN) convey information from the input layer to the output layer without cycles or loops. Neurons within the same layer do not have direct connections to each other. A standard FNN architecture consists of fully connected or dense layers. Fig. 5.2 illustrates a simple representation of a feedforward neural network layer.

### 5.2.1.2 Recurrent neural networks (RNN)

In a recurrent neural network (RNN), neurons in the hidden layer are connected to themselves, allowing information to persist and be updated over time. This recurrent connection introduces a loop in the network, enabling the capture of sequential dependencies in data. Fig. 5.3 provides a simplified representation of an RNN layer.

RNNs address the challenge of learning long-term dependencies in sequential data by incorporating memory cells. One common type of memory cell is the long short-term memory (LSTM) cell, which utilizes peephole connections to enhance its ability to capture long-term depen-

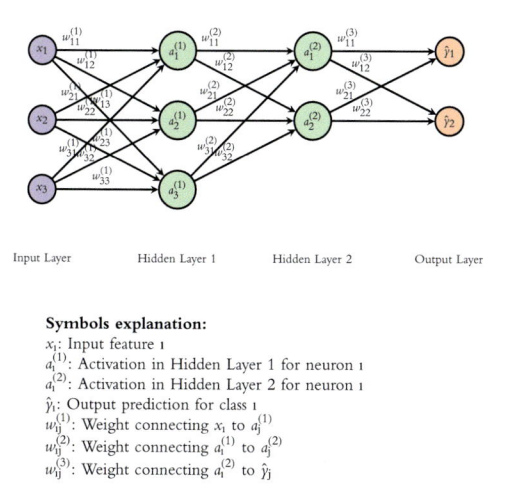

Figure 5.2  Feedforward neural network architecture.

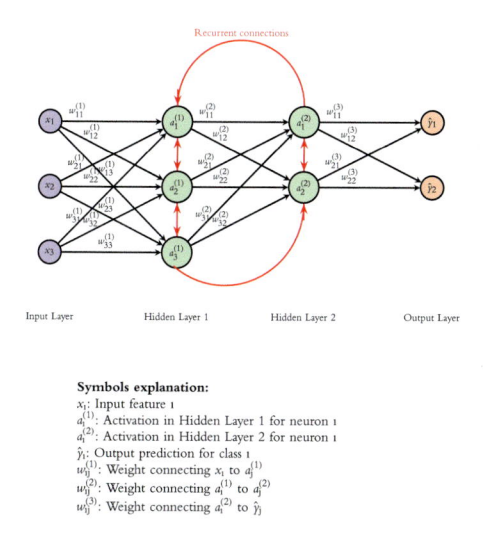

Figure 5.3  Recurrent neural networks architecture.

dencies. These connections allow the cell to consider its internal state history when making decisions. The three types of peephole connections are

1. Connection to the forget gate: The cell state from the previous time step influences the decision of the forget gate in determining what information to discard from the cell state.

2. Connection to the input gate: The input gate, which chooses what fresh data to store in the cell state, is likewise influenced by the cell state.

3. Connection to the output gate: The cell state influences the output gate, which regulates the information that gets passed to the next layer of the network.

In Fig. 5.4, we present a simple textual representation of a feedforward neural network layer.

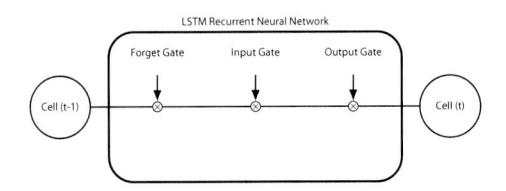

**Figure 5.4** An example of LSTM memory cell architecture.

## 5.2.2 Time series forecasting with neural networks

Time series data, a fundamental element in many fields, represents a sequence of observations ordered chronologically. This type of data inherently captures the temporal dependencies and patterns that evolve over time. Understanding and predicting the future values of a time series is a crucial aspect, and this chapter delves into the application of neural networks for effective time series forecasting using neural networks, which involves harnessing historical observations to predict future values. Neural networks, with their capacity to capture intricate patterns and relationships, offer a robust approach to modeling time-dependent data. Before we embark on the intricacies of applying neural networks to time series forecasting, let's first establish a fundamental understanding of the basic components that constitute time series data.

### 5.2.2.1 Basic components of time series data

Time series data consists of four primary components:

1. Trend: The trend represents the long-term movement or pattern in a time series. It indicates the general direction in which the time series is moving. Trends can be upward, downward, or stable over time.

2. Cyclical: Cyclical patterns in a time series are fluctuations that occur over a long period. Cyclical movements are not as regular as seasonal patterns and can last for an extended period.

3. Seasonality: Seasonality refers to regular, periodic fluctuations in a time series that occur at fixed intervals. These patterns are associated with seasonal factors, such as weather, holidays, etc. Seasonality repeats in a predictable manner, such as monthly or quarterly.

4. Irregular fluctuations, also known as random or residual components, represent the unpredictable and random variations in a time series. They are not attributed to any specific trend, cyclical, or seasonal factors. Irregular fluctuations can be caused by unpredictable events, noise, or other external factors.

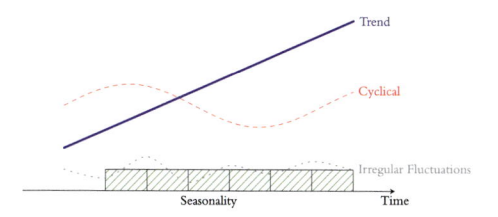

**Figure 5.5** Basic components of time series data.

In Fig. 5.5, we present the basic components of time series data. The blue line in the plot represents the trend, showing a general upward movement over time. The red dashed line illustrates cyclical fluctuations, which demonstrate periodic movements with no fixed pattern. The green shaded areas represent seasonality, with recurring patterns at regular intervals. The gray, loosely dotted line represents irregular fluctuations, indicating random and unpredictable variations. These components collectively contribute to the complex dynamics observed in time series data, making their accurate modeling and forecasting a challenging yet essential task. Neural networks, with their capacity to capture intricate patterns, offer a versatile and powerful approach to uncovering the underlying structures within time series data for effective forecasting.

Having elucidated the basic components of time series data in Fig. 5.5, we now delve into the methodologies employed for their effective modeling and forecasting. Understanding the intricacies of trends, cyclical fluctuations, seasonality, and irregularities sets the stage for exploring how neural networks navigate the temporal intricacies inherent in time series data.

### 5.2.2.2 Input representation

In the context of time series forecasting, the input to the neural network consists of past observations. The temporal sequence of these data points is crucial for capturing the underlying patterns and trends within the time series.

### 5.2.2.3 Sliding windows

To facilitate the training of neural networks for time series forecasting, the concept of sliding windows is often employed. A sliding window involves selecting a fixed-size subset of consecutive past observations as input. The corresponding future value, occurring immediately after the window, becomes the target output. This sliding window approach allows the network to learn from different temporal contexts and adapt to changing patterns over time.

### 5.2.2.4 Training process

During the training process, the neural network adjusts its internal weights based on the error between predicted and actual values. The sequential nature of time series data necessitates models that can discern patterns across multiple time steps. Convolutional neural networks (CNNs) and long short-term memory (LSTM) networks, which inherently capture temporal dependencies, are commonly employed for this task.

### 5.2.2.5 Generalization to unseen data

Neural networks trained on historical data can generalize their learning to make accurate predictions for future time steps. This adaptability is crucial for capturing the dynamic nature of time series data. The models learn to recognize patterns, trends, and seasonality, enabling them to provide meaningful forecasts, even when faced with previously unseen data.

### 5.2.2.6 Challenges and considerations

Though neural networks offer powerful capabilities for time series forecasting, delving into detailed discussions of challenges, such as ensuring an adequate amount of historical data for training, overfitting, choosing appropriate network architectures, and hyperparameter tuning is beyond the scope of this chapter. However, it is crucial to address these challenges in practical applications.

Instead, we will briefly present two general applications of neural networks in the context of astronomical data. These applications will showcase the adaptability and effectiveness of neural networks in addressing real-world challenges in time series forecasting for celestial dynamics.

## 5.2.3 Applicability to celestial dynamics

Analyzing the complex and dynamic nature of celestial bodies involves grappling with intricate patterns and relationships embedded in time-varying data. In this context, neural networks emerge as a compelling choice due to their inherent capacity to capture nonlinear dependencies and adapt to diverse data structures. Celestial dynamics, which encompass the movements and behaviors of planets, stars, and galaxies, benefit from the abilities of neural networks, particularly architectures such as convolutional neural networks (CNNs) or long short-term memory (LSTM) networks, to uncover subtle and complex interactions. Despite the promise, challenges such as data scarcity, interpretability, and the risk of overfitting need thoughtful consideration. This section explores the fitting choice of neural networks for celestial body dynamics analysis, delves into their advantages in capturing intricate patterns, and addresses potential challenges, setting the stage for specific applications that showcase their effectiveness in uncovering meaningful insights within celestial dynamics data.

### 5.2.3.1 Predicting the solar flux

Small bodies in the solar system, such as asteroids and comets, frequently navigate through regions with varying solar flux conditions. Accurate predictions of solar flux play a crucial role in refining models for orbital dynamics around these small bodies. In this application, we establish a connection between the macro-scale challenges of orbiting satellites and the micro-scale dynamics of small bodies, highlighting the interdisciplinary nature of our research.

In recent years, a significant upsurge in the population of objects orbiting our planet, particularly in low-Earth orbit (LEO), has been observed. The mitigation of manmade objects in orbit is greatly impacted by the natural environment of low-Earth orbit. This is because atmospheric-satellite interaction, which is mathematically described as the perturbation owing to drag, which affects the loss of orbit mechanical energy. For satellites in low-Earth orbit, a reduced model is typically used to lower computing costs during propagations. The Earth's Keplerian gravitational field, disturbance from the center body's nonsphericity, and air drag are the main factors affecting motion. At altitudes lower than 400 km, other disturbances, such as solar radiation pressure, tides, albedo, and third-body interactions from the Sun and Moon, are often insignificant (Vallado, 2007). Within an inertial system situated at the center of mass of the Earth, the equation of motion for a satellite in low-Earth orbit is represented as

$$\ddot{\vec{r}} = -g_{4\times4}\vec{r} + \vec{a}_D. \tag{5.1}$$

The Earth's gravitational model (EGM-08) of order $4 \times 4$ is represented by $g_{4\times4}$ in this case. The drag acceleration vector, $a_D$, acts in opposition to the airflow vector, $\vec{V}_\infty$, and $\ddot{\vec{r}}$ indicates the inertial acceleration vector. The difference between the atmospheric velocity caused by the Earth's rotation and the inertial velocity vectors, including winds, is known as the airflow. Many drag models have been used to explain atmosphere-satellite interaction and lower uncertainty, such as the one put out by Mostaza Prieto et al. (2014). The fundamental model of drag acceleration is written as

$$\vec{a}_D = -\rho\left(\frac{C_D A}{2m}\right)V_\infty \vec{V}_\infty. \tag{5.2}$$

The satellite's mass is $m$, the coefficient of drag is $C_D$, and the mean area normal to its velocity vector is $A$. The ballistic coefficient is a widely used term to describe the value $m/AC_D$. The drag forces on a satellite with a low ballistic coefficient will be significant. The atmospheric density, $\rho$, is a difficult metric to calculate as it depends on solar activity, altitude, local time, and geographic coordinates.

$C_D$ is estimated with a mean value reported in the scientific literature as 2.2 for satellites in the upper atmosphere in free molecular flow (FMF) (Vallado, 2007). This estimation accounts for uncertainties arising from satellite geometry, materials, and attitude. Though Mostaza Prieto et al. (2014); Rafano Carná and Bevilacqua (2019); Tewari (2009) demonstrate the feasibility of implementing a high-fidelity drag model for FMF and/or rarefied flow, such details are beyond the scope of our current study.

The precision of the drag perturbation poses a significant challenge to orbital determination and propagation in low-Earth orbit. The drag model is intricately linked to atmospheric density and is simultaneously influenced by space weather, as indicated by Eq. (5.2). Space weather, being a stochastic model, introduces various uncertainties, including atmospheric conditions resulting from solar and magnetic activity, predictions of atmospheric density derived from empirical models, and atmospheric dynamics (including winds). In this context, the ability to reasonably forecast the behavior of weather data, to predict the behavior of weather data, and calculate atmospheric density in the future becomes crucial.

Dealing with this complex topic, various authors have approached it differently. For example, Lean et al. (2009) employed a linear autoregressive method with lags dependent on the autocorrelation function. This method utilizes the highest correlations of each day to anticipate the next, similar to the straightforward naïve forecasting method we will employ later in this work. Additionally, Henney et al. (2012) predicted the solar 10.7 cm (2.8 GHz) radio flux using the global solar magnetic field, whereas Warren et al. (2017) used a different forecasting model to predict the solar flux from 1 to 45 days by linearly combining the previous 81 measurements.

In our study, Aljbaae et al. (2023), we will leverage Earth orientation parameters (EOP) and Space Weather Data[1] to access historical solar activity data spanning from January 10, 1957, to January 11, 2021. Deep learning techniques will be applied to perform time-series forecasting in our analysis.

### Methodology and results

The EOP and Space Weather Data provide daily data for the solar radio flux (**F10.7 OBS**) since 1/10/1957. The top panel of Fig. 5.6 displays a univariate series, and from this plot, a possible seasonality is discerned.

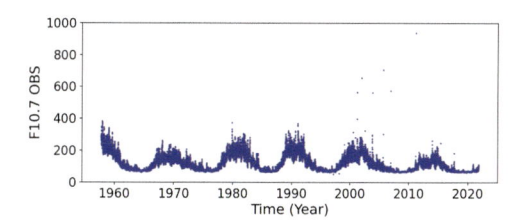

**Figure 5.6** Space weather data vs timestep from the EOP and Space Weather Data.

We used one-step time-series forecasting to anticipate the values of the aforementioned metrics **F10.7 OBS** as a preliminary step towards a more thorough investigation. Using all of the remaining observations, a basic naïve forecasting algorithm is fitted across the past 365 time steps, or one year, as a test set. Using this forecasting method, the preceding era is extrapolated to the current one. The model's performance is assessed using the walk-forward validation approach, which applies a single, one-step forecast for each of the test observations before adding the real data to the training set for the subsequent forecast. To compare our findings with the actual test set, we employed the mean absolute percentage error regression loss (MAPE) method. **scikit-learn** documentation[2] defines this measure as fol-

---

[1] https://celestrak.com/SpaceData/, accessed on November 2021.
[2] https://scikit-learn.org/stable/modules/model_evaluation.html#mean-absolute-percentage-error

lows:

$$\text{MAPE} = \frac{1}{n} \sum_{i=0}^{n-1} \frac{|y_i - \hat{y}_i|}{max(\epsilon, |y_i|)}, \qquad (5.3)$$

where $y, \hat{y}$ are the real and predicted values, respectively. The number of samples, $n$, and an arbitrarily small positive value, $\epsilon$, are used to prevent undefinable outcomes when y equals zero. An excellent performance with a MAPE of 0.025% is noticed in the top panel of Fig. 5.7, where our data are displayed.

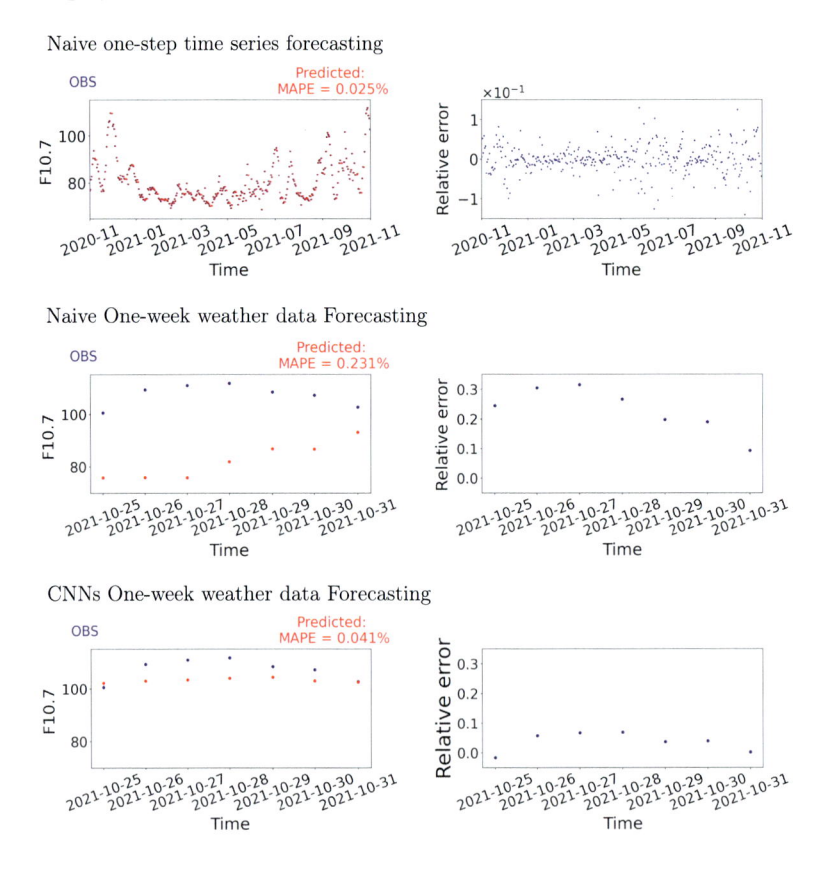

**Figure 5.7** Weather data Forecasting.

We then tried to create deep learning models for projections that were one week ahead of time. To do this, we employed a convolutional neural network (CNN) after first using a straightforward naïve technique. We have

23,406 days total in the weather data, or 3,343 complete weeks. A training set of 3,008 weeks and a test set of 335 weeks each were created from the data. In this case, the models were also assessed using a walk-forward validation technique, in which the training set is expanded to include the actual data for the week after the model makes a prediction. Throughout every week in the test set, the procedure is repeated. The findings are displayed in Fig. 5.7's center panel, where they demonstrate a comparatively poor performance with a relative error of less than 0.3 and a MAPE of 0.231%.

Our results were enhanced by employing an advanced CNN technique. Although CNNs were first developed for picture data, time-series prediction tasks may be fitted to a number of CNN models. To return the positive portion of the input, we employed a CNN model with two convolutional layers and the rectified linear unit activation function (ReLU) specified. A convolutional operation with a kernel size of three reads the tested week three times in each layer; this procedure is repeated 32 times (32 filters). Next, a pooling layer is used to determine the highest value inside a two-size window. Then 300 nodes are added to the linked layers that interpret the characteristics. The model was trained by subjecting it to the whole training dataset (100 epochs) 100 times. For every epoch, the model's weights are modified for every 12 samples (a batch size of 12). We direct the reader to Brownlee (2020) for further information about our methodology. The findings, which are displayed in Fig. 5.7's bottom panel, demonstrate that CNN can significantly lower the MAPE for the solar flux (**F10.7 OBS**) to 0.041% and a relative error of less than 0.1.

### 5.2.4 Motion relative to (99942) Apophis

Tumbling asteroids, characterized by an angular velocity vector unaligned with their principal axes of inertia, present challenges in modeling spacecraft trajectories for orbiting these objects. In a specific application (Aljbaae et al. (2021a,b)), time series prediction using recurrent neural networks (RNNs) was employed to classify orbits around the asteroid (99942) Apophis. The classification aimed to establish a relationship between prediction difficulty and orbit stability, effectively identifying the most regular orbits. The study demonstrated a strong correlation between the time-series prediction approach and established metrics, such as *MEGNO* (Cincotta and Simó (2000)) or the *perturbation map of type II* (Sanchez and Prado (2017, 2019)).

### Dynamical model

In this application, we investigate the orbital motion of a spacecraft orbiting Apophis, taking into account the Souchay et al. (2018) estimated fluctuations in the target's spin axis during near encounters. The origin of the equations of motion, which align with the J2000 ecliptic plane, is at the center of mass of the asteroid in an inertial frame. Brozović et al. (2018) models Apophis as a cloud of 3996 point masses, which match to the faces in the form model. This model is used to determine the gravitational potential; see Aljbaae et al. (2021b) for validation information.

We include the gravitational effects of the Sun, eight planets, the Moon, Pluto, and three big asteroids (Ceres, Pallas, and Vesta) to achieve accuracy similar to JPL's HORIZONS ephemerides. The starting conditions for all bodies are determined from JPL's HORIZONS ephemerides on March 1, 2029. For 60 days, we recorded the orbit every 30 seconds using the Runge-Kutta 7/8 integrator. The encounter minimum distance is found to be 37723 km, which is in close agreement with HORIZONS' result of 37728 km, with a 0.01% difference.

By comparing the orbits of every planet in the solar system with HORIZON, we were able to validate our integrator and get results that were suitable for brief integration times. Longer periods, however, should result in bigger inaccuracies because of unaccounted-for disturbances in our model, such as solar system perturbations and relativistic effects. For our study's brief 60-day integration period, this constraint is acceptable.

Spherical harmonics are used to increase the gravitational potential of the Earth and Moon up to degree and order 4. Furthermore, we apply solar radiation pressure (SRP) only to the spacecraft, following the description found in Beutler (2005). Taking into account the spacecraft's mass-to-area ratio ($60 \ \text{kg.m}^{-2}$) and reflectance (0.4), the dynamical model incorporates an OSIRIS-REx-like object. Nongravitational disturbances on the asteroid are insignificant because of the short integration time.

To sum up, our research offers a comprehensive dynamical model for a spacecraft orbiting Apophis, taking into account many gravitational effects and verifying against HORIZONS ephemerides over brief durations.

$$\ddot{\mathbf{r}} = U_r + \sum_{i=1}^{14} \mathcal{G} m_i \left( \frac{\mathbf{r}_i - \mathbf{r}}{|\mathbf{r}_i - \mathbf{r}|^3} - \frac{\mathbf{r}_i}{|\mathbf{r}_i|^3} \right) +$$
$$P_E + P_M + \nu P_R, \qquad\qquad (5.4)$$

where r represents the spacecraft's position vector in the inertial frame of reference, and $\mathcal{G}m_i$ and $r_i$ denote the $i^{\text{th}}$ body's position vector and gravitational parameter, respectively, with $\mathcal{G} = 6.67259 \times 10^{-20} \text{ km}^3 \text{ s}^{-2} \text{ kg}^{-1}$. The acceleration resulting from the Earth's and the Moon's gravitational potential is represented by $P_E$ and $P_M$, respectively, and is defined by spherical harmonics up to degree and order 4. According to Aljbaae et al. (2021a), $\nu P_R$ denotes the acceleration brought on by direct radiation pressure, while taking the shadowing phenomena into account. The gradient of the asteroid's gravitational potential, $U_r$, is computed from a total of 3996 points after the shape (Brozović et al. (2018)) is rotated about the origin in terms of both longitude and obliquity. As stated in Brozović et al. (2018), we took into consideration a constant rotation period $P_\omega = 27.38$ h and the precession period $P_\psi = 263$ h in this case. Future research should take into account the calculation of these times during the near approach with our planet, as it remains an unanswered subject.

To align Apophis with our frame of reference, we implement a series of rotations that may be expressed as follows:

$$R_z = \left( \frac{2\pi}{p_\psi} t + \lambda_0 + \Delta\psi \right),$$
$$R_x = (\varepsilon_0 + \Delta\varepsilon), \tag{5.5}$$
$$R_z = \left( \frac{2\pi}{p_\omega} t + \Delta\omega \right),$$

where the rotation matrices about the x- and z-axes are denoted by the symbols $R_x$ and $R_z$, respectively. $\Delta\psi$, $\Delta\varepsilon$, and $\Delta\omega$ are the variations of the Apophis axis, as computed in the research of Souchay et al. (2018) to identify the lowest and maximum values of the amplitudes of changes in the orientation of the spin axis. We studied how these modifications affected the dynamics of a spacecraft during the near approach when it was orbiting Apophis. As seen in Fig. 5.8, the pairings $(\lambda_0, \varepsilon_0) = (19.7°, 60.9°)$ and $(96.4°, 20.6°)$ respectively, have the lowest and maximum values of the variations. Here, we assume that throughout the Apophis/Earth near approach, the asteroid's form and the periods $p_\omega$ and $p_\psi$ do not vary considerably. This is one of our work's limits, though. There is still more work to be done on the estimate of these changes.

### Study of the orbital stability

In this section, we qualitatively analyze the orbital stability of a spacecraft around Apophis, focusing on two extreme cases of initial spin orienta-

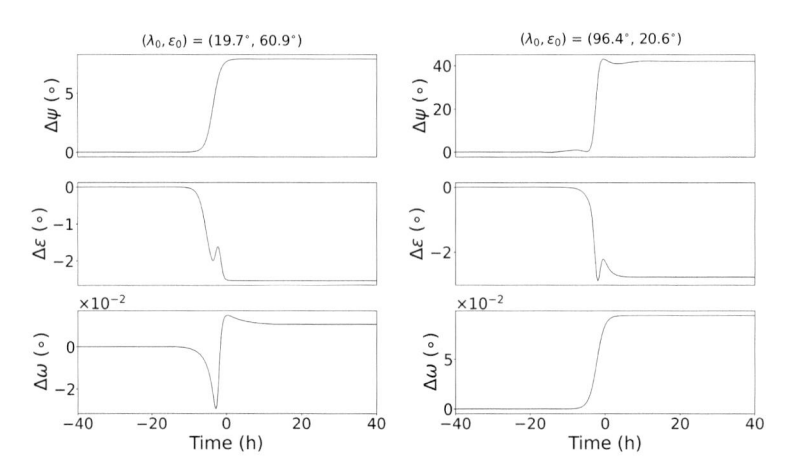

**Figure 5.8** Apophis's spin axis orientation's minimum (left column) and maximum (right column) variations.

tions: Spin-1 (19.7°, 60.9°) and Spin-2 (96.4°, 20.6°). Using Eq. (5.4), we model the spacecraft's 60-day motion around Apophis, factoring in various perturbations. The initial conditions for the planets are derived from HORIZONS on March 1, 2029, covering 43 days before and 16 days after the close encounter with Earth, allowing sufficient time for spacecraft maneuvers.

Fig. 5.9 presents the final states of orbits integrated for 40 and 60 days. Orbits are classified as escaping if the distance becomes three times larger than the Apophis Hill sphere (34 km) or colliding if they cross a 3D ellipsoid around the central body. About 95% of orbits exhibit collisions or escapes postclose encounter, confirming previous findings (Aljbaae et al., 2021a,b). The initial spin orientation subtly influences the distribution of colliding and escaping orbits, underscoring its computational significance.

The perturbation levels after 40 days, as indicated by the peak-to-peak amplitude ($\Delta a$) of semimajor axis fluctuations, are shown in Fig. 5.10. In comparison to Spin-2, the Spin-1 instance displays a longer and less disturbed zone; this might be due to the uneven form projection on the spacecraft's orbital plane. Fig. 5.11 displays the least perturbed orbits, which correspond to $\Delta a \approx 2$ m for Spin-1 and $\Delta a \approx 35$ m for Spin-2.

In summary, the study highlights the impact of Apophis' initial spin orientation on spacecraft orbits, showcasing the prevalence of chaotic orbits postclose encounter. Further investigations employ three different methods to analyze the phase space structure of these orbits.

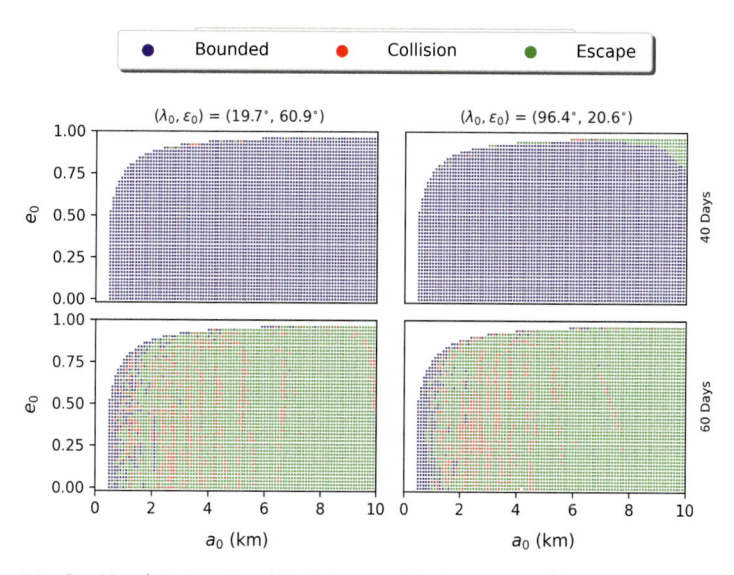

**Figure 5.9** On March 1, 2029, orbital characterization around (99942) Apophis for 40 days (top) and 60 days (bottom).

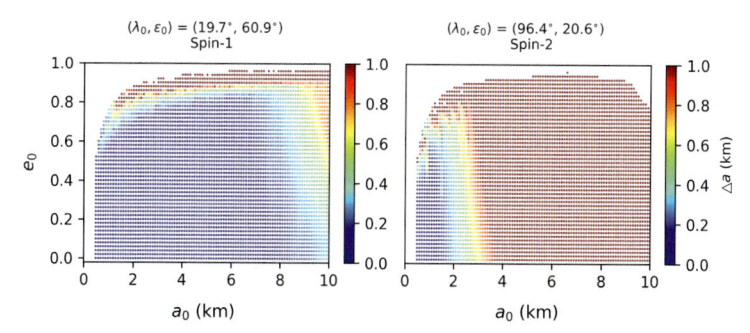

**Figure 5.10** Semimajor axis variation maps from the Apophis system's ensemble perturbations before to the system's close encounter with Earth.

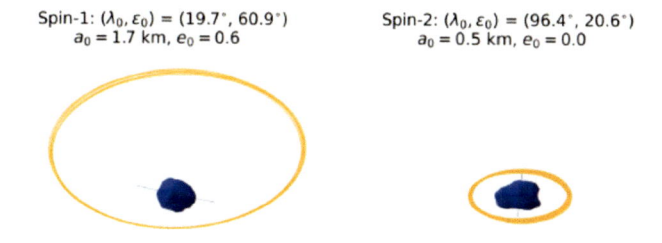

**Figure 5.11** Less agitated orbits around Apophis prior to the near-Earth encounter.

## Use of the MEGNO algorithm

In this subsection, we apply the discrete-time version of the mean exponential growth factor of nearby orbits (MEGNO), as developed by Cincotta
and Simó (2000). This method provides a global perspective on dynamics
by calculating the average relative divergence of the orbit, as expressed by
Mestre et al. (2011):

$$\text{MEGNO} = \frac{2}{T} \sum_{k=1}^{T} k \ln \left( \frac{\delta(k)}{\delta(k-1)} \right). \tag{5.6}$$

The deviation vector in phase space is denoted by $\delta(k)$ in this case,
and the total integration time is $T$. Every 30-second iteration adds up to a
maximum of 60 days.

The results are depicted in the top panel of Fig. 5.13. Larger MEGNO
values signify increased chaos and instability. Notably, isolated regions (in
dark blue) suggest spacecraft maintaining quasiperiodic orbits.

It's crucial to note that, in practice, MEGNO values may not converge
to the typical range of 2.0 to a large number, as observed in many dynamical systems. This stems from the specific nature of our system, where
a majority of orbits experience collisions with the central body or escape
within approximately 40 days. In such cases, the MEGNO indicator may
not converge to the expected value of 2, as highlighted by Cincotta and
Simó (2000).

Despite this, our analysis provides valuable insights into the system's dynamics. The rapid collision of orbits within a short timeframe is a distinctive
feature, impacting the typical behavior of MEGNO. It is important to note
that, according to the definition in Cincotta's expression, orbits that escape will be chaotic, and technically, orbits that collide may or may not be
chaotic; this depends on the specific dynamical system and its considerations. For instance, assuming a defined radius and considering the collision
of a sphere, the orbit may not be chaotic, but it could collide due to the
body's dimensions. In such cases, it is common for a higher MEGNO to be
returned to indicate this information on a single map. However, discussing
this further will be beyond the scope of this chapter.

Though MEGNO may not fully converge within the limited integration time, the observed behaviors offer a qualitative understanding of the
system's dynamics. The presented figures vividly depict the unique characteristics of our system, even if MEGNO values deviate from standard
convergence patterns. We acknowledge these limitations and emphasize

that our study offers a meaningful exploration of the system's behavior under these specific conditions.

### Perturbation map of type II (PMap)

We compute the type II perturbation maps using this approach, as described in Sanchez and Prado (2017, 2019). The following equation is used to measure the energy disturbances that the spacecraft experiences:

$$PI_{ii} = \frac{1}{T} \int_0^T \langle a, \frac{v}{|v|} \rangle \mathrm{dt}, \tag{5.7}$$

where $v$ is the spacecraft's velocity, $T$ is the end time of the numerical integration, and $a$ is the acceleration owing to all perturbations to Keplerian motion about Apophis. With this method, the energy fluctuation brought on by the perturbations is well-represented by the value of $PI_{ii}$. As an example, the blue zone in Fig. 5.13's middle panel corresponds to a negative value of the integral, suggesting a potential collision with the asteroid due to energy loss and a resulting decreasing semimajor axis. Nevertheless, this integral does not always have a negative value for collisional orbits. As we shall see later, the integral value is positive if the orbit collides after approximately 10 days of integration. It is evident that the top panel of Fig. 5.13 displays the lowest MEGNO values, which are consistent with the zone with negative values of $PI_{ii}$. As the PMap method demonstrates, even if the trajectories are severely perturbed after 10 days of integration, MEGNO—a technique primarily dedicated to identifying chaos—may not exhibit chaotic behavior. Stated differently, we might draw the conclusion that, in the vicinity of the central body, the PMap approach can yield more information than the MEGNO method.

### Time series prediction

Time series prediction does not make use of adjacent orbits, in contrast to the other two techniques. It uses time series forecasting and computational intelligence to forecast orbit coordinate distribution behavior based only on historical trends. The PYTHON language, KERAS (Chollet et al., 2015), and TENSORFLOW frameworks (Abadi et al., 2016) were utilized in the construction of our model. The process is fitting a model to past data using a series of random variables, and then utilizing the model to forecast future events.

TSix characteristics (velocity and location) are captured every 30 seconds and make up the dataset for each orbit. Every characteristic has values that vary in range from one another. Thus we normalize all the feature values between 0 and 1 before training a neural network. The model is trained using the first 90% of the points in each orbit, or 54 days, to forecast the spacecraft's location for the final 6 days of the orbit. After 43 days of integration, the Earth contact occurred; thus 11 days following the encounter were included in our training. By doing this, we can ensure that our training data includes perturbations of the close encounter.

We have 172800 points in each orbit; 155520 of those points are in our training set. That is more than sufficient to forecast each orbit's remaining 18280 points. Our model gathers data for the first 12 hours (1440 observations), sampling every 2.5 minutes to maximize performance. The places following fifteen observations serve as the label. When the validation loss is no longer improving, our training is stopped. Additional information on the procedure may be found on the Keras documentation pages at https://keras.io/. Next, we compute $\mathcal{A}$, the area between the actual and anticipated data. The orbit becomes more predictable as the region decreases, which facilitates better mapping and planning of the spacecraft mission. We illustrate a limited orbit and an orbit experiencing an escape after about 45 days in Fig. 5.12, taking into account both the Spin-1 and Spin-2 requirements.

In the bottom panel of Fig. 5.13, the forecasting map utilizing the area $\mathcal{A}$ normalized between 0 and 1 is displayed. This chart shows the conceptual compatibility of the three approaches examined here. Though the findings remain very comparable beyond $a_0 = 5.0$ km, PMap appears to display more information for the orbits that are closest to the asteroid. Taking the two neighboring orbits ($a_0 = 0.5$, $e_0 = 0.12$) and ($a_0 = 0.6$, $e_0 = 0.12$), for example, we can see that the second orbit survives the 60-day integration, whereas the first orbit collides with the asteroid after 21 days (Fig. 5.14). These orbits have values of 0.8 m/s and -0.45 m/s per year, respectively, and are included in a separate category on the PMap. On the other hand, the MEGNO and time-series prediction maps show them as near values.

### Comparison of the three methods

Lastly, we use the Pearson correlation coefficient (Pearson, 1895), which quantifies a linear relationship between two provided variables, to assess the coherence between the three approaches discussed previously. The standard formula is shown in Carruba et al. (2021), where the authors found similarities between four chaos indicators: MEGNO, the frequency analysis technique (Laskar, 1990), the auto-correlation function (Carruba et al., 2021), and the fast Lyapunov exponents (Froeschlé and Lega, 2000).

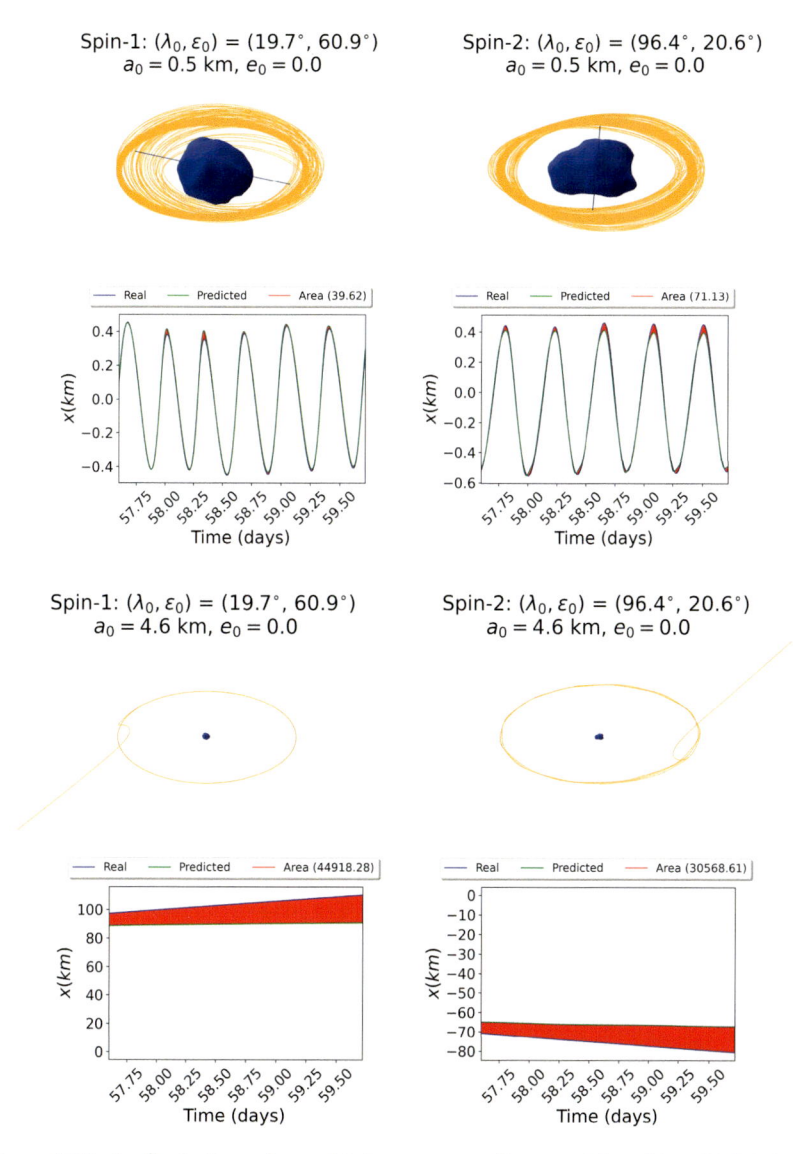

**Figure 5.12** An illustration of an orbit for a spacecraft around Apophis, which is both regular (top panel) and irregular (bottom panel), using the discrepancy between the actual and anticipated data.

There are always two possible values for the Pearson correlation coefficient: anticorrelation (–1.0) and correlation (+1.0). The variables exhibit independence as the value gets closer to zero. Fig. 5.15 exhibits with our

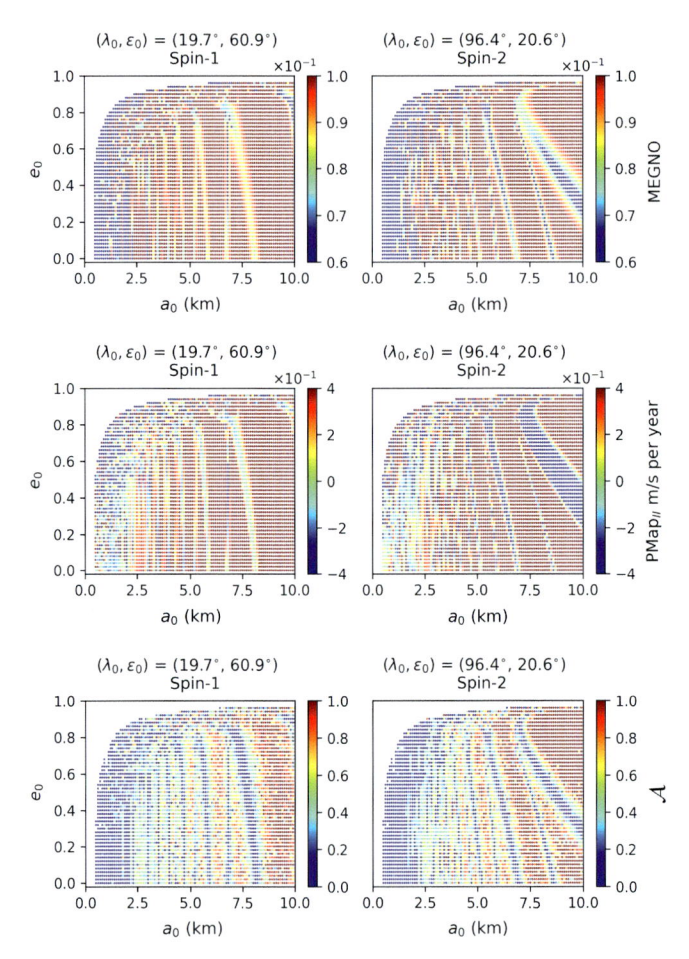

**Figure 5.13** Spacecraft circling around Apophis employing MEGNO dynamical maps, type II perturbation maps (PMap), and time-series prediction-based forecasting maps.

results. We discovered a strong correlation between the time series and PMap approaches.

## 5.3. Conclusions

In this exploration at the intersection of machine learning and space science, we've embarked on two compelling journeys: motion relative to (99942) Apophis and predicting solar flux:

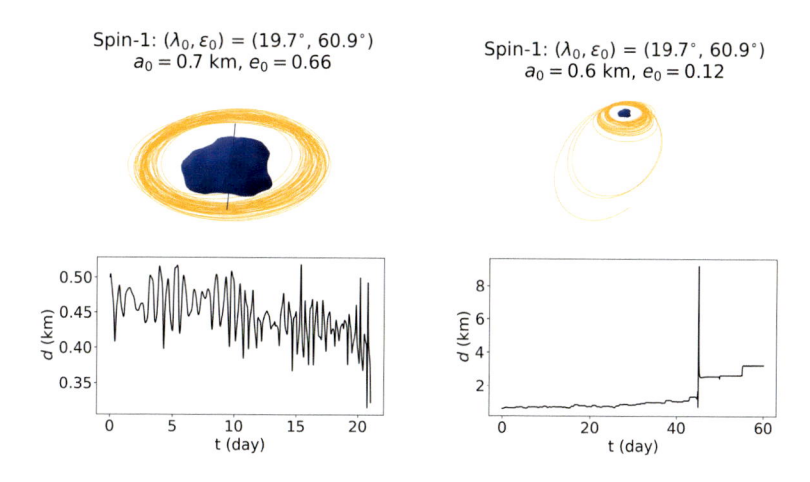

**Figure 5.14** An illustration of two neighboring orbits in the Apophis system.

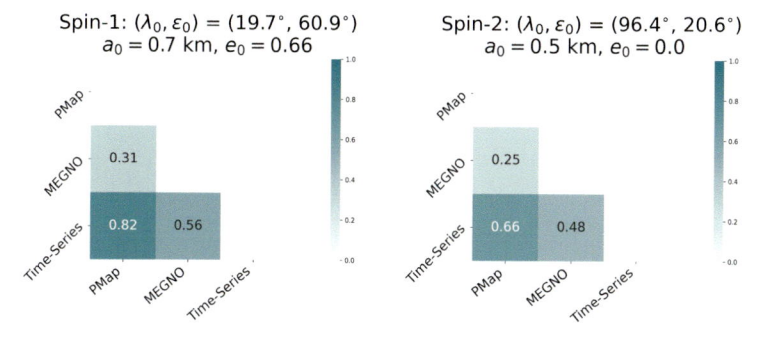

**Figure 5.15** The correlation matrix pertaining to the three techniques employed in examining the phase space structure linked to Apophis.

## Orbital dynamics

We explored how the rotation of asteroid (99942) Apophis during its 2029 close encounter with Earth impacts spacecraft orbits. Utilizing a model based on our previous work, we found that the initial spin orientation of Apophis significantly influences the spacecraft's orbital behavior. Over a 60-day integration period surrounding the close encounter, we observed that most orbits either collided or escaped due to encounter-induced perturbations, despite exhibiting stability beforehand. Employing MEGNO, PMap, and time series forecasting, we delved deeper into orbit analysis. Time series forecasting, our analytical guide, not only classified orbits based on predictability but also identified the most stable ones. The method aligned well with MEGNO and PMap, providing consistent celestial insights. Our focus was on a realistic analysis of the celestial mechanics challenge posed by a spacecraft orbiting an asteroid experiencing gravitational effects from a close encounter with Earth, rather than developing a mission plan around Apophis.

### Predicting solar flux

We presented a machine learning method to forecast daily fluctuations in solar activity using a time series forecasting model. This is a first step toward a more thorough dynamical study that aims to reduce the uncertainty associated with drag perturbation caused by air density estimate. Utilizing historical solar activity data from 1957 to the present day, obtained from Space Weather Data and the EOP, our projections showed a strong correlation with empirical data. We were able to optimize forecasts at each time step when new data became available by utilizing a convolutional neural network (CNN) within a walk-forward validation framework in conjunction with a basic naïve technique. The CNN model significantly outperformed these findings, decreasing the error to less than 0.3, while the analytical model was able to achieve a relative error of less than 0.15 for single one-step predictions and less than 0.3 for one-week forecasts.

Looking ahead, the future of space science and machine learning integration holds exciting possibilities. As technology advances, we anticipate the development and refinement of specialized algorithms and packages tailored for space applications. Enhanced deep learning models, perhaps leveraging more complex architectures or hybrid approaches, could provide even more accurate predictions for solar activity and atmospheric conditions. Furthermore, advancements in data assimilation techniques, integrating real-time observational data with predictive models, may lead to more robust and adaptive space weather forecasting systems. Collaborations between the space science and machine learning communities are likely to yield innovative solutions, enabling a deeper understanding of celestial dynamics and facilitating more precise spacecraft navigation in challenging environments.

## 5.4. Code availability

All data on the studied asteroid and code used in the work are available from the author upon reasonable request.

# References

Abadi, M., Barham, P., Chen, J., Chen, Z., Davis, A., Dean, J., Devin, M., Ghemawat, S., Irving, G., Isard, M., et al., 2016. Tensorflow: a system for large-scale machine learning.

Aljbaae, S., Murcia-Pineros, J., Prado, A.F.B.A., Moraes, R.V., Carruba, V., Carita, G.A., 2023. Machine learning to predict the solar flux and geomagnetic indices to model density and Drag in Satellites. https://doi.org/10.48550/arXiv.2307.05002. arXiv e-prints, arXiv:2307.05002, 2023.

Aljbaae, S., Sanchez, D.M., Prado, A.F.B.A., Souchay, J., Terra, M.O., Negri, R.B., Marchi, L.O., 2021a. First approximation for spacecraft motion relative to (99942) Apophis. Romanian Astronomical Journal 31, 241–264. arXiv:2012.06781.

Aljbaae, S., Souchay, J., Carruba, V., Sanchez, D.M., Prado, A.F.B.A., 2021b. Influence of Apophis' spin axis variations on a spacecraft during the 2029 close approach with Earth. Romanian Astronomical Journal 31, 317–337. arXiv:2105.14001.

Beutler, G., 2005. Methods of celestial mechanics. Vol. II: Application to planetary system geodynamics and satellite geodesy.

Brownlee, J., 2020. Deep Learning for Time Series Forecasting. Ed. Machine Learning Mastery, San Juan, PR, USA.

Brozović, M., Benner, L.A.M., McMichael, J.G., et al., 2018. Goldstone and Arecibo radar observations of (99942) Apophis in 2012-2013. Icarus 300, 115–128.

Carruba, V., Aljbaae, S., Domingos, R.C., Huaman, M., Barletta, W., 2021. Chaos identification through the autocorrelation function: the ACFI indicator. Celestial Mechanics & Dynamical Astronomy 133 (8), 38. https://doi.org/10.1007/s10569-021-10036-6.

Chollet, F., et al., 2015. Keras. https://keras.io.

Cincotta, P.M., Simó, C., 2000. Simple tools to study global dynamics in non-axisymmetric galactic potentials - I. The AAPS Journal 147, 205–228.

Dachwald, B., 2004. Optimization of interplanetary solar sailcraft trajectories using evolutionary neurocontrol. Journal of Guidance, Control, and Dynamics 27, 66–72.

Froeschlé, C., Lega, E., 2000. On the structure of symplectic mappings. The fast Lyapunov indicator: a very sensitive tool. Celestial Mechanics & Dynamical Astronomy 78, 167–195.

Hennes, D., Izzo, D., Landau, D., 2016. Fast approximators for optimal low-thrust hops between main belt asteroids, pp. 1–7.

Henney, C.J., Toussaint, W.A., White, S.M., Arge, C.N., 2012. Forecasting f10.7 with solar magnetic flux transport modeling. Space Weather 10. https://agupubs.onlinelibrary.wiley.com/doi/abs/10.1029/2011SW000748.

Laskar, J., 1990. The chaotic motion of the solar system: a numerical estimate of the size of the chaotic zones. Icarus 88, 266–291.

Lean, J.L., Picone, J.M., Emmert, J.T., 2009. Quantitative forecasting of near-term solar activity and upper atmospheric density. Journal of Geophysical Research: Space Physics 114. https://agupubs.onlinelibrary.wiley.com/doi/abs/10.1029/2009JA014285.

Mereta, A., Izzo, D., Wittig, A., 2017. Machine learning of optimal low-thrust transfers between near-earth objects, pp. 543–553.

Mestre, M.F., Cincotta, P.M., Giordano, C.M., 2011. Analytical relation between two chaos indicators: FLI and MEGNO. Monthly Notices of the Royal Astronomical Society 414, L100–L103.

Mostaza Prieto, D., Graziano, B.P., Roberts, P.C., 2014. Spacecraft drag modelling. Progress in Aerospace Sciences 64, 56–65. https://www.sciencedirect.com/science/article/pii/S0376042113000754.

Pearson, K., 1895. Note on regression and inheritance in the case of two parents. Proceedings of the Royal Society of London. Series I 58, 240–242.

Peng, H., Bai, X., 2018. Artificial neural network-based machine learning approach to improve orbit prediction accuracy. Journal of Spacecraft and Rockets 55, 1248–1260.

Rafano Carná, S.F., Bevilacqua, R., 2019. High fidelity model for the atmospheric re-entry of CubeSats equipped with the Drag De-Orbit Device. Acta Astronautica 156, 134–156.

Sanchez, D.M., Prado, A.F.B.A., 2017. On the use of mean motion resonances to explore the Haumea system. In: AAS/AIAA Astrodynamics Specialist Conference, vol. 162, pp. 1507–1524.

Sanchez, D.M., Prado, A.F.B.A., 2019. Searching for less-disturbed orbital regions around the near-Earth asteroid 2001 SN263. Journal of Spacecraft and Rockets 56, 1775–1785.

Sánchez-Sánchez, C., Izzo, D., 2018. Real-time optimal control via deep neural networks: study on landing problems. Journal of Guidance, Control, and Dynamics 41, 1122–1135.

Souchay, J., Lhotka, C., Heron, G., Hervé, Y., Puente, V., Folgueira Lopez, M., 2018. Changes of spin axis and rate of the asteroid (99942) Apophis during the 2029 close encounter with Earth: a constrained model. Astronomy & Astrophysics 617, A74.

Tewari, A., 2009. Entry trajectory model with thermomechanical breakup. Journal of Spacecraft and Rockets 46, 299–306.

Vallado, D.A., 2007. Fundamentals of Astrodynamics and Applications.

Viavattene, G., Ceriotti, M., 2020. Artificial neural network design for tours of multiple asteroids.

Viavattene, G., Devereux, E., Snelling, D., Payne, N., Wokes, S., Ceriotti, M., 2022. Design of multiple space debris removal missions using machine learning. Acta Astronautica 193, 277–286.

Warren, H.P., Emmert, J.T., Crump, N.A., 2017. Linear forecasting of the f10.7 proxy for solar activity. Space Weather 15, 1039–1051.

# CHAPTER SIX

# Asteroid spectro-photometric characterization

Dagmara Oszkiewicz[a], Antti Penttilä[b], and Hanna Klimczak-Plucińska[a]

[a]Astronomical Observatory Institute, Faculty of Physics and Astronomy, Adam Mickiewicz University, Poznań, Poland
[b]Department of Physics, University of Helsinki, Helsinki, Finland

## 6.1. Introduction

Understanding the formation and evolution of the Solar System (and by extension, other planetary systems) requires mineralogical and taxonomical information for a large number of asteroids across the Solar System. The spatial distribution of the different compounds is key in decoding the different stages of Solar System formation and evolution, and is fundamental in understanding how water and life formed on Earth (DeMeo and Carry, 2014). Several ways of obtaining this compositional information with various levels of precision have been used over the past decades.

Sample return missions are the most effective way to link asteroid spectral types with detailed mineralogical content measured in a laboratory environment. To date, samples from several celestial bodies have been collected via crewed and robotic space missions. Moon samples were brought to Earth during Apollo space missions (Johnston and Hull, 1975; Jerde, 2021). Asteroids (25143) Itokawa, (162173) Ryugu, and (101955) Bennu were probed by Hayabusa (JAXA), Hayabusa2 (JAXA), and OSIRIS-Rex (NASA) space missions, respectively (Yoshikawa et al., 2014, 2021; Watanabe et al., 2017; Carvano et al., 2010; Bierhaus et al., 2018; Lauretta et al., 2021). NASA Stardust mission collected samples of comet 81P/Wild and interstellar dust particles (Brownlee, 2014; Sandford et al., 2021). Solar wind samples have been delivered to Earth by NASA Genesis mission (Wiens et al., 2021; Burnett et al., 2003). These space missions have already provided a wealth of information. For example, fragments of (25143) Itokawa directly proved that S-type asteroids are the parent bodies of ordinary chondrites (specifically LL chondrites) (Nakamura et al., 2011; Yurimoto et al., 2011). Before the mission, there were no reliable spectral matches linking the meteorites with S-type asteroids, although the S-type

*Machine Learning for Small Bodies in the Solar System*
https://doi.org/10.1016/B978-0-44-324770-5.00011-8
147

asteroids OC meteorites link was broadly established already in the early 1970s Chapman and Salisbury (1973). Furthermore, the effect of space weathering alternating spectra of airless Solar System bodies was only well understood for Moon soil samples. Due to the Hayabusa mission, it is now known that the regolith on the surfaces of asteroids is much less affected by space weathering than the Moon's soil, which explains the spectral mismatch between S-type asteroids and ordinary chondrites (OCs) (Noguchi et al., 2011). Samples from the primitive C-complex asteroid (162173) Ryugu provided a direct link to CI chondrites (Pilorget et al., 2022; Yada et al., 2022); fragments of the B-type (101955) Bennu have not yet been analyzed.

Observed asteroid falls and, in particular, observations of the so-called imminent impactors producing "fresh" meteorites constitute another direct method of providing links between asteroid taxonomic types and meteorite types. For example, an F-type asteroid 2008 TC was observed prior to its impact in 2008, and was identified as the parent body of Urelite meteorites based on recovered samples of the Almahata Sitta meteorite (Jenniskens et al., 2010). The asteroid was discovered just 19 hours before the impact. Several other small asteroids were discovered hours before impacting Earth (2023 CX1, 2022 WJ1, 2022 EB5, 2019 MO, 2018 LA, 2014 AA, 2023 CX1, 2024 BX1). However, because of the short response time, no spectroscopic measurements were obtained for these objects prior to the fall. Preliminary reports of the remains of 2024 BX1 indicate that the recovered meteorites are aubrites, and fragments of 2023 CX1 recovered in France are ordinary chondrites.

Finding the closest spectral matches (using curve matching) between asteroids and meteorites is the most accessible and most common method to link the two groups of objects. The best-established connection is between V-type asteroids and howardite, eucrite, and diogenite (HED) meteorites (McCord et al., 1970; Consolmagno and Drake, 1977; Binzel and Xu, 1993; DeMeo et al., 2022), although a small sample of these objects could also be related to mesosiderites (DeMeo et al., 2022; Wadhwa et al., 2003). The S-complex asteroids were linked to OC meteorites (Chapman and Salisbury, 1973; Sasaki et al., 2001; Chapman, 2004; Strazzulla et al., 2005). However, asteroids of the S complex are very diverse, and other types of meteorites may also originate from this group (Gaffey et al., 1993; Rudnick, 2005; Vernazza et al., 2014). A recent study by DeMeo et al. (2022) confirms that most of the S-complex asteroid spectra can be matched with OC, but some matches were also found with howardites, eucrites, and R chondrites. The asteroids of the C complex are matched with carbonaceous chondrites (Hiroi et al., 1993; Burbine, 1998). Hydrated Ch and Cgh types were partially matched to CI and CM chondrites (Hiroi et al., 1993; Rivkin, 2012; Bland and Travis, 2017; Vernazza et al., 2017). Vernazza et al. (2014) also suggest interplanetary dust particles (IDPs) as a possible link to C-complex asteroids. CK and CO chondrites were also found to be possible matches (Carvano et al., 2003). Interestingly, DeMeo et al. (2022) found additional connections of the C-complex asteroids to urelites. This link was explained by shock darkening, which decreases the depth of the absorption bands (Kohout et al., 2014). About 14% of ordinary chondrites have been estimated to be shock-darkened (Britt and Pieters, 1991). However, since it would not be possible to shock the entire surface of large bodies, it is expected that the shock-darkened S-types are less common among the main belt objects (DeMeo et al., 2022). The X-complex asteroids are mainly paired with carbonaceous and enstatite chondrites and some with iron meteorites and ordinary chondrites (DeMeo et al., 2022). Previous links to mesosiderites have also been pointed out (Vernazza et al., 2009). D-type asteroids were found to be related to carbonaceous and ordinary chondrites (Hiroi et al., 2001; Marrocchi et al., 2021; Kebukawa et al., 2019), and even lunar materials (Yamamoto et al., 2018). DeMeo et al. (2022) found spectral matches to CM and CI carbonaceous chondrites, irons, and enstatite chondrites. The A- and Sa-type olivine-rich asteroids are matched with pure olivine mineral spectra, pallasites, bracchinites, R chondrites, diagonites, urelites, and OC (Cruikshank and Hartmann, 1984; Sunshine et al., 2007; Zolensky et al., 2010; DeMeo et al., 2022). The K-type asteroids have been matched with various carbonaceous chondrites and R chondrites (Bell, 1988; Burbine et al., 2001; Clark et al., 2009; DeMeo et al., 2022).

Although asteroid-meteorite connections are not entirely exclusive, they remain highly valuable in contextualizing theories of the Solar System evolution. In particular, the current observed structure and architecture of the Solar System has to be matched by dynamical formation and evolution models (e.g., the Grand Tack and Nice models Walsh et al. (2012); Tsiganis et al. (2005); Gomes et al. (2005); Morbidelli et al. (2005); Bottke et al. (2006, 2012); Levison et al. (2009)). For instance, the observed gradient from "redder" objects in the inner main belt to "bluer" in the outer might be attributed to a remnant of the original thermal and compositional gradient of the solar nebula (Gradie and Tedesco, 1982; Zellner et al., 1985; Gaffey et al., 1989; DeMeo and Carry, 2014). This broad compositional gradient is interrupted with implanted objects, e.g., primitive asteroids in the inner main belt, thermamorphosed objects in the outer main belt beyond the snow line, or reddish objects across the whole asteroid belt (Clark et al., 1995; Hardersen et al., 2004; Burbine and Binzel, 2002; Mothé-Diniz et al., 2003; DeMeo et al., 2014). This significant compositional diversity of asteroids is interpreted as turbulent mixing through planet migration and other processes (DeMeo and Carry, 2014). The properties and distribution of basaltic V-type asteroids in the Solar System have suggested that their parent bodies (differentiated planetesimals) likely formed close to the Sun in the terrestrial planet region and were then scattered into the main asteroid belt (Bottke et al., 2006; Morbidelli et al., 2022; Oszkiewicz et al., 2023a,b). Jupiter trojans have unique spectral properties consistent with objects in the outer main belt, possibly indicating they may have been implanted and captured from the Kuiper belt (Bottke et al., 2023; Emery et al., 2014). As additional taxonomic classifications become available for a broader range of asteroids, a more intricate understanding of the asteroid population emerges, enabling the development of more nuanced evolutionary models.

The number of taxonomical classifications available for small Solar System objects has grown exponentially in recent years, mostly due to the availability of spectro-photometric measurements from various sky surveys. Among the most important early large spectro-photometric surveys is the Eight-Color Asteroid Survey (ECAS, Zellner et al. (1985); Tedesco et al. (1982)), which led to a taxonomic system developed by (Tholen, 1984). This system was based on observations of almost 1,000 asteroids in broad bands covering wavelength ranges from 0.31 μm to 1.06 μm, as well as albedo measurements. It resulted in the derivation of 14 taxonomic types divided into broader categories and a few other types that do not match these categories. Spectroscopic surveys, such as S3OS2 (Lazzaro et al., 2004) and the MIT SMASS surveys (Xu et al., 1995; Bus and Binzel, 2002), conducted low-resolution spectroscopic measurements in visible wavelengths for hundreds of asteroids. The MIT data were utilized to create a taxonomic system at visible wavelengths (Bus, 1999), which was then extended to near-infrared observations by DeMeo et al. (2009). The extended scheme was based on almost 400 spectra in the 0.45–2.5 μm wavelength region and contained 24 classes divided into three broad complexes and a group of end members. Subsequently, Binzel et al. (2019) added a new Xn class for objects with a narrow 0.9 μm band and otherwise a relatively flat spectrum. The scheme was obtained by applying principal component analysis (PCA) to spectra of 371 objects. Recently, a collection of about 3,000 spectra from the literature was used by Mahlke (2022); Mahlke et al. (2022) to create a taxonomic system capable of classifying partial and incomplete spectra. This system identified 17 classes compatible with the previous taxonomies and defined a new Z class for extremely red objects in the main belt. The Deep European Near-Infrared Southern Sky Survey (DENIS, Epchtein et al. (1994)) performed observations in the $I, J$, and $K$ near-infrared filters for about 1,200 asteroids Baudrand et al. (2004). The Two Micron All-Sky Survey (2MASS, Skrutskie et al. (2006); Huchra et al. (2012)) conducted a survey from 1997 to 2001 in the $J, H, K_s$ bands, collecting data for several thousand asteroids (Sykes et al., 2000).

A significant step into the "big data" era for asteroids was made by the Sloan Digital Sky Survey (SDSS, Kent (1994); York et al. (2000)). The fourth SDSS data release contained measurements in the $u, g, r, i, z$ filters for about 470,000 asteroids, a huge increase in the number of taxonomic classifications compared to the number of previously classified objects (Ivezić et al., 2001). Subsequently, additional observations of about 370,000 asteroids were extracted from the SDSS images by Sergeyev and Carry (2021). Carvano et al. (2010) derived taxonomic templates based on SDSS spectrophotometric data for 9 classes. Multifilter observations of about 200,000 asteroids were extracted from the SkyMapper survey frames by Sergeyev et al. (2022). The SkyMapper filter set in the $u, v, g, r, i, z$ bands is similar to that of the SDSS. The Visible and Infrared Survey Telescope for Astronomy (VISTA, Emerson and Sutherland (2002)) is mapping the southern sky using filters $Y, J, H, K_s$. This effort has already resulted in the extraction of photometry for approximately 53,500 asteroids (Popescu et al., 2016, 2018a). The Javalambre Photometric Local Universe Survey (J-PLUS, Morate et al. (2021)) provided the first catalog of asteroid colors in 12 narrow and intermediate band filters for 3,122 minor bodies. The Gaia space mission (Gaia Collaboration et al., 2016), in its third data release (DR3), provided spectro-photometric measurements in 16 channels for about 60,500 asteroids (Gaia Collaboration et al., 2023). Recently Carruba et al. (2024) classified over 68000 asteroids observed in the course of the Dark Energy Survey (DES) into C- and S-complexes and V-types. Future surveys will continue, and even exceed, this trend in the number of observed asteroids in multiple filters and over various wavelength ranges. The recently launched Euclid mission (Laureijs et al., 2012) is expected to provide observations of about 150,000 asteroids in the $VIS_E$, $Y_E, J_E, H_E$ bands (Carry, 2018). The Legacy Survey of Space and Time (LSST) of the

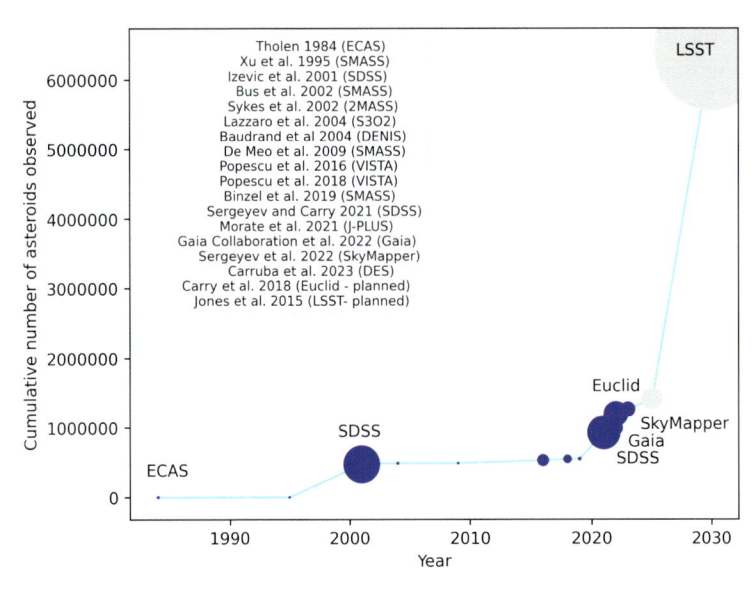

**Figure 6.1** Progressive accumulation of spectral and spectro-photometric observations of asteroids over the past decades. Grey denotes future and blue past surveys, whereas point size corresponds to the contribution of each survey. Notably, the SDSS survey significantly expanded the number of characterized asteroids, and the forthcoming LSST survey is poised to mark another substantial advancement. The plot does not take into account the overlap in the observed asteroids between the different surveys.

Vera C. Rubin Observatory is expected to discover and characterize (in the $u, g, r, i, z, y$ filters) more than 5 million asteroids during its ten-year survey lifetime (Jones et al., 2015; Olivier et al., 2012). The growth of spectral and spectrophotometric data over the past decades and planned into the future is shown in Fig. 6.1. This huge amount of data will not only provide a deeper understanding of the compositional distribution in the Solar System, but will also present computing challenges, in which machine learning algorithms might prove to be valuable tools.

## 6.2. Traditional classification methods

Depending on the data type, several methods of classifying asteroids into different taxonomic schemes have been developed. For asteroids that have visible (VIS) and near-infrared (NIR) measurements, linking observations to taxonomy by applying the PCA directions derived by DeMeo et al. (2009) in the Bus–DeMeo taxonomic scheme is possible. DeMeo et

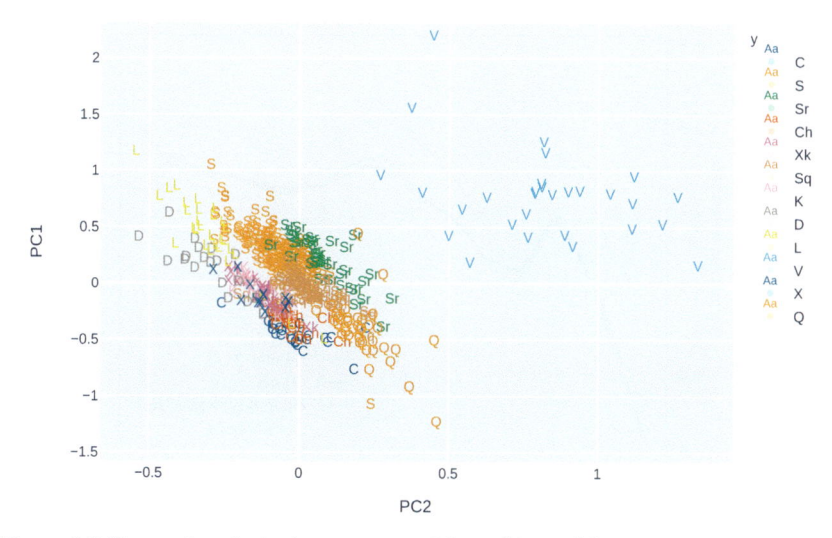

**Figure 6.2** Two main principal components PC1 and PC2 of the spectra. The data consist of 371 objects from DeMeo et al. (2009) and 195 objects from Binzel et al. (2019). The letters denote the taxonomic types, whereas colors denote the complexes. Source: Klimczak et al. (2021).

al. (2009) developed an online classifier,[1] which requires a traditional dense spectrum covering the full range from 0.45 μm to 2.5 μm. The tool maps the spectroscopic data into the PCA directions and assigns a taxonomic type based on location in the PCA component space and on the overall spectral slope (see Fig. 6.2). This space is divided into polygonal areas defining the classes.

Linking observations to taxonomy by curve matching is an alternative approach for objects with spectra that do not cover full wavelength ranges. A method based on curve–matching algorithms is used in the M4Ast tool,[2] developed by Popescu et al. (2012). The tool requires a spectrum in any range of visual or near-infrared wavelengths. The spectrum can be assigned a type in three different taxonomic schemes (Bus–DeMeo, G13, G9 (DeMeo et al., 2009; Birlan et al., 1996; Fulchignoni et al., 2000)) using different distance metrics. Several closest-matching taxonomic types are given along with a distance measure and a reliability score based on the wavelength range covered. An example of this technique is shown in Fig. 6.3. The spectrum of asteroid (7) Iris was matched with an S-

[1] http://smass.mit.edu/cgi-bin/busdemeoclass-cgi
[2] https://spectre.imcce.fr/m4ast/index.php/index/home

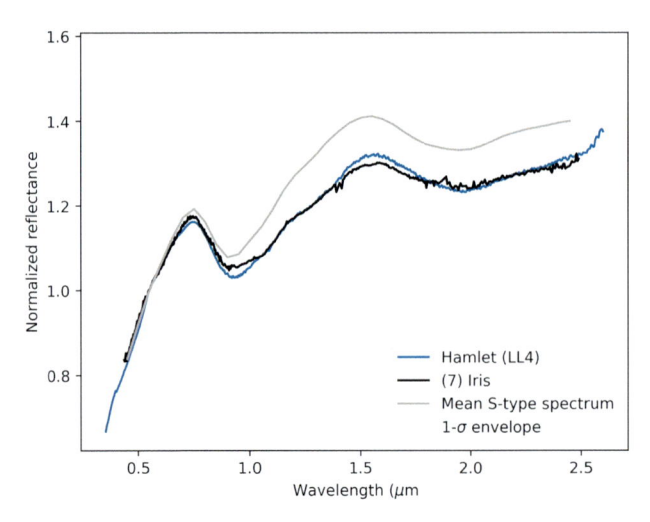

**Figure 6.3** An example of spectral curve matching. The spectrum of asteroid (7) Iris is matched with the mean S-type spectrum and with ordinary chondrite Hamlet using a curve-matching algorithm implemented in the M4Ast service (Popescu et al., 2012). The asteroid spectrum is sourced from Xu et al. (1995); Bus and Binzel (2002); Burbine and Binzel (2002), whereas the meteorite spectrum is obtained from the RELAB database (Pieters and Hiroi, 2004).

type template using the mean squared error metrics and the reliability was estimated as 100%. The closest meteorite match was the LL4 ordinary chondrite Hamlet. The tool also encompasses other functionalities, such as band and mineralogical analysis. Curve matching was recently extensively used by DeMeo et al. (2022) to identify spectral similarities between asteroid and meteorite spectra. Based on this technique, the authors confirmed known linkages between asteroid types and meteorite groups (e.g., ordinary chondrites with S-complex asteroids; pristine CM carbonaceous chondrites with Ch-type asteroids and heated CMs with C-type asteroids; HED meteorites with V types; enstatite chondrites with Xc-type asteroids; CV meteorites with K-type asteroids; Brachinites, Pallasites, and R chondrites with olivine-dominated A-type asteroids) and identified new matches (e.g., X complex with ordinary chondrites, carbonaceous chondrites with slopes matching D types).

    Matching color indices has been another popular technique, especially for objects difficult to follow with spectroscopy, such as near-Earth objects with fast sky motions Licandro et al. (2017). It is not always possible even for fast-slewing telescopes to keep very fast-moving objects in the slit for extended periods to obtain the usual spectral measurements. Therefore obtaining color photometry and color indices for those objects has become a conventional technique. Appropriate corrections for changing geometry and rotational modulations are typically applied to these measurements, and then taxonomic types are assigned based on how well the derived colors match the average colors of various taxonomic classes (Dandy et al., 2003; Koleńczuk et al., 2022; Hromakina et al., 2021, 2022).

Similar techniques based directly on matching color indices or related parameters have also been widely applied to survey data by various authors. Ivezić et al. (2002) classified the SDSS data into C-, S-, and V-types based on colors and the first principal component $a^*$ of the $r - i$ and $r - g$ space. Nesvorný et al. (2005) also utilized the PCA applied to SDSS to identify S-type objects and study the trends of space weathering among them. Parker et al. (2008) used a Gaussian distribution model in both orbital and color space to identify asteroid families in SDSS spectrophotometry. Masiero et al. (2013) utilized proper elements and geometric albedos of asteroids observed in the course of the NEOWISE survey to identify 76 distinct asteroid families, and Carruba et al. (2013) performed the task in the domain of proper elements, SDSS colors, and albedos. Moskovitz et al. (2008) identified V-type objects in the SSDS data based on the similarity of the observed colors with the colors of members of the Vesta family. Carvano et al. (2010) derived templates for the main taxonomic classes based on SDSS data and derived a new SSDS-based taxonomic system. DeMeo and Carry (2013) matched the SDSS data with the Bus–DeMeo taxonomic templates to study the distribution of asteroids throughout the Solar System. A similar approach was applied by Sergeyev and Carry (2021) to the extracted SDSS data and by Sergeyev et al. (2022) to the SkyMapper data. Sergeyev et al. (2023) used visible colors of near-Earth objects (NEOs) from the SDSS and SkyMapper surveys to study the properties of planet-crossing asteroids.

## 6.3. Overview of ML applications to asteroid spectroscopy and spectro-photometry

The recent advancements in the field of machine learning have inspired researchers to experiment with the application of these automatic methods in various tasks in astronomy. Among these methods, two major groups can be distinguished: supervised and unsupervised learning, both successfully used on spectrometric and photometric data and further discussed in this chapter.

Supervised learning can be applied to classification problems, i.e., problems where the expected outcome of the processing is known beforehand at least in a training dataset (see Chapter 1 for more details). The user must define the desired outcomes of the model, and only correctly labeled datasets must be used. This drastically reduces the pool of available data. In cases when the large amounts of existing data have not been analyzed and there are no labels present, an alternative is unsupervised learning. In this methodology, only the input features are provided to the model, and the data is automatically analyzed by the model itself. The model outputs it's perception of patterns and main themes present in the data. The user can then draw conclusions based on that processed data. Clustering is one example of unsupervised leaning.

Spectro-photometry was among the first data sources used for training neural networks in asteroid studies. In Howell et al. (1994), one of the first publications on the application of neural networks to asteroid classification, the authors tested the network's capability in classification according to the eight-color taxonomy proposed by Tholen. For this classification task, two color surveys have been used, the Eight Color Asteroid Survey (Zellner et al., 1985) and the 52 Color Asteroid Survey (Bell et al., 1988). The proposed architecture is a hybrid of a two-dimensional self-organizing map (Kohonen, 1989) and a classification output layer, with the Widrow–Hoff learning rule (Pao, 1989). This architecture works in two phases. The first phase creates a two-dimensional topologically-ordered map that clusters the objects into similar groups. The similarity, in this case, is defined as the Euclidean distance, which is compatible with the methodology from Tholen (1984). This approach does not make assumptions about the organization of the topological map or the number of clusters. In the second stage, the network is trained to classify objects according to the given label. Only the classification layer changes at this stage. The authors used the Tholen class labels as the starting point, later using subsets of the data with some examples unlabeled to check the generalization capabilities of the model. The properties of this architecture provide that only the classes that are compatible with the previously generated topological structure can be learned by the model. When the network had trouble learning to classify some objects, the examination of the Kohonen map was performed and alterations to the existing taxonomy were made. A new taxonomy was proposed as a result of this work, generally compatible with that of Tholen (1984), but with some rearrangements within a few classes and defining two new subclasses in the

S complex. The neural architecture proposed in this work obtains around 90% average prediction score.

The use of neural networks after Howell et al. (1994) was not particularly active in spectral studies, but Merényi et al. (1996) used neural networks along with linear mixture modeling and supervised spectral pattern classification to map the spectral variations on the surface of Mars and De León et al. (2010) to compare near-Earth asteroids with ordinary chondrites. Recently, there has been an increase in the applications of neural networks applied to taxonomic classification. Hietala (2020) studied the performance of a neural network in classifying the observations used for the Bus–DeMeo taxonomy (DeMeo et al., 2009) and made in the MITH-NEOS survey (Binzel et al., 2019). Penttilä et al. (2021) studied how a neural network classifier would perform in classifying Gaia spectral data to the Bus–DeMeo system, and Penttilä et al. (2022) researched how LSST observations could be classified. A similar study was done by Luo et al. (2023) for the spectral data from the Chinese Space Survey telescope.

Another popular method, the $k$-nearest neighbor algorithm, was utilized in Popescu et al. (2018b), using the NIR colors of the MOVIS catalog of the VISTA-VHS survey to group observations. This classification was found to align with that of the Bus–DeMeo taxonomy. The authors provided taxonomical classifications for 18,265 asteroids, of which the final classification was validated for 6,496 objects. Interestingly, the authors derived size-frequency distributions for V- and A-type asteroids and found that the distributions were identical, with V-type asteroids being five times more common than A types. This observation led the authors to indicate that the two most common scenarios for explaining the apparent lack of olivine material in the main asteroid belt are unlikely. These scenarios include the "battered to bits" hypothesis (Burbine et al., 1996) and the suggestion that the classical differentiation mechanism may be uncommon (Elkins-Tanton et al., 2011). The analysis of V-type asteroids was also performed by Oszkiewicz et al. (2014) using the naive Bayes method on data from the Sloan Digital Sky Survey and the Wide-field Infrared Survey explorer. The data were further enhanced with phase curve data. In Oszkiewicz et al. (2023a), more powerful methods, such as support vector machines (SVMs), gradient boost, and neural networks, were employed to detect V-types in Gaia data.

One of the challenges in ML applications is the selection of features, parameters, and methods that optimize the classification accuracy. This issue, particularly concerning the spectroscopic and spectrophotometric data, was explored by Klimczak et al. (2021) and Klimczak et al. (2023). The authors discovered that a relatively small number of features (four to five) accounted for the majority of the accuracy score, and adding additional classification features did not significantly influence prediction accuracy. This finding mirrors the classical PCA by DeMeo et al. (2009), who also observed that the top four to five dimensions contain the major part of the variance. Additionally, Klimczak et al. (2021) and Klimczak et al. (2023) utilized forward feature selection methods to quantify which photometric bands and colors provided the most accuracy for taxonomic classification (Whitney, 1971). These bands could prove instrumental in planning future spectrophotometric surveys aimed at optimizing the science output for planetary science.

A similar challenge of selecting ML methods to optimize the classification accuracy in various surveys was investigated by Klimczak et al. (2022). The authors utilized simulated spectro-photometric data from 12 diverse surveys. They found that among the methods analyzed, the multilayer perceptron and the gradient boosting yielded the highest accuracy, whereas the naive Bayes achieved the lowest accuracy (Fig. 6.4). In the most extreme case, selecting the appropriate ML algorithm doubled the success rate. The most successful surveys were those covering the 1-μm and 2-μm olivine/pyroxene absorption bands, which, along with spectral slope, are the most prominent features of asteroid spectra.

Mommert et al. (2016) analyzed several ML methods (nearest neighbor, SVM, and Gaussian naive Bayes) in classifying data from the United Kingdom infrared telescope (UKIRT). Initially, they utilized the MIT-UH-IRTF Joint Campaign for NEO spectral reconnaissance to derive synthetic colors in the UKIRT bands. These simulated data served as the training dataset, and the trained algorithms were then applied to actual UKIRT observations. Mommert et al. (2016) employed a version of the Bus–DeMeo taxonomy reduced to four broad categories (C/X, S, V, D) and compared the classification accuracy using varying numbers of color indices. This resulted in prediction accuracies ranging from approximately 50% to 95%, depending on the method and number of features used. Application to the UKIRT data revealed that the fraction of S-complex asteroids in the whole NEO population is lower than the fraction of OCs in meteorite fall statistics.

The ML methods have also been applied to asteroid spectra to provide insights directly into their mineralogical composition. For instance, Korda et al. (2023b) trained a neural network with 306 spectra of silicate samples and their mixtures for application for spectra of asteroids rich in dry silicates. The authors demonstrated that the abundance and mineral chemistry of common silicates (olivine and pyroxene) can be derived with an accuracy better than 10%.

Later, Korda et al. (2023a) investigated the performance of neural networks in studying the spatially-resolved taxonomy and surface composition of asteroids (433) Eros and (25143) Itokawa. They utilized spectra of Eros collected by the near-infrared spectrometer (NIS) onboard the NEAR Shoemaker spacecraft and spectra of Itokawa from the near-infrared spectrometer of the Hayabusa spacecraft. The two models were trained with spectra of asteroids obtained from the MITHNEOS and SMASS MIT surveys (DeMeo et al., 2009; Binzel et al., 2019) and with laboratory spectra of olivine/pyroxene mixtures and meteorites. The trained models were then applied to asteroid spacecraft measurements. The authors found that the surface of Eros is mostly homogeneous, whereas Itokawa exhibits more local variations in space-weathered surfaces. The composition model for Eros showed consistency with L/LL chondrites and for Itokawa with LL chondrites. Interestingly, the accuracy of the taxonomic classification models indicated a worsening of predictions with decreasing resolution, whereas the error model was found not to significantly depend on the resolution of reflectance spectra.

In continuation to the mineralogical studies of silicates, Korda and Kohout (2024) applied their neural network classifier to optimize the retrieved information versus the data volume in hyperspectral images of the ASPECT instrument onboard the Milani/Hera CubeSat. Hera is an ESA mission to the NASA DART mission target, the binary asteroid system (65803) Didymos.

Furthermore, Dyar et al. (2023) compared four different methods (SVM, logistic regression, kernel Fisher discriminant analysis, quadratic discriminant analysis) to determine the best approach for mapping asteroid spectra to meteorite classes based on laboratory measurements. The authors utilized 1,422 laboratory meteorite spectra divided into eight categories based on petrologic studies. Finally, they employed the best-performing model—logistic regression—to classify spectra of 605 asteroids into existing meteorite classes, uncovering several links between asteroid types and meteorite classes. A similar topic was explored by Saito et al. (2020), who used $K$-means and similarity matrices to study matches between asteroids and meteorites.

Unsupervised methods have been less commonly utilized in spectral and spectro-photometric studies. In Villmann et al. (2003), the authors investigated self-organizing maps (SOMs) for the analysis of spectral images obtained from airborne and satellite-based imaging spectrometers. Colazo et al. (2022) employed the fuzzy $c$-means algorithm to classify data from the Sloan Moving Objects Catalog into the three main taxonomic complexes (S, C, and X) and V-types, and Roh et al. (2022) classified SDSS data using Gaussian mixture models. Furthermore, Mahlke et al. (2022) introduced a new taxonomic classification for asteroids using a clustering method called mixture of common factor analyzers (MCFA). This taxonomy was established based on 2,983 observations of 2,125 individual asteroids. The authors derived 17 classes divided into the three complexes C, M, and S, including the novel Z class for extremely red objects in the main belt.

## 6.4. Comparison of ML and traditional methods

Modern ML methods can offer robustness and ability to process large amounts of data over traditional methods. The current de facto method in asteroid taxonomic classification is the Bus–DeMeo taxonomy and the MIT web-based tool implementing the classification.[3] The classification is based on the PCA transformation of the spectra and on the location of the object in the coordinate space given by the three first PCA components, and on the overall spectral slope. Finally, to resolve the taxonomic subtype, a visual inspection of absorption bands is sometimes needed. In fact, Burbine et al. (2019) estimated needing visual inspection in about 40% of the cases when classifying meteorite spectra from the NASA RELAB database. In the tool, the wavelength range of the spectral input must match the wavelengths used when building the classification system, that is, from 0.45 to 2.45 μm.

From ML methods, neural networks are extremely flexible and can, in theory, learn to mimic any linear or nonlinear classification process. In the case of classification into Bus–DeMeo taxonomy, for example, it is reasonable to assume that neural network could learn the whole process, including the final visual inspection, and thus be able to be used autonomously on large amounts of data. Furthermore, neural networks can be tuned to be robust against missing data on some wavelengths and can learn possible bi-

---

[3] http://smass.mit.edu/busdemeoclass.html

ases present in the data (e.g., systematics in Gaia DR3 asteroid spectral data Oszkiewicz et al. (2023a)).

The downside of neural networks is that they typically need a larger amount of labeled training data than more traditional methods to be properly trained and validated. This is understandable since the number of free parameters in even a small neural network can be magnitudes larger than with traditional classification methods. With the high granularity of information carried by the spectra, effectively training an ML system on such data still poses a challenge. According to the curse of dimensionality, a common phenomenon existing in the AI field, with each additional dimension of input data, the required number of observations drastically increases. Therefore it is important to gather as much data as possible to ensure the ML methods can have the best performance. Though the quantities of spectra currently available are much smaller than for photometry, there are many ongoing efforts aiming to increase the scope of available data, one recent being the Gaia DR3 effort.

The ability of neural networks and other ML methods to learn an existing classification method, such as the Bus–DeMeo taxonomy, can be used to apply taxonomy to data that do not meet the wavelength range and spectral resolution of the original system. This feature can enable the classification of observations from present and upcoming large surveys producing large spectral databases. This was studied by Penttilä et al. (2021) for Gaia data in the wavelength range of 0.33–1.05 μm; Penttilä et al. (2022) for LSST data with six broadband filters; Luo et al. (2023) for the Chinese Space Survey telescope's wavelength range of 0.255–1.0 μm, and by Klimczak et al. (2022) for 12 different surveys. All studies showed reasonably good results in obtaining Bus–DeMeo classification for wavelength ranges different from the nominal 0.45–2.45 μm, or even for multiple broadband colors. Typically, a classification accuracy of 80–90% was achieved; see the example from Klimczak et al. (2022) in Fig. 6.4. It should be noted, however, that in all the studies mentioned, the Bus–DeMeo taxonomy was simplified to some extent, and not all the subclasses of the complexes were used.

Another approach to new surveys with new wavelength ranges or broadband filter collections is to create a completely new taxonomy for the particular survey. The benefit of this approach is that if the wavelengths of the new survey cover ranges not included in the present taxonomy, there can be unseen features that can enable new insights into differences between asteroid compositions. For example, the Gaia taxonomy is already planned

**Figure 6.4** Balanced accuracy measure for five ML classification methods and 12 surveys. The figure is replicated from Klimczak et al. (2022); please see details from there.

to be formed at the end of the mission using a clustering method based on minimal spanning tree (Delbò et al., 2012; Galluccio et al., 2008). Gaia spectra are extended further down in UV than Bus–DeMeo, to 0.33 μm. Before Bus–DeMeo, the UV part was employed by the Tholen taxonomy and was diagnostic, for example, to Tholen F-type distinguishing them from the B-type.

A more general approach was taken by Mahlke et al. (2022), where they developed a new taxonomy based on a clustering approach with a latent-variable data model that can input missing wavelengths. Their taxonomy also takes the albedo of the asteroid into account in addition to the normalized spectra.

The benefits of using spectro-photometric studies as opposed to spectral data include the high availability of large data volumes, crucial for effectively training ML models. On the other hand, the actual number of features in one of these studies is much smaller than when using the entire spectra. This property can have two-fold effects. On the one hand, providing less detailed information to the methods can lead to omitting important characteristics of the object's composition, therefore making it impossible for the methods to gather a deep understanding of key features. However, if the photometric survey was designed with enough detail to cover those characteristics, this could increase the signal-to-noise ratio of the features, acting in a way as a preprocessing step. It is important to take careful consideration of the input features so that the training process is effectively supplied with essential information.

## 6.5. Conclusions and future directions

It is evident that with the expected massive increase in available asteroid spectra and spectro-photometric data, any method used to derive compositional information via taxonomic classification must be fully automated. Particularly, the Large Synoptic Survey of the Vera C. Rubin Observatory is expected to discover over 5 million main-belt asteroids, almost 100,000 near-Earth objects (NEOs), around 280,000 Jupiter trojans, and about 40,000 trans-Neptunian objects over a 10-year operation (Eggl et al., 2019). In comparison, the current number of discovered small bodies is about 1.3 million (as of February 21, 2024), out of which only a very small fraction is taxonomically characterized. The LSST will provide not only astrometry, but also filtered observations (in the $u$, $g$, $r$, $i$, $z$, and $y$ filters) with an accuracy of approximately 10 millimagnitudes (Jones et al., 2015). A typical main belt object will be observed 200–300 times (Jones et al., 2015). This is likely the first time in history when the number of physical characterizations will be similar to the number of known orbital parameters. This enormous amount of data will provide a taxonomical view of the asteroid population down to much smaller sizes and, in terms of numbers, for a few magnitudes more objects than previously characterized. This wealth of taxonomical characterizations will allow for more complex and detailed models of solar system formation and evolution, better identification of asteroid families, analysis of the source regions of NEOs, and the tackling of many other problems in planetary science. At this scale and with this enormous amount of data, ML classifications will clearly become necessary.

Furthermore, it should be acknowledged that different surveys and their output (dense spectra vs broadband filters) can produce different depths or details in classification. With only a few broadband filters, it might be possible to distinguish only siliceous types (S) from carbonaceous (C), whereas dense UV-VIS-NIR spectra might distinguish down to subtypes in the complexes and allow for mineralogical analysis. Moreover, a probability estimate for the assigned taxonomic class would help to judge the degree of certainty of the classification result.

Preliminary classifications provided based on spectro-photometry may also become increasingly useful in pinpointing objects for further analysis.

For example, DeMeo et al. (2019) followed up on olivine-dominated A-type objects suggested based on the analysis of SDSS spectro-photometry with the goal of identifying the missing mantle material of differentiated planetesimals. Oszkiewicz et al. (2023a) listed peculiar basaltic V–type asteroids selected from the Gaia DR3 data for follow-up to identify several of different differentiated parent bodies. Thanks to the SDSS data, DeMeo et al. (2014) were able to follow up D-type interlopers in the inner main belt and found that these objects have relatively high albedos compared to other D-types, which are typically considered to have formed in the outer Solar System far from the Sun. The presence of these objects has yet to be explained by dynamical models.

Clearly, there is a need for both novel machine learning techniques and traditional modeling in the coming decades of big surveys and discoveries.

## 6.6. Code availability

A Jupyter notebook containing an example model for classifying spectro-photometry data is accessible via the following links:

Docs: https://solar-system-ml.github.io/book/chapter6/
Repository: https://github.com/solar-system-ml/book/tree/main/docs/chapter6

## References

Baudrand, A., Bec-Borsenberger, A., Borsenberger, J., 2004. Asteroidal I, J, K magnitudes recovered in the DENIS survey: second release. Astronomy & Astrophysics 423, 381–383.

Bell, J., Owensby, P., Hawke, B.R., Gaffey, M., 1988. The 52-Color Asteroid Survey: final results and interpretation. In: Abstracts of the Lunar and Planetary Science Conference, p. 57.

Bell, J.F., 1988. A probable asteroidal parent body for the CO or CV chondrites. Meteoritics 23, 256–257.

Bierhaus, E., Clark, B., Harris, J., Payne, K., Dubisher, R., Wurts, D., Hund, R., Kuhns, R., Linn, T., Wood, J., et al., 2018. The OSIRIS-REx spacecraft and the touch-and-go sample acquisition mechanism (TAGSAM). Space Science Reviews 214, 1–46.

Binzel, R., DeMeo, F., Turtelboom, E., Bus, S., Tokunaga, A., Burbine, T., Lantz, C., Polishook, D., Carry, B., Morbidelli, A., et al., 2019. Compositional distributions and evolutionary processes for the near-Earth object population: results from the MIT-Hawaii Near-Earth Object Spectroscopic Survey (MITHNEOS). Icarus 324, 41–76.

Binzel, R.P., Xu, S., 1993. Chips off of asteroid 4 Vesta: evidence for the parent body of basaltic achondrite meteorites. Science 260, 186–191.

Birlan, M., Barucci, M., Fulchignoni, M., 1996. G-mode analysis of the reflection spectra of 84 asteroids. Astronomy & Astrophysics 305, 984.

Bland, P.A., Travis, B.J., 2017. Giant convecting mud balls of the early solar system. Science Advances 3, e1602514.

Bottke, W.F., Marschall, R., Nesvorný, D., Vokrouhlický, D., 2023. Origin and evolution of Jupiter's trojan asteroids. Space Science Reviews 219, 83.

Bottke, W.F., Nesvorný, D., Grimm, R.E., Morbidelli, A., O'Brien, D.P., 2006. Iron meteorites as remnants of planetesimals formed in the terrestrial planet region. Nature 439, 821–824.

Bottke, W.F., Vokrouhlický, D., Minton, D., Nesvorný, D., Morbidelli, A., Brasser, R., Simonson, B., Levison, H.F., 2012. An Archaean heavy bombardment from a destabilized extension of the asteroid belt. Nature 485, 78–81.

Britt, D.T., Pieters, C.M., 1991. Black ordinary chondrites: an analysis of abundance and fall frequency. Meteoritics 26, 279–285.

Brownlee, D., 2014. The Stardust mission: analyzing samples from the edge of the solar system. Annual Review of Earth and Planetary Sciences 42, 179–205.

Burbine, T., Binzel, R., Bus, S., Clark, B., 2001. K asteroids and CO3/CV3 chondrites. Meteoritics & Planetary Science 36, 245–253.

Burbine, T., Wallace, S., Dyar, M., 2019. Applying the Bus-deMeo asteroid taxonomy to meteorite spectra. In: Proceedings of the 50th Lunar and Planetary Science Conference. The Woodlands, Texas, USA, p. 2132.

Burbine, T.H., 1998. Could G-class asteroids be the parent bodies of the CM chondrites? Meteoritics & Planetary Science 33, 253–258.

Burbine, T.H., Binzel, R.P., 2002. Small main-belt asteroid spectroscopic survey in the near-infrared. Icarus 159, 468–499.

Burbine, T.H., Meibom, A., Binzel, R.P., 1996. Mantle material in the main belt: battered to bits? Meteoritics & Planetary Science 31, 607–620.

Burnett, D., Barraclough, B., Bennett, R., Neugebauer, M., Oldham, L., Sasaki, C., Sevilla, D., Smith, N., Stansbery, E., Sweetnam, D., et al., 2003. The Genesis Discovery mission: Return of solar matter to Earth. Space Science Reviews 105, 509–534.

Bus, S.J., 1999. Compositional structure in the asteroid belt: Results of a spectroscopic survey. Ph.D. thesis. Massachusetts Institute of Technology, USA.

Bus, S.J., Binzel, R.P., 2002. Phase II of the Small Main-Belt Asteroid Spectroscopic Survey: the observations. Icarus 158, 106–145.

Carruba, V., Camargo, J.I., Aljbaae, S., Ferreira, F., Lin, E., Figueiredo-Peixoto, V., Banda-Huarca, M., Pieres, A., Boufleur, R., Da Costa, L., et al., 2024. Main belt asteroids taxonomical information from dark energy survey data. Monthly Notices of the Royal Astronomical Society 527, 6495–6505.

Carruba, V., Domingos, R., Nesvorný, D., Roig, F., Huaman, M., Souami, D., 2013. A multidomain approach to asteroid families' identification. Monthly Notices of the Royal Astronomical Society 433, 2075–2096.

Carry, B., 2018. Solar system science with ESA Euclid. Astronomy & Astrophysics 609, A113.

Carvano, J., Hasselmann, P., Lazzaro, D., Mothé-Diniz, T., 2010. SDSS-based taxonomic classification and orbital distribution of main belt asteroids. Astronomy & Astrophysics 510, A43.

Carvano, J., Mothé-Diniz, T., Lazzaro, D., 2003. Search for relations among a sample of 460 asteroids with featureless spectra. Icarus 161, 356–382.

Chapman, C.R., 2004. Space weathering of asteroid surfaces. Annual Review of Earth and Planetary Sciences 32, 539–567.

Chapman, C.R., Salisbury, J.W., 1973. Comparisons of meteorite and asteroid spectral reflectivities. Icarus 19, 507–522.

Clark, B.E., Bell, J.F., Fanale, F.P., O'Connor, D.J., 1995. Results of the seven-color asteroid survey: infrared spectral observations of ~50-km size S-, K-, and M-type asteroids. Icarus 113, 387–402.

Clark, B.E., Ockert-Bell, M.E., Cloutis, E.A., Nesvorny, D., Mothé-Diniz, T., Bus, S.J., 2009. Spectroscopy of K-complex asteroids: parent bodies of carbonaceous meteorites? Icarus 202, 119–133.

Colazo, M., Alvarez-Candal, A., Duffard, R., 2022. Zero-phase angle asteroid taxonomy classification using unsupervised machine learning algorithms. Astronomy & Astrophysics 666, A77.

Consolmagno, G.J., Drake, M.J., 1977. Composition and evolution of the eucrite parent body: evidence from rare earth elements. Geochimica Et Cosmochimica Acta 41, 1271–1282.

Cruikshank, D.P., Hartmann, W.K., 1984. The meteorite-asteroid connection: two olivine-rich asteroids. Science 223, 281–283.

Dandy, C., Fitzsimmons, A., Collander-Brown, S., 2003. Optical colors of 56 near-Earth objects: trends with size and orbit. Icarus 163, 363–373.

De León, J., Licandro, J., Serra-Ricart, M., Pinilla-Alonso, N., Campins, H., 2010. Observations, compositional, and physical characterization of near-Earth and Mars-crosser asteroids from a spectroscopic survey. Astronomy & Astrophysics 517, A23.

Delbò, M., Gayon-Markt, J., Busso, G., Brown, A., Galluccio, L., Ordenovic, C., Bendjoya, P., Tanga, P., 2012. Asteroid spectroscopy with Gaia. Planetary and Space Science 73, 86–94.

DeMeo, F., Carry, B., 2013. The taxonomic distribution of asteroids from multi-filter all-sky photometric surveys. Icarus 226, 723–741.

DeMeo, F.E., Binzel, R.P., Carry, B., Polishook, D., Moskovitz, N.A., 2014. Unexpected D-type interlopers in the inner main belt. Icarus 229, 392–399.

DeMeo, F.E., Binzel, R.P., Slivan, S.M., Bus, S.J., 2009. An extension of the bus asteroid taxonomy into the near-infrared. Icarus 202, 160–180.

DeMeo, F.E., Burt, B.J., Marsset, M., Polishook, D., Burbine, T.H., Carry, B., Binzel, R.P., Vernazza, P., Reddy, V., Tang, M., et al., 2022. Connecting asteroids and meteorites with visible and near-infrared spectroscopy. Icarus 380, 114971.

DeMeo, F.E., Carry, B., 2014. Solar System evolution from compositional mapping of the asteroid belt. Nature 505, 629–634.

DeMeo, F.E., Polishook, D., Carry, B., Burt, B.J., Hsieh, H.H., Binzel, R.P., Moskovitz, N.A., Burbine, T.H., 2019. Olivine-dominated A-type asteroids in the main belt: distribution, abundance and relation to families. Icarus 322, 13–30.

Dyar, M.D., Wallace, S.M., Burbine, T.H., Sheldon, D.R., 2023. A machine learning classification of meteorite spectra applied to understanding asteroids. Icarus 406, 115718.

Eggl, S., Jones, L., Juric, M., 2019. LSST asteroid discovery rates. https://dmtn-109.lsst.io/. Version 2, published at 2019-12-06.

Elkins-Tanton, L.T., Weiss, B.P., Zuber, M.T., 2011. Chondrites as samples of differentiated planetesimals. Earth and Planetary Science Letters 305, 1–10.

Emerson, J.P., Sutherland, W., 2002. Visible and infrared survey telescope for astronomy: overview. In: Survey and Other Telescope Technologies and Discoveries, International Society for Optics and Photonics, pp. 35–42.

Emery, J.P., Marzari, F., Morbidelli, A., French, L.M., Grav, T., et al., 2014. The complex history of Trojan asteroids. In: Bottke, W.F., DeMeo, F.E., Michel, P. (Eds.), Asteroids IV. University of Arizona Press, pp. 203–220.

Epchtein, N., De Batz, B., Copet, E., Fouque, P., Lacombe, F., Le Bertre, T., Mamon, G., Rouan, D., Tiphène, D., Burton, W., et al., 1994. Denis: a deep near-infrared survey of the southern sky. In: Science with Astronomical Near-Infrared Sky Surveys. Springer, pp. 3–9.

Fulchignoni, M., Birlan, M., Barucci, M.A., 2000. The extension of the G-mode asteroid taxonomy. Icarus 146, 204–212.

Gaffey, M., Bell, J., Cruikshank, D., 1989. Reflectance spectroscopy and asteroid surface mineralogy. In: Binzel, R., Gehrels, T., Matthews, M. (Eds.), Asteroids II. The University of Arizona Press, pp. 98–127.

Gaffey, M.J., Bell, J.F., Brown, R.H., Burbine, T.H., Piatek, J.L., Reed, K.L., Chaky, D.A., 1993. Mineralogical variations within the S-type asteroid class. Icarus 106, 573–602.

Gaia Collaboration, Galluccio, L., Delbo, M., De Angeli, F., Pauwels, T., Tanga, P., Mignard, F., Cellino, A., Brown, A., Muinonen, K., et al., 2023. Gaia Data Release 3: reflectance spectra of Solar System small bodies. Astronomy & Astrophysics 674, A35.

Gaia Collaboration, Prusti, T., de Bruijne, J.H.J., Brown, A.G.A., Vallenari, A., Babusiaux, C., Bailer-Jones, C.A.L., Bastian, U., Biermann, M., Evans, D.W., et al., 2016. The Gaia mission. Astronomy & Astrophysics 595, A1.

Galluccio, L., Michel, O., Bendjoya, P., Slezak, E., 2008. Unsupervised clustering on astrophysics data: asteroids reflectance spectra surveys and hyperspectral images. AIP Conference Proceedings 1082, 165–171.

Gomes, R., Levison, H.F., Tsiganis, K., Morbidelli, A., 2005. Origin of the cataclysmic Late Heavy Bombardment period of the terrestrial planets. Nature 435, 466–469.

Gradie, J., Tedesco, E., 1982. Compositional structure of the asteroid belt. Science 216, 1405–1407.

Hardersen, P.S., Gaffey, M.J., Abell, P.A., 2004. Mineralogy of asteroid 1459 Magnya and implications for its origin. Icarus 167, 170–177.

Hietala, H., 2020. Asteroid Spectra and Machine Learning. Master's thesis. University of Helsinki, Finland. http://urn.fi/URN:NBN:fi:hulib-202003241601.

Hiroi, T., Pieters, C.M., Zolensky, M.E., Lipschutz, M.E., 1993. Evidence of thermal metamorphism on the C, G, B, and F asteroids. Science 261, 1016–1018.

Hiroi, T., Zolensky, M.E., Pieters, C.M., 2001. The Tagish Lake meteorite: a possible sample from a D-type asteroid. Science 293, 2234–2236.

Howell, E., Merenyi, E., Lebofsky, L., 1994. Classification of asteroid spectra using a neural network. Journal of Geophysical Research: Planets 99, 10847–10865.

Hromakina, T., Birlan, M., Barucci, M., Fulchignoni, M., Colas, F., Fornasier, S., Merlin, F., Sonka, A., Petrescu, E., Perna, D., Dotto, E., the NEOROCKS Team, 2021. Photometric survey of 55 near-earth asteroids. Astronomy & Astrophysics 656, A89.

Hromakina, T., Birlan, M., Barucci, M.A., Fulchignoni, M., Colas, F., Fornasier, S., Merlin, F., Sonka, A., Anghel, S., Perna, D., Dotto, E., the NEOROCKS Team, 2022. Updated dataset of neos surface colors obtained within the neorocks project. In: Europlanet Science Congress. Granada, Spain. pp. EPSC2022–364.

Huchra, J.P., Macri, L.M., Masters, K.L., Jarrett, T.H., Berlind, P., Calkins, M., Crook, A.C., Cutri, R., Erdoğdu, P., Falco, E., et al., 2012. The 2MASS redshift survey—description and data release. The Astrophysical Journal. Supplement Series 199, 26.

Ivezić, Ž., Lupton, R.H., Jurić, M., Tabachnik, S., Quinn, T., Gunn, J.E., Knapp, G.R., Rockosi, C.M., Brinkmann, J., 2002. Color confirmation of asteroid families. The Astronomical Journal 124, 2943.

Ivezić, Ž., Tabachnik, S., Rafikov, R., Lupton, R.H., Quinn, T., Hammergren, M., Eyer, L., Chu, J., Armstrong, J.C., Fan, X., et al., 2001. Solar system objects observed in the Sloan Digital Sky Survey commissioning data. The Astronomical Journal 122, 2749.

Jenniskens, P., Vaubaillon, J., Binzel, R.P., DeMEO, F.E., Nesvorný, D., Bottke, W.F., Fitzsimmons, A., Hiroi, T., Marchis, F., Bishop, J.L., et al., 2010. Almahata Sitta (=asteroid 2008 TC$_3$) and the search for the ureilite parent body. Meteoritics & Planetary Science 45, 1590–1617.

Jerde, E.A., 2021. The Apollo program. In: Longobardo, A. (Ed.), Sample Return Missions: The Last Frontier of Solar System Exploration. Elsevier, pp. 9–36.

Johnston, R.S., Hull, W.E., 1975. Apollo missions. In: Johnston, R., Dietlein, L., Berry, C. (Eds.), Biomedical Results of Apollo. National Aeronautics and Space Administration, Washington, DC, pp. 9–40.

Jones, R.L., Jurić, M., Ivezić, Ž., 2015. Asteroid discovery and characterization with the Large Synoptic Survey Telescope. Proceedings of the International Astronomical Union 10, 282–292.

Kebukawa, Y., Ito, M., Zolensky, M.E., Greenwood, R.C., Rahman, Z., Suga, H., Nakato, A., Chan, Q.H., Fries, M., Takeichi, Y., et al., 2019. A novel organic-rich meteoritic clast from the outer solar system. Scientific Reports 9, 3169.

Kent, S.M., 1994. Sloan Digital Sky Survey. Astrophysics and Space Science 217, 27–30.

Klimczak, H., Kotłowski, W., Oszkiewicz, D., DeMeo, F., Kryszczyńska, A., Wilawer, E., Carry, B., 2021. Predicting asteroid types: importance of individual and combined features. Frontiers in Astronomy and Space Sciences 8, 216.

Klimczak, H., Wilawer, E., Kwiatkowski, T., Kryszczyńska, A., Oszkiewicz, D., Kotłowski, W., DeMeo, F., 2023. Optimization of future multifilter surveys toward asteroid characterization. The Astronomical Journal 166, 230.

Klimczak, H., Oszkiewicz, D., Carry, B., Penttilä, A., Kotlowski, W., Kryszczyńska, A., Wilawer, E., 2022. Comparison of machine learning algorithms used to classify the asteroids observed by all-sky surveys. Astronomy & Astrophysics 667, A10.

Kohonen, T., 1989. Self-Organization and Associative Memory, 3rd edition. Springer-Verlag, Berlin, Heidelberg.

Kohout, T., Gritsevich, M., Grokhovsky, V.I., Yakovlev, G.A., Haloda, J., Halodova, P., Michallik, R.M., Penttilä, A., Muinonen, K., 2014. Mineralogy, reflectance spectra, and physical properties of the Chelyabinsk LL5 chondrite–insight into shock-induced changes in asteroid regoliths. Icarus 228, 78–85.

Koleńczuk, P., Kwiatkowski, T., Kamińska, M., Kamiński, K., Colas, F., Klotz, A., Kim, T., Birlan, M., 2022. Colours and taxonomy of 2022 ab: a super fast rotating near-Earth asteroid. In: Europlanet Science Congress. Granada, Spain, pp. EPSC2022–1161.

Korda, D., Kohout, T., 2024. Silicate mineralogy from Vis–NIR reflectance spectra. The Planetary Science Journal 5, 85.

Korda, D., Kohout, T., Flanderová, K., Vincent, J.B., Penttilä, A., 2023a. (433) Eros and (25143) Itokawa surface properties from reflectance spectra. Astronomy & Astrophysics 675, A50.

Korda, D., Penttilä, A., Klami, A., Kohout, T., 2023b. Neural network for determining an asteroid mineral composition from reflectance spectra. Astronomy & Astrophysics 669, A101.

Laureijs, R., Gondoin, P., Duvet, L., Criado, G.S., Hoar, J., Amiaux, J., Auguères, J.L., Cole, R., Cropper, M., Ealet, A., et al., 2012. Euclid: ESA's mission to map the geometry of the dark universe. In: Space Telescopes and Instrumentation 2012: Optical, Infrared, and Millimeter Wave, International Society for Optics and Photonics, p. 84420T.

Lauretta, D.S., Enos, H.L., Polit, A.T., Roper, H.L., Wolner, C.W., 2021. OSIRIS-REx at Bennu: overcoming challenges to collect a sample of the early Solar System. In: Longobardo, A. (Ed.), Sample Return Missions: The Last Frontier of Solar System Exploration. Elsevier, pp. 163–194.

Lazzaro, D., Angeli, C., Carvano, J., Mothé-Diniz, T., Duffard, R., Florczak, M., 2004. S3OS2: the visible spectroscopic survey of 820 asteroids. Icarus 172, 179–220.

Levison, H.F., Bottke, W.F., Gounelle, M., Morbidelli, A., Nesvorný, D., Tsiganis, K., 2009. Contamination of the asteroid belt by primordial trans-Neptunian objects. Nature 460, 364–366.

Licandro, J., Popescu, M., Morate, D., de León, J., 2017. V-type candidates and Vesta family asteroids in the Moving Objects VISTA (MOVIS) catalogue. Astronomy & Astrophysics 600, A126.

Luo, N., Wang, X., Gu, S., Penttilä, A., Muinonen, K., Liu, Y., 2023. Taxonomic analysis of asteroids with artificial neural networks. The Astronomical Journal 167, 13.

Mahlke, M., 2022. Asteroid taxonomy: a probabilistic synthesis of spectrometry and albedo from complete and partial observations. Ph.D. thesis. Université Côte d'Azur.

Mahlke, M., Carry, B., Mattei, P.A., 2022. Asteroid taxonomy from cluster analysis of spectrometry and albedo. Astronomy & Astrophysics 665, A26.

Marrocchi, Y., Avice, G., Barrat, J.A., 2021. The Tarda meteorite: a window into the formation of D-type asteroids. The Astrophysical Journal Letters 913, L9.

Masiero, J.R., Mainzer, A., Bauer, J., Grav, T., Nugent, C., Stevenson, R., 2013. Asteroid family identification using the hierarchical clustering method and wise/neowise physical properties. The Astrophysical Journal 770, 7.

McCord, T.B., Adams, J.B., Johnson, T.V., 1970. Asteroid Vesta: spectral reflectivity and compositional implications. Science 168, 1445–1447.

Merényi, E., Singer, R.B., Miller, J.S., 1996. Mapping of spectral variations on the surface of Mars from high spectral resolution telescopic images. Icarus 124, 280–295.

Mommert, M., Trilling, D.E., Borth, D., Jedicke, R., Butler, N., Reyes-Ruiz, M., Pichardo, B., Petersen, E., Axelrod, T., Moskovitz, N., 2016. First results from the rapid-response spectrophotometric characterization of near-Earth objects using UKIRT. The Astronomical Journal 151, 98.

Morate, D., Carvano, J.M., Alvarez-Candal, A., De Prá, M., Licandro, J., Galarza, A., Mahlke, M., Solano-Márquez, E., Cenarro, J., Cristóbal-Hornillos, D., et al., 2021. J-PLUS: a first glimpse at the spectrophotometry of asteroids–the MOOJa catalog. Astronomy & Astrophysics 655, A47.

Morbidelli, A., Baillie, K., Batygin, K., Charnoz, S., Guillot, T., Rubie, D.C., Kleine, T., 2022. Contemporary formation of early Solar System planetesimals at two distinct radial locations. Nature Astronomy 6, 72–79.

Morbidelli, A., Levison, H.F., Tsiganis, K., Gomes, R., 2005. Chaotic capture of Jupiter's Trojan asteroids in the early Solar System. Nature 435, 462–465.

Moskovitz, N.A., Jedicke, R., Gaidos, E., Willman, M., Nesvorný, D., Fevig, R., Ivezić, Ž., 2008. The distribution of basaltic asteroids in the main belt. Icarus 198, 77–90.

Mothé-Diniz, T., árcio Carvano, J.M., Lazzaro, D., 2003. Distribution of taxonomic classes in the main belt of asteroids. Icarus 162, 10–21.

Nakamura, T., Noguchi, T., Tanaka, M., Zolensky, M.E., Kimura, M., Tsuchiyama, A., Nakato, A., Ogami, T., Ishida, H., Uesugi, M., et al., 2011. Itokawa dust particles: a direct link between S-type asteroids and ordinary chondrites. Science 333, 1113–1116.

Nesvorný, D., Jedicke, R., Whiteley, R.J., Ivezić, Ž., 2005. Evidence for asteroid space weathering from the Sloan Digital Sky Survey. Icarus 173, 132–152.

Noguchi, T., Nakamura, T., Kimura, M., Zolensky, M., Tanaka, M., Hashimoto, T., Konno, M., Nakato, A., Ogami, T., Fujimura, A., et al., 2011. Incipient space weathering observed on the surface of Itokawa dust particles. Science 333, 1121–1125.

Olivier, S.S., Riot, V.J., Gilmore, D.K., Bauman, B., Pratuch, S., Seppala, L., Ku, J., Nordby, M., Foss, M., Antilogus, P., et al., 2012. LSST camera optics design. In: Ground-Based and Airborne Instrumentation for Astronomy IV, International Society for Optics and Photonics, p. 84466B.

Oszkiewicz, D., Klimczak, H., Carry, B., Penttilä, A., Popescu, M., Krüger, J., Aron Keniger, M., 2023a. Spectral analysis of basaltic asteroids observed by the Gaia space mission. Monthly Notices of the Royal Astronomical Society 519, 2917–2928.

Oszkiewicz, D., Kwiatkowski, T., Tomov, T., Birlan, M., Geier, S., Penttilä, A., Polińska, M., 2014. Selecting asteroids for a targeted spectroscopic survey. Astronomy & Astrophysics 572.

Oszkiewicz, D., Troianskyi, V., Galád, A., Hanuš, J., Ďurech, J., Wilawer, E., Marciniak, A., Kwiatkowski, T., Koleńczuk, P., Skiff, B.A., et al., 2023b. Spins and shapes of basaltic asteroids and the missing mantle problem. Icarus 397, 115520.

Pao, Y.H., 1989. Adaptive Pattern Recognition and Neural Networks. Addison-Wesley Longman Publishing Co., Inc., USA.

Parker, A., Ivezić, Ž., Jurić, M., Lupton, R., Sekora, M.D., Kowalski, A., 2008. The size distributions of asteroid families in the SDSS Moving Object Catalog 4. Icarus 198, 138–155.

Penttilä, A., Fedorets, G., Muinonen, K., 2022. Taxonomy of asteroids from the legacy survey of space and time using neural networks. Frontiers in Astronomy and Space Sciences 9, 816268.

Penttilä, A., Hietala, H., Muinonen, K., 2021. Asteroid spectral taxonomy using neural networks. Astronomy & Astrophysics 649, A46.

Pieters, C.M., Hiroi, T., 2004. RELAB (Reflectance Experiment Laboratory): a NASA multiuser spectroscopy facility. In: Lunar and Planetary Science Conference, p. 1720.

Pilorget, C., Okada, T., Hamm, V., Brunetto, R., Yada, T., Loizeau, D., Riu, L., Usui, T., Moussi-Soffys, A., Hatakeda, K., et al., 2022. First compositional analysis of Ryugu samples by the MicrOmega hyperspectral microscope. Nature Astronomy 6, 221–225.

Popescu, M., Birlan, M., Nedelcu, D., 2012. Modeling of asteroid spectra - M4AST. Astronomy & Astrophysics 544, A130.

Popescu, M., Licandro, J., de Leon, J., Morate, D., Boaca, I.L., 2018a. MOVIS catalog: near-infrared colors and taxonomy of asteroids observed by VISTA-VHS survey. In: European Planetary Science Congress. pp. EPSC2018–273.

Popescu, M., Licandro, J., Morate, D., de León, J., Nedelcu, D., Rebolo, R., McMahon, R., Gonzalez-Solares, E., Irwin, M., 2016. Near-infrared colors of minor planets recovered from VISTA-VHS survey (MOVIS). Astronomy & Astrophysics 591, A115.

Popescu, M., Licandro, J., Carvano, J.M., Stoicescu, R., de Leo'n, J., Morate, D., Boacau, I.L., Cristescu, C.P., 2018b. Taxonomic classification of asteroids based on MOVIS near-infrared colors. Astronomy & Astrophysics 617, A12.

Rivkin, A.S., 2012. The fraction of hydrated C-complex asteroids in the asteroid belt from SDSS data. Icarus 221, 744–752.

Roh, D.G., Moon, H.K., Shin, M.S., DeMeo, F.E., 2022. A new approach to feature-based asteroid taxonomy in 3D color space. I. SDSS photometric system. Astronomy & Astrophysics 664, A51.

Rudnick, R.L., 2005. The Crust, vol. 3. Elsevier.

Saito, Y., Hong, P.K., Niihara, T., Miyamoto, H., Fukumizu, K., 2020. Data-driven taxonomy matching of asteroid and meteorite. Meteoritics & Planetary Science 55, 193–206.

Sandford, S.A., Brownlee, D.E., Zolensky, M.E., 2021. The Stardust sample return mission. In: Longobardo, A. (Ed.), Sample Return Missions: The Last Frontier of Solar System Exploration. Elsevier, pp. 79–104.

Sasaki, S., Nakamura, K., Hamabe, Y., Kurahashi, E., Hiroi, T., 2001. Production of iron nanoparticles by laser irradiation in a simulation of lunar-like space weathering. Nature 410, 555–557.

Sergeyev, A., Carry, B., Marsset, M., Pravec, P., Perna, D., DeMeo, F., Petropoulou, V., Lazzarin, M., La Forgia, F., Di Petro, I., 2023. Compositional properties of planet-crossing asteroids from astronomical surveys. Astronomy & Astrophysics 679, A148.

Sergeyev, A., Carry, B., Onken, C., Devillepoix, H., Wolf, C., Chang, S.W., 2022. Multifilter photometry of Solar System objects from the SkyMapper Southern Survey. Astronomy & Astrophysics 658, A109.

Sergeyev, A.V., Carry, B., 2021. A million asteroid observations in the Sloan Digital Sky Survey. Astronomy & Astrophysics 652, A59.

Skrutskie, M., Cutri, R., Stiening, R., Weinberg, M., Schneider, S., Carpenter, J., Beichman, C., Capps, R., Chester, T., Elias, J., et al., 2006. The two micron all sky survey (2MASS). The Astronomical Journal 131, 1163.

Strazzulla, G., Dotto, E., Binzel, R., Brunetto, R., Barucci, M.A., Blanco, A., Orofino, V., 2005. Spectral alteration of the Meteorite Epinal (H5) induced by heavy ion irradiation: a simulation of space weathering effects on near-Earth asteroids. Icarus 174, 31–35.

Sunshine, J.M., Bus, S.J., Corrigan, C.M., McCoy, T.J., Burbine, T.H., 2007. Olivine-dominated asteroids and meteorites: distinguishing nebular and igneous histories. Meteoritics & Planetary Science 42, 155–170.

Sykes, M.V., Cutri, R.M., Fowler, J.W., Tholen, D.J., Skrutskie, M.F., Price, S., Tedesco, E.F., 2000. The 2MASS asteroid and comet survey. Icarus 146, 161–175.

Tedesco, E.F., Tholen, D.J., Zellner, B., 1982. The eight-color asteroid survey-standard stars. Astronomical Journal 87, 1585–1592.

Tholen, D., 1984. Asteroid taxonomy from cluster analysis of photometry. Ph.D. thesis. University of Arizona, USA.

Tsiganis, K., Gomes, R., Morbidelli, A., Levison, H.F., 2005. Origin of the orbital architecture of the giant planets of the Solar System. Nature 435, 459–461.

Vernazza, P., Brunetto, R., Binzel, R., Perron, C., Fulvio, D., Strazzulla, G., Fulchignoni, M., 2009. Plausible parent bodies for enstatite chondrites and mesosiderites: implications for Lutetia's fly-by. Icarus 202, 477–486.

Vernazza, P., Castillo-Rogez, J., Beck, P., Emery, J., Brunetto, R., Delbo, M., Marsset, M., Marchis, F., Groussin, O., Zanda, B., et al., 2017. Different origins or different evolutions? Decoding the spectral diversity among C-type asteroids. The Astronomical Journal 153, 72.

Vernazza, P., Zanda, B., Nakamura, T., Scott, E., Russell, S., 2014. The formation and evolution of ordinary chondrite parent bodies. In: Bottke, W.F., DeMeo, F.E., Michel, P. (Eds.), Asteroids IV. University of Arizona Press, pp. 617–634.

Villmann, T., Mere'nyi, E., Hammer, B., 2003. Neural maps in remote sensing image analysis. Neural Networks 16, 389–403.

Wadhwa, M., Shukolyukov, A., Davis, A., Lugmair, G., Mittlefehldt, D., 2003. Differentiation history of the mesosiderite parent body: constraints from trace elements and manganese-chromium isotope systematics in Vaca Muerta silicate clasts. Geochimica Et Cosmochimica Acta 67, 5047–5069.

Walsh, K.J., Morbidelli, A., Raymond, S.N., O'brien, D., Mandell, A., 2012. Populating the asteroid belt from two parent source regions due to the migration of giant planets — "The Grand Tack". Meteoritics & Planetary Science 47, 1941–1947.

Watanabe, S.i., Tsuda, Y., Yoshikawa, M., Tanaka, S., Saiki, T., Nakazawa, S., 2017. Hayabusa2 mission overview. Space Science Reviews 208, 3–16.

Whitney, A.W., 1971. A direct method of nonparametric measurement selection. IEEE Transactions on Computers 100, 1100–1103.

Wiens, R.C., Reisenfeld, D., Jurewicz, A., Burnett, D., 2021. The Genesis Solar-Wind mission: first deep-space robotic mission to return to Earth. In: Longobardo, A. (Ed.), Sample Return Missions: The Last Frontier of Solar System Exploration. Elsevier, pp. 105–122.

Xu, S., Binzel, R.P., Burbine, T.H., Bus, S.J., 1995. Small main-belt asteroid spectroscopic survey: initial results. Icarus 115, 1–35.

Yada, T., Abe, M., Okada, T., Nakato, A., Yogata, K., Miyazaki, A., Hatakeda, K., Kumagai, K., Nishimura, M., Hitomi, Y., et al., 2022. Preliminary analysis of the Hayabusa2 samples returned from C-type asteroid Ryugu. Nature Astronomy 6, 214–220.

Yamamoto, S., Watanabe, S., Matsunaga, T., 2018. Space-weathered anorthosite as spectral D-type material on the Martian satellites. Geophysical Research Letters 45, 1305–1312.

York, D.G., Adelman, J., Anderson Jr, J.E., Anderson, S.F., Annis, J., Bahcall, N.A., Bakken, J., Barkhouser, R., Bastian, S., Berman, E., et al., 2000. The Sloan Digital Sky Survey: technical summary. The Astronomical Journal 120, 1579.

Yoshikawa, M., Kawaguchi, J., Fujiwara, A., Tsuchiyama, A., 2014. Hayabusa sample return mission. In: Bottke, W.F., DeMeo, F.E., Michel, P. (Eds.), Asteroids IV. University of Arizona Press, pp. 397–418.

Yoshikawa, M., Kawaguchi, J., Fujiwara, A., Tsuchiyama, A., 2021. The Hayabusa mission. In: Longobardo, A. (Ed.), Sample Return Missions: The Last Frontier of Solar System Exploration. Elsevier, pp. 123–146.

Yurimoto, H., Abe, K.i., Abe, M., Ebihara, M., Fujimura, A., Hashiguchi, M., Hashizume, K., Ireland, T.R., Itoh, S., Katayama, J., et al., 2011. Oxygen isotopic compositions of asteroidal materials returned from Itokawa by the Hayabusa mission. Science 333, 1116–1119.

Zellner, B., Tholen, D., Tedesco, E., 1985. The eight-color asteroid survey: results for 589 minor planets. Icarus 61, 355–416.

Zolensky, M., Herrin, J., Mikouchi, T., Ohsumi, K., Friedrich, J., Steele, A., Rumble, D., Fries, M., Sandford, S., Milam, S., et al., 2010. Mineralogy and petrography of the Almahata Sitta ureilite. Meteoritics & Planetary Science 45, 1618–1637.

# CHAPTER SEVEN

# Machine learning-assisted dynamical classification of trans-Neptunian objects

**Kathryn Volk[a] and Renu Malhotra[b]**
[a]Planetary Science Institute, Tucson, AZ, United States
[b]The University of Arizona, Tucson, AZ, United States

## 7.1. Introduction to dynamical classification of TNOs

Trans-Neptunian objects (TNOs) are small solar system bodies with semimajor axes ($a$) in the range 30–2000 au, beyond Neptune but interior to the Oort cloud. Their orbits are perturbed by Neptune and other solar system giant planets, but relatively unperturbed by external forces, such as those of passing stars and galactic tides. Astronomical surveys for TNOs have thus far sampled only a very small fraction of the whole population. The observed set of approximately 4000 TNOs have complex distributions in $a$, eccentricity ($e$), and inclination ($i$) that reveal multiple dynamical subclasses (described in Section 7.1.1). The distribution of TNOs in these subclasses has revealed important details about the dynamical history of the outer solar system's giant planets and early planetesimal disk, though many open questions remain (see, e.g., a recent review by Gladman and Volk 2021). The Vera Rubin Observatory's Legacy Survey of Space and Time (LSST) is expected to increase the number of known TNOs to $\sim 40,000$ (Ivezić et al., 2019; Schwamb et al., 2019). We have thus far heavily relied on manual dynamical classification of TNOs, but leveraging the full LSST TNO dataset to further constrain dynamical models of the early solar system will require automated approaches.

The rest of subsection 7.1 provides an overview of the dynamical classes of TNOs, the challenges faced in classifying TNOs, and the need for improved machine learning classifiers. Subsection 7.2 describes an improved classifier based on a supervised learning approach with a large and diverse labeled training/testing dataset. Subsection 7.3 summarizes our work and describes anticipated future applications.

*Machine Learning for Small Bodies in the Solar System*
https://doi.org/10.1016/B978-0-44-324770-5.00012-X

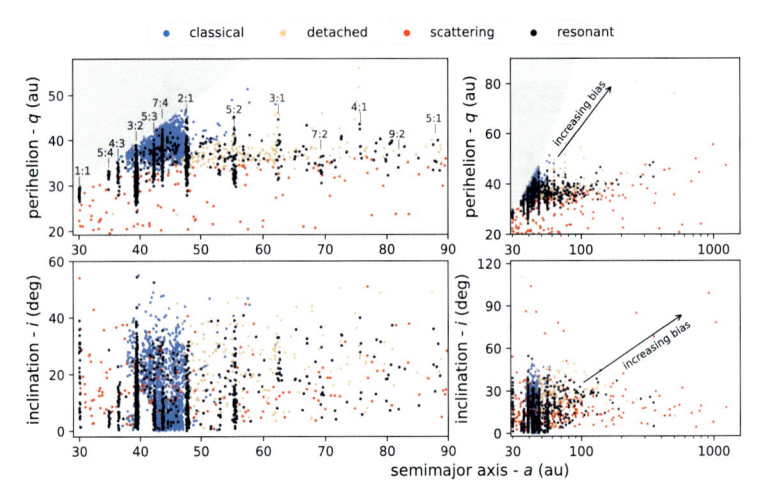

**Figure 7.1** Current inventory of 3357 multiopposition TNOs with orbits sufficiently well-constrained to classify using the Gladman et al. (2008) scheme. The top panels show perihelion distance vs semimajor axis (the grayed out areas are unphysical), whereas the bottom panels show ecliptic inclination vs semimajor axis. The right panels show a zoomed-out view over a larger parameter space range; note the log-scale for semimajor axis. The left panels show a zoomed-in view of the closer-in TNO populations; various prominent mean motion resonances with Neptune are labeled. The data underlying this plot is published in Volk and Van Laerhoven (2024), and a version of this plot with a smaller sample of TNOs was published in Gladman and Volk (2021).

## 7.1.1 The dynamical classes of TNOs and current approaches to classification

Fig. 7.1 shows the current census of TNOs divided into four dynamical classes by Volk and Van Laerhoven (2024): resonant, scattering, classical, and detached. We describe the exact definitions of these classes below, but it is useful to review their broad features and implications. The resonant TNOs are objects that librate in Neptune's exterior mean motion resonances (described in detail in subsection 7.1.2). They are prominent both in the biased observational sample (e.g., Elliot et al., 2005; Kavelaars et al., 2009; Bannister et al., 2018; Bernardinelli et al., 2022; Smotherman et al., 2023) as well as intrinsically (e.g., Gladman et al., 2012; Adams et al., 2014; Volk et al., 2016; Chen et al., 2019; Crompvoets et al., 2022). The present-day large stable resonant TNO populations imply the capture of TNOs into Neptune's resonances during the epoch of giant planet migration (see, e.g., Malhotra 1995, 2019). The resonant TNOs are thus of particular interest to identify, because they provide an important test for dynamical models (dis–

cussed in Gladman and Volk 2021). The classical TNOs appear in Fig. 7.1 as a collection of relatively lower-$e$ (higher perihelion distance $q$) orbits concentrated at semimajor axes between Neptune's 3:2 and 2:1 resonances. The lowest-$e$ and lowest-$i$ members of the classical belt (often called "cold" classicals) are thought to be the only in-situ remnant of the outer solar system's original planetesimal disk, whereas the more dynamically excited "hot" classical population was transplanted into this region from closer-in portions of the disk, which was dispersed during planet migration (see discussion in recent reviews by Morbidelli and Nesvorný 2020 and Gladman and Volk 2021). The scattering TNOs are objects with low-enough perihelion distances to experience strong perturbations from Neptune, which can change their orbital energy (and thus semimajor axis) on timescales much shorter than the age of the solar system. In Fig. 7.1, only resonant TNOs have perihelion distances that overlap with the scattering population, but they are phase protected from encountering Neptune at perihelion by their resonant orbits. Detached TNOs have larger $q$ than the scattering population and correspondingly more stable orbits; these are TNOs outside the classical belt region that are neither resonant with Neptune nor experiencing significant changes in $a$ due to planetary encounters. Both the scattering and detached TNOs are remnants of the original planetesimal disk that was scattered outward when the giant planets migrated to their present orbits (see, e.g., Gomes et al. 2008; Dones et al. 2015). The distributions of TNOs amongst these different dynamical classes provide critical constraints on the exact details of the planet migration era. These distributions are affected by many observational biases that must be accounted for when comparing observations to theoretical models. The first step in this comparison is dynamical classification of the observed orbits.

In this chapter, we will define TNO dynamical classes using the scheme presented by Gladman et al. (2008). In this nomenclature system, the observed orbit of a TNO is numerically integrated under the influence of the Sun and all four giant planets for 10 Myr, and the integration is used to classify the TNOs as follows:

- Resonant TNOs: the resonant angle for one of Neptune's external $p$:$q$ mean motion resonances shows libration for at least 50% of the integration
- Scattering TNOs: a nonresonant orbit that exhibits a barycentric semimajor axis change $\Delta a > 1.5$ au during the integration
- Detached TNOs: a nonresonant, nonscattering orbit with $e > 0.24$ (which tends to coincide with $a \gtrsim 50$ au)

- Classical TNOs: a nonresonant, nonscattering orbit with $e \leq 0.24$ (these are most concentrated between Neptune's 3:2 and 2:1 resonances; $39.4 < a < 47.8$ au).

Gladman et al. (2008) further divide the classical TNOs into inner, main, and outer based on semimajor axis, but we do not consider these divisions here. Note that the order of operations in implementing this scheme is important.

The dynamical behavior of TNOs is a spectrum rather than being discrete, so the above scheme represents just one set of possible dividing lines, with a number of specific choices to divide the spectrum of dynamical evolution. The first choice is the timescale over which the dynamical behavior is classified. For the majority of observed TNOs, 10 Myr is a timescale that will capture a reasonable picture of their present dynamics. Resonant libration periods are typically $10^4 - 10^5$ years, so this timescale covers many libration cycles. For close-in TNOs ($a \lesssim 50$ au), secular variations in $e$ and $i$ have timescales of a few Myr, so 10 Myr will capture their full range. For large-$a$ TNOs (especially those with $a \gtrsim 100$ au), 10 Myr is no longer sufficient to capture this range. As we discover more large-$a$ TNOs, it will be worth reconsidering appropriate classification timescales, because longer-term $e$ variations can cause objects to switch between the scattering and detached populations.

The next specific choice is the $\Delta a > 1.5$ au definition for the scattering population. This is a reasonable dividing line between stable and unstable orbits for the moderate-$a$ ($a \lesssim 100$ au) scattering population, which dominates the observed set of TNOs (scattering objects with smaller $a$ are more likely to be observed than those with larger $a$ in brightness limited surveys). Some unstable classical objects do not meet this threshold to qualify as scattering objects, because their tightly-bound orbits require larger energy changes, and some very distant objects with large perihelia would likely be better described as diffusing due to smaller perturbations on their weakly bound orbits (e.g., Bannister et al., 2017). This is also a case for which changes in the criterion, such as defining a relative $\Delta a/a$ threshold and/or adding an additional class of diffusing objects, should be considered in the future as more TNOs are discovered.

The division between the classical and detached TNOs might similarly evolve with more discoveries. The motivation for the eccentricity cut between these populations is to accommodate the possibility of a second or extended low-eccentricity classical belt beyond the currently known one. A more distant population of TNOs on nearly circular orbits has not yet been ruled out by observations (see discussion in Gladman and Volk 2021), and currently unseen TNO populations have been proposed based on dust measurements (e.g., Doner et al., 2024). If discovered, such a population would likely not have the same dynamical emplacement history as the known detached population, a motivation for providing a separation. However, for the known TNOs, the classical and detached populations share very similar dynamics, even if they have different semimajor axis ranges. We will test our classifier with their current, separate class labels as well as combining them into a single class in subsection 7.2.3.

The last choice of note in the scheme above is the 50% threshold for resonant libration over 10 Myr. This places many resonance-interacting objects in the classical and detached populations and many intermittently resonant objects in the resonant population. We will discuss the resonant TNOs in detail in the next section, including the challenges presented by TNOs with intermittent resonant libration.

Thus far, the most accurate way to divide TNOs into the four classes described above is to rely heavily on manual inspection of the 10 Myr integration outputs. This is essentially entirely due to how difficult it can be to unambiguously identify resonant behavior in automated time-series analyses. Though scattering objects are relatively easy to identify through their large changes in $a$, it is more difficult to classify the objects that remain relatively stationary in $a$ over long timespans. A lack of mobility in $a$ can mean an object is not experiencing significant perturbations (i.e., a classical or detached object), or it can mean an object is resonant with Neptune; typically, resonant angles are calculated and checked for libration to identify resonant objects. However, as we describe in the next section, determining resonant status based solely on the analysis of resonant angles is both challenging and inefficient. In contrast, it is remarkably easy to visually identify most resonant behavior by examining plots of $a$ vs time. Humans are very good at pattern recognition, and after examining thousands of plots of TNO orbits, a human finds it very easy to distinguish resonant librations in $a$ from nonresonant variations in $a$. This visual inspection combined with simple codes that can apply the other criteria above results in the most accurate classifications; but this approach is not scalable to very large numbers of TNOs.

## 7.1.2 The inherent challenges of identifying resonant TNOs

Neptune's external mean motion resonances have resonant angles with the following form:

$$\phi = p\lambda - q\lambda_N - m\varpi - r\varpi_N - n\Omega - s\Omega_N, \qquad (7.1)$$

where $\lambda$ is the mean longitude, $\varpi$ is the longitude of perihelion, and $\Omega$ is the longitude of ascending node; in each case, the subscript N refers to the orbit of Neptune, and the nonsubscripted elements are for the TNO. The integers $p$ and $q$ $(p > q > 0)$ describe the period ratio of the TNO and Neptune such that the TNO completes $q$ orbits for every $p$ orbits of Neptune. The integers $m$, $r$, $n$, and $s$ (all $\geq 0$) must sum to equal $p - q$, and the integers $n$ and $s$ must be zero or even. The value of $p - q$ is often referred to as the "order" of a resonance, because in classical analyses of mean motion resonances of low-$e$, low-$i$ orbits, the strength of the perturbation associated with a particular resonant argument is proportional to $e$ and/or $\sin i$ raised to the power of their corresponding integers $m$, $r$, $n$, and $s$ (see, e.g., Murray and Dermott 1999 for a full discussion of resonant angles in the context of the disturbing function). The eccentricities and inclinations of TNOs are typically much larger than those of Neptune, so resonances with $r, s \neq 0$ are much weaker than those with $m, n \neq 0$. We will thus only be considering a simplified range of Neptune's resonances in this chapter:

$$\phi = p\lambda - q\lambda_N - m\varpi - n\Omega. \qquad (7.2)$$

We refer to resonances with $n = 0$ as eccentricity-type resonances and term those with $n \neq 0$ as mixed-type resonances. In principle, inclination-type resonances ($m = 0$) can also occur, but there are not yet any observed TNOs confirmed in such resonances; in our entire training/testing set of orbit integrations (subsection 7.2.1), there is only one case for which the inclination-type resonant argument librates, while the mixed and eccentricity-type arguments do not. Eccentricity-type resonances are, by far, the most common resonances in the observed set of TNOs (see Table 7.1).

    An immediate challenge presented in the search for resonant TNOs is that there are an infinite number of possible combination of integers $p$ and $q$ that could describe an object's period ratio with Neptune. We must somehow limit that infinite set. As noted above, for low-$e$, low-$i$ orbits, one could limit the choices of $p$ and $q$ by some maximum difference between the integers based on a resonance strength argument. However, the eccentricities of many TNOs are so large that the traditional $p - q$ resonance order is not particularly useful for predicting the strength of an eccentricity-type resonance. The strength of a resonance is only predicted by $p - q$ if all of the conjunctions between the two resonant bodies are dynamically meaningful. Bodies with a period ratio of $p{:}q$ will experience $p - q$ conjunctions over one resonant cycle (the time it takes to return to their starting configuration, $q$ orbits of the TNO). For low-$e$ orbits, the separation between the two bodies at conjunction does not vary hugely over the resonant cycle, so each conjunction is a significant perturbation; having many conjunctions over a resonant cycle tends to "smear" out the effects of the resonance, making it dynamically irrelevant for large-enough $p - q$. For the relatively lower-$e$ TNOs in the classical belt region ($a \lesssim$ 50 au), the low $p - q$ resonances do dominate in strength (see, e.g., Gallardo 2006); though some observed TNOs do display libration in the classical belt region resonances with $p - q > 10$.

However, when we consider a more distant resonant TNO on a highly-eccentric orbit with a large semimajor axis, $p - q$ becomes a less useful proxy for resonance strength. The vast majority of such a resonant TNO's conjunctions with Neptune will occur when the TNO is far from perihelion and at such large separations that they have very little effect on the TNO's orbit. Only the interactions between the TNO and Neptune when the TNO is at or near perihelion can happen at small-enough separations that they can affect its evolution. Thus for high-$e$ TNO orbits, large values of $q$ (the number of TNO perihelion passages over one resonant cycle), rather than large $p - q$, will cause Neptune's perturbations to "smear" out and make the resonance weaker. The resonances with smaller values of $q$ will be stronger regardless of $p - q$ (see, e.g., Gallardo 2006; Lan and Malhotra 2019). We will hereafter refer to $q$ as the resonance order when considering Neptune's resonances (following, e.g., Pan and Sari 2004). It is the best proxy for resonance strength for high-$a$ TNOs, and because of the radial distribution of low-$e$ TNOs, the resonances relevant for low-$e$ orbits that have large $p - q$ also have large $q$; so they are high-order resonances in either definition of "resonance order."

The above considerations still leave us with the question: what cutoff value of $p - q$ and/or $q$ should we adopt for the practical task of dynamical classification? We will not have a conclusive answer to this question in this chapter, but the development and performance of our machine learning classifier (detailed in subsection 7.2) offers some insights. Though all of the resonant TNO classifications (described in subsection 7.2.1) in the training and testing dataset for the classifier are based on visual confirmation of a librating resonant angle (the check of which is often triggered by visual inspection of the semimajor axis evolution), we do not feed any information about resonant angles to the machine learning classifier. This is because of the difficulties described above with the number of potential resonant angles that would need to be calculated and tested for libration. Even if we limit ourselves to a set of $p:q$ resonances with previously identified members, the number of resonant angles in the classical belt region can easily become impractically large to calculate and check (every variation of $m$ and $n$ in Eq. (7.2) must be considered for every $p$ and $q$ combination). An arbitrary limit would be premature in the face of the many new TNO discoveries to come, which might occupy new resonances. To construct our labeled training and testing set, we did in fact calculate a *very* large number of potential resonant angles to very high resonance orders for every particle and perform an automated simple running-window analysis to assess if any of those angles librated (i.e., were confined to a range of $\Delta\phi < 355°$). This extensive computer-based search often took more computational time than the 10 Myr orbital integration, and it often missed borderline cases of intermittent, or merely very high-amplitude resonant libration. These missed resonant cases were almost always immediately apparent to the human eye during visual inspection of plots of the semimajor axis time series for the particle. In these cases, a manual search for a librating resonant angle was performed to confirm resonance. A typical example is shown in the left panels of Fig. 7.2, where the resonant nature of the orbit is very apparent to an experienced human from the time evolution of $a$, but the libration of the resonant angle is more difficult to characterize using any simple computer-based analyses, such as testing for confinement in a series of time windows; the right panels in Fig. 7.2 show an example of well-behaved libration for comparison. Though the $\phi$ evolution could be characterized with more complex analyses (including machine learning approaches!), doing such analyses for a large enough set of potential resonant angles for any given particle becomes computationally prohibitive.

When classifying TNOs manually, visual inspection of the semimajor axis evolution by an experienced evaluator is essentially always sufficient to identify resonant behavior, with inspection of the resonant angle used only to confirm the classification of the $a$ evolution and to identify the specific resonance. This and the computational cost described above for calculating resonant angles, motivates us to discard resonant angles for our machine learning classifier. We will instead rely solely on characterizations

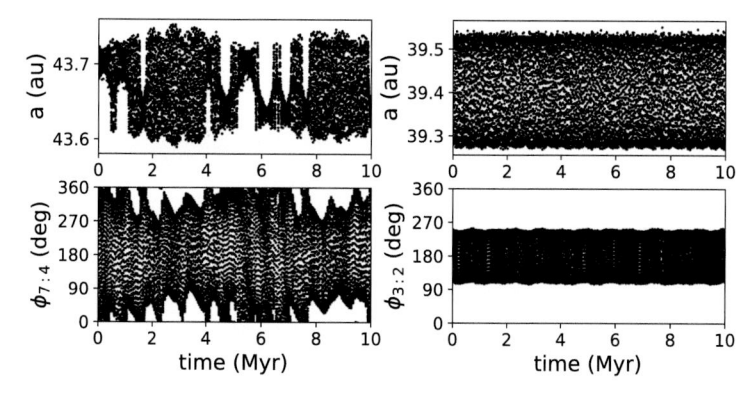

**Figure 7.2** Evolution of semimajor axis (top panels) and resonant angles (bottom panels) of the best-fit orbits of TNO 2013 UG17 in Neptune's 7:4 resonance (left panels) and TNO 534161 in Neptune's 3:2 resonance (right panels). The resonance angles are $\phi_{7:4} = 7\lambda - 4\lambda_N - 3\varpi$ and $\phi_{3:2} = 3\lambda - 2\lambda_N - \varpi$. TNO 534161's evolution is an example of clean, relatively low-amplitude resonant libration, which is easily characterized using simple bounds on $\phi$. TNO 2013 UG17's evolution is a very typical example of intermittent libration, which is more difficult to characterize, but its resonant nature is very readily recognized by an experienced human inspecting the evolution of $a$.

of the standard set of orbital elements and position data, including calculating parameters designed to mimic the resonant characteristics and patterns that are apparent by visual examination by a human (see subsection 7.2.2). However, this approach also has some downsides. Sometimes an extensive resonant angle search reveals libration of very high-order resonant arguments, which would not be readily apparent from visual inspection of the orbital evolution. Fig. 7.3 shows the semimajor axis evolution from a short, high-resolution-output integration and a longer, lower-resolution-output integration for two slightly different orbits within the observational uncertainties of TNO 2014 UJ299. We find that the orbit shown in the left panels librates in the high-order mixed-$e$-$i$-type resonance characterized by $\phi = 21\lambda - 13\lambda_N - 4\varpi - 4\Omega$, whereas the orbit in the right panels shows no libration of any resonant angle that we tested. Aside from the behavior of that resonant angle and the short-term semimajor axis evolution of the resonant clone, which appears *slightly* different (left panels), the longer-timescale evolution is nearly indistinguishable between the two clones, and the resonance has no effect on their $e$ or $i$ evolution. A comparison of periodograms in $a$ for these two particles does not provide substantially different information than this visual assessment of Fig. 7.3. As we will find in subsection 7.2.3, a significant fraction of the misclassifications made by our

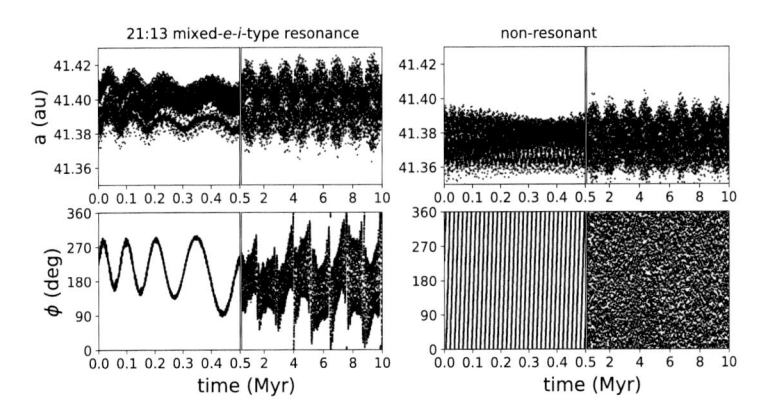

**Figure 7.3** Evolution of two clones of TNO 2014 UJ299 in semimajor axis (top panels) and resonant angle $\phi$ (bottom panels). The left panels show a clone librating in a high-order mixed-$e$-$i$-type resonance with resonant angle $\phi = 21\lambda - 13\lambda_N - 4\varpi - 4\Omega$, whereas the right panels show a nearby nonresonant clone. In each panel the time-axis is discontinuous with the left portion showing the high-resolution output from 0–0.5 Myr and the right portion showing lower-resolution output from 0.5–10 Myr. While the high-resolution output does reveal some differences in the semimajor axis behavior between the two clones, they look nearly identical over 10 Myr timescales. Their inclinations and eccentricities also evolve nearly identically. So we are left to ponder about the dynamical significance of the libration of this very high-order resonant argument.

machine learning classifier are failures to identify cases of very high-order ($q \geq 10$) resonances, such as the one in Fig. 7.3. But perhaps the fact that these resonant orbits are so difficult to distinguish from nearby nonresonant orbits, both by eye and by analyses of their orbital evolution, hints at a limit to the true dynamical relevance of such high-order resonances.

Even after a resonance has been identified, one challenge remains: deciding how and where to draw the line between mostly resonant behavior and mostly nonresonant behavior. The Gladman et al. (2008) scheme chooses to label cases that librate in $\phi$ for at least 50% of a 10 Myr simulation as resonant and those librating less than 50% as nonresonant. But even that determination can be challenging for objects such as the one shown in the left panels of Fig. 7.2. When trying to automate classification, whether the libration meets the 50% threshold depends strongly on the window size over which the time series is examined, the time-sampling of $\phi$, the exact tolerances on the maximum range in $\phi$, etc. Visual examination of $\phi_{7:4}$ in Fig. 7.2 places this object in the resonant category, though even that is a bit of a judgment call as it is very close to the 50% threshold; and as we will discuss in the next section, visual examination and human judgment is not a sustainable approach to this problem. We will show in subsection 7.2.3 that many of the misclassifications by the machine learning classifier involve objects that display intermittent resonant interactions. It is unsurprising that these edge cases remain a challenge whether being classified by humans or via machine learning.

### 7.1.3 The need to improve automated classification

The strongest motivation for automating TNO classification is the expected $\sim 40,000$ new TNOs that will be discovered by LSST (Ivezić et al., 2019; Schwamb et al., 2019). Even considering just best-fit orbits, manual classification of such a large number of TNOs would be daunting. The Gladman et al. (2008) scheme for determining whether a real TNO's dynamical classification is secure or insecure requires classifying three variants on each TNO's orbit (the best-fit orbit, and a minumum- and maximum-semimajor axis orbit), tripling the number of integrations to be visually inspected. Though the three-clone approach is quite useful, it does not yield probabilities of dynamical class membership for a TNO. Additionally, for objects with large orbital uncertainties, the gaps between the best-fit and extreme clones can be too large to adequately sample the range of possible dynamical classes. We ideally would like to run many more clones (100 or more) of each observed TNO orbit to fully sample the dynamical classes consistent with the observations and provide meaningful probabilities for the most likely dynamical class. For the expected LSST TNO sample, this aspiration means classifying millions of orbits. At this scale, human assessment must be removed from the classification process.

Machine learning classifiers are well-suited for the task of replacing human inspection in the TNO classification process. Smullen and Volk (2020) described a first pass at training and testing a gradient boosting classifier for assigning dynamical classes according to the Gladman et al. (2008) scheme based on a simple set of data features in short integrations. This classifier achieved 97% accuracy (the percentage of all predictions that are correct) on a set of TNOs that met the Gladman et al. (2008) criteria for secure dynamical classifications. The accuracy dropped to 75% when considering only insecurely classified TNOs, which is unsurprising as those are often insecurely classified because they exhibit a complicated mix of resonant and nonresonant behavior. As discussed above, these are the most challenging cases to classify. When the Smullen and Volk (2020) classifier was applied to a new set of TNOs (with a typical mix of secure and insecure classifications as determined by the manual classification approach), the accuracy was

92.4%. Given the relatively small training and testing set available, and the limited set of data features used for the classifier, this was a very promising result. However, that classifier would still result in thousands of misclassified TNOs from the LSST discoveries. In the next section, we describe an updated and improved machine learning classifier based on a training and testing set an order of magnitude larger than that in Smullen and Volk (2020) and employing an improved set of data features. This new classifier provides the same classification as a human 97–98% of the time and dynamically relevant classifications $> 99\%$ of the time (see subsection 7.2.3), making it a reasonable replacement for human classification in the LSST era.

## 7.2. Building a machine learning classifier for TNOs

Building on the success of Smullen and Volk (2020), we will take a supervised learning approach to TNO classification. This requires constructing a sample of correctly labeled TNO orbits, deciding what data from those orbits will be provided to the classifier, and then training, testing, and optimizing the classifier. We describe our training/testing set of TNO orbits and the dynamical labels we assign to those orbits in subsection 7.2.1. Subsection 7.2.2 describes how we turn the integrations of those TNO orbits into a discrete set of data features to use in the classifier. Finally, we train and test a variety of classifiers in subsection 7.2.3 based on those dynamical labels and data features.

### 7.2.1 Building and labeling an adequately large and diverse training dataset

We built an initial training and testing set based on the observed set of multiopposition TNOs pulled from the Minor Planet Center as of December 2023. These TNOs were integrated and classified based on the Gladman et al. (2008) scheme, which involves integrating 3 cloned particles representing variations of an observed TNO's orbit. The details of this set of observed objects and their resulting dynamical classifications are given in Volk and Van Laerhoven (2024). For the purposes of this chapter, the important details are that we have orbital time series data for 9477 test particles representing 3159 real, observed TNOs, and that these time series data have been examined visually to confirm that the evolution of each test particle is accurately labeled with the following information:

1. the Gladman et al. (2008) classification as resonant, scattering, classical, or detached;

2. if resonant, whether the resonant argument is for the typical $e$-type resonance or if it is for a mixed $e$-$i$-type resonance;

3. if resonant, whether the test particle's resonant libration is intermittent or remains clearly bounded for the entire 10 Myr integration;

4. if classical or detached, whether the test particle experiences significant resonant interactions, even though it is dominantly nonresonant; if a particle experiences short periods ($< 5$ Myr total) of resonant libration or clearly crosses back and forth across the separatrix of a resonance, it is labeled as "resonant-interacting."

The integration outputs include the typical orbital elements (barycentric $a$, $e$, $i$, $\Omega$, $\varpi$, argument of perihelion $\omega$, and mean anomaly $M$) for the particle and the giant planets. We have high-resolution output for the first 0.5 Myr (outputs every 50 yr; 10,000 outputs per particle) and lower-resolution outputs over the 10 Myr timespan (outputs every 1000 yr; 10,000 outputs per particle). Plots of both timescales were examined to assign the Gladman et al. (2008) classifications and additional labels listed above. Table 7.1 provides details on the number of particles in this dataset divided into the different dynamical categories.

Some of the dynamical categories are not well-sampled in the set of observed TNOs, notably high-order resonances ($p$:$q$ MMRs with $q \geq 5$) and mixed-$e$-$i$ resonances. Scattering and detached particles are also less numerous than classical and resonant ones; scattering, detached, and distant resonant TNOs spend most of their time very far from the Sun compared to classical and closer-in resonant TNOs, and thus suffer from stronger observational biases (see, e.g., Gladman and Volk 2021 for a discussion of observational biases in the TNO populations). To augment the training and testing set, we generated synthetic TNO orbits targeting these under-sampled populations (see Appendix 7.4 for details of how these orbits were generated) and integrated and classified them just as above. Table 7.1 lists the number of synthetic TNOs of each dynamical type that were added to the dataset. We increased the number of mixed-$e$-$i$ resonant particles and high-order resonant particles by an order of magnitude, more than doubling the total resonant training set. We also nearly doubled the number of scattering particles and increased the number of detached particles by 60%. This enhanced training set still does not fully sample the range of orbits possible for TNOs, but it provides significantly more examples of orbits that are underrepresented in the observationally based data; this en-

**Table 7.1** Classifications of the 15375 particles in the training set. The top four rows show the number of test particles classified into each of the four Gladman et al. (2008) dynamical classes. Below that are divisions by additional dynamical detail. For each category, we indicate how many of the test particles are generated directly from observed TNO orbits and how many are generated from synthetic TNO orbits.

| | number of real TNO particles | number of synthetic TNO particles | total |
|---|---|---|---|
| resonant | 3203 | 4840 | 8043 |
| scattering | 712 | 658 | 1370 |
| detached | 618 | 372 | 990 |
| classical | 4944 | 28 | 4972 |
| $e$-type resonant | 3085 | 3762 | 6847 |
| mixed-$e$-$i$ resonant | 119 | 1078 | 1197 |
| cleanly librating | 2877 | 3226 | 6103 |
| intermittently librating | 327 | 1614 | 1941 |
| $p$:$q$ resonant, $q \geq 5$ | 336 | 3023 | 3359 |
| $p$:$q$ resonant, $q \geq 10$ | 63 | 325 | 388 |
| non-resonance-interacting classical/detached | 4897 | 104 | 4793 |

sures that there are enough of these rarer object types to both train and test the classifier. As we will discuss in subsection 7.2.3, the inclusion of these synthetic TNOs significantly decreases the rate at which the classifier provides dynamically irrelevant classifications.

The total testing and training dataset comprises integrations of 15,375 particles with human-assigned dynamical labels. Unlike the training set previously used in Smullen and Volk (2020), this set is not limited to just the observed set of TNOs, and it is not limited to only TNOs with secure dynamical classifications. We have also labeled each particle with information not contained in the Gladman et al. (2008) classifications, which will enable us to test the ability to divide TNOs into different classes besides the four standard ones. The next section describes the analyses performed on the integration outputs of our training and testing set orbits to prepare them for the classifier.

## 7.2.2 Choosing appropriate and useful time series data features for the classifier

The type of classifier we employ in subsection 7.2.3 is not given a set of rules by which to classify TNOs, and it does not ingest the entire time series data of each particle it classifies. It is instead given a set of so–called data features, which are parameters calculated to summarize important aspects the time series data. It uses a training set of data features that are labeled by class to construct a set of rules for how to use those data features to predict the correct class. The rules constructed from the training set are then tried out on a testing set of data features, and the predicted classes are compared to the known correct classes to determine the classifier's accuracy. See Chapter 1 for a more detailed description of this kind of machine learning classifier.

We use data features calculated from both the short, high-resolution orbital integration time series and the long, lower-resolution time series. We begin by considering the same set of data features as in Smullen and Volk (2020), which are the following: 1) the minimum, maximum, mean, and standard deviation of each of the particle's orbital elements; 2) the minimum, maximum, mean, and standard deviation of the rate of change of each orbital element from one simulation output to the next; and 3) the maximum range of each of the orbital elements and its rate of change. In Smullen and Volk (2020), the orbital elements considered for these parameters were $a$, $e$, $i$, $\Omega$, and $\omega$, and their values were all calculated from just a short $10^5$ year integration. In addition to calculating these features from both short and long timescale integrations (to accommodate some of the additional data features we describe below), we also make some modifications to this set of simple features. First, we add the longitude of perihelion ($\varpi = \Omega + \omega$), the perihelion distance ($q$), and the Tisserand parameter with respect to Neptune ($T_N = a_N/a + 2\sqrt{a/a_N(1 - e^2)} \cos i$) to the list of time series variables considered. We also discard most of the features based on the absolute values of the orbital angles $\Omega$ and $\varpi$, because the initial values of these angles are subject to strong observational biases based on where in the sky surveys have looked for TNOs (see, e.g., Shankman et al. 2017). Resonant objects are subject to additional, epoch-dependent observational biases in $\varpi$ (e.g., Gladman et al., 2012; Volk and Malhotra, 2020). Even though these biases might wash out in the longer integrations as these angles precess, and there is some important dynamical information in their overall distribution (e.g., JeongAhn and Malhotra, 2014), we do not wish to potentially bias the classifier. We keep the features calculated based on the rate of change for each of these angles. We also keep the minimum, maximum, mean, and standard deviation for $\omega$, because the dynamical information contained in the $\omega$ evolution outweighs the potential for introducing biases. Many resonant TNOs undergo so-called Kozai libration of $\omega$ (see, e.g.,

Gomes et al. 2005), meaning specific values for the average and range of $\omega$ could be an important marker for resonant behavior. For $a$ and $q$, we also include values of the standard deviation and maximum range normalized to their average values.

Next we consider a range of new data features. Following Volk and Malhotra (2020), we calculate a spectral fraction for the evolution of each particle's semimajor axis, components of the eccentricity and inclination vectors ($e \sin \varpi$ and $\sin i \sin \Omega$), and the angular momentum deficit ($\mathrm{AMD} = a(1 - \sqrt{1 - e^2}) \cos i$). The spectral fraction is a parameter that captures whether the evolution of a time series parameter is dominated by just a few frequencies or contains many, potentially overlapping frequencies. We define the spectral fraction of a time-series variable by taking a fast Fourier transform (FFT) of that time-series and determining what fraction of the frequencies in that FFT have an associated power (defined as the amplitude squared at that frequency) greater than 5% of the highest-amplitude frequency's power. A small spectral fraction means the single, highest-amplitude frequency dominates the evolution, whereas a large spectral fraction means there are many frequencies affecting the evolution. Volk and Malhotra (2020) found that for simulated multi-planet systems, the spectral fraction could be used as an indicator of long-term stability or instability. It is likely that the same holds true for TNO orbits as those with many overlapping frequencies are more likely to experience chaotic evolution. Additionally, the semimajor axis evolution of resonant and nonresonant TNOs are very different in the frequency domain; the resonant libration of $a$ adds a powerful, lower-frequency term to the $a$ evolution of resonant objects. For $a$ and the AMD, we calculate a spectral fraction from both the short and long data series, because both simulations cover timescales relevant to the dynamics. The dominant timescales for the eccentricity and inclination evolution are typically on Myr timescales, so we only calculate the spectral fraction for those parameters from the 10 Myr time series. For each parameter, we also include as data features the power of the highest-amplitude FFT frequency, the summed power of the top three FFT frequencies, and the values of the top three FFT frequencies.

We also consider data features related to the spatial distribution of a particle's path in a frame rotating at Neptune's instantaneous azimuthal rate. Particles in resonance with Neptune follow distinct paths in this rotating frame as they come to their resonant perihelion locations relative to Neptune (see, e.g., Malhotra 1996; Gladman et al. 2012; Gladman and Volk 2021). For each particle, we transform the simulation outputs to this rotating frame with Neptune and consider the distribution of the barycentric distance $r_b$ and the projected angle with respect to Neptune in the x-y plane, an angle we will denote as $\theta_N$. We then parameterize the distribution of points in this reference frame by dividing a particle's range of barycentric distances into 10 bins, and $\theta_N$ into 20 bins from 0–360°, yielding a 200 space grid. Fig. 7.4 illustrates the distribution of $r_b$ and $\theta_N$ across this grid for two resonant orbits and one classical belt orbit. For the distance bin that includes the particle's perihelion, we calculate a Rayleigh parameter (e.g., Fisher et al., 1993) to describe how uniformly $\theta_N$ is distributed. This parameter is given by

$$R = \sqrt{(< \sin \theta_N >)^2 + (< \cos \theta_N >)^2}, \tag{7.3}$$

where the averages are taken over all values of $\sin \theta_N$ and $\cos \theta_N$ with barycentric distances in the closest $r_b$ bin. A particle with a uniform distribution in $r_b$ and $\theta_N$ would have $R = 0$, whereas one that was perfectly concentrated at a single value of $\theta_N$ would have $R = 1$. For the classical belt object in Fig. 7.4, $R = 0.008$. Particles in $p{:}q$ resonances, where $q \neq 1$, also have small values of $R$, because they come to perihelion at more than one $\theta_N$ value; the 3:2 resonant particle in Fig. 7.4 has $R = 0.007$. So we must generalize our calculation of $R$ for $p{:}q$ resonances to

$$R_q = \sqrt{(< \sin q\theta_N >)^2 + (< \cos q\theta_N >)^2} \tag{7.4}$$

for $q = 1, 2, 3, .., 10$; we do not consider $R_q$ for higher-order resonances as the values become less statistically distinguishable from uniform in this parameter with additional perihelion locations. We take the maximum value of $R_{1-10}$ as the data feature for the classifier ($R_{max,peri}$). For our example 3:2 particle, $R_{max,peri} = R_2 = 0.65$; similarly, our example 12:7 particle in Fig. 7.4 has $R_{max,peri} = R_7 = 0.80$. For every particle, we also calculate an analogous feature using the maximum $r_b$ bin, $R_{max,apo}$.

We calculate several features based on the grid in $r_b$ and $\theta_N$. We determine the number of empty grid spaces across the minimum and maximum $r_b$ bins ($n_{empty,peri}$ and $n_{empty,apo}$) and the standard deviation in the number of visits per grid space in these bins ($\sigma_{n,peri}$ and $\sigma_{n,apo}$); all of these parameters will be larger for resonant particles than nonresonant ones. For the minimum $r_b$ bins, we determine how many of the bins on either side of Neptune's location at $\theta_N = 0$ are empty ($n_{empty,N}$); this will be highest for low-order resonances, smaller for high-order resonances, and close to zero for nonresonant particles. We also determine the standard deviation in the

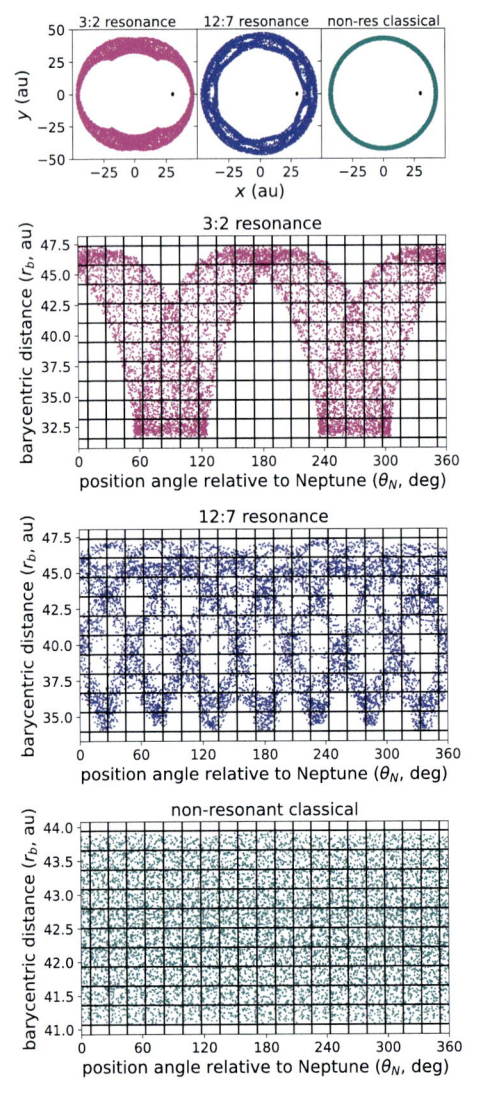

**Figure 7.4** Top panel: Position of three TNO orbits (colored) and Neptune (black) over 10 Myr in a frame rotating at Neptune's instantaneous azimuthal rate. Bottom three panels: Barycentric distance, $r_b$ vs longitude angle relative to Neptune, $\theta_N$, at every output over the 10 Myr integrations for the three TNOs in the top panel. Neptune would be centered at the point (0,30.06) in these plots. We divide the evolution of the particle in this plane into a 10 by 20 grid, with the 10 bins in $r_b$ bounding the particle's minimum and maximum barycentric distances; the grid in $\theta_N$ starts and ends with half a bin so that Neptune is centered in the wrapped bin. We then calculate data features based on this grid, including the number of empty grid spaces overall as well as in the smallest distance range (near perihelion) and the largest distance range (near aphelion); the average and standard deviation in the number of points in all the grid spaces as well as in the perihelion and aphelion grid spaces.

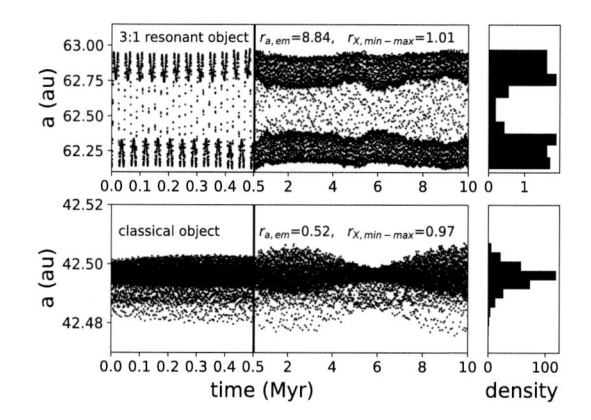

**Figure 7.5** Semimajor axis time-series for a 3:1 resonant object (top panels) and a nonresonant classical object (bottom panels). In the left panels, the time-axis is discontinuous with the first portion showing the high-resolution output over 0–0.5 Myr and the remaining portion showing lower-resolution output over 0.5–10 Myr. The right panels show the histogram of $a$ values across the 10 Myr integration, highlighting the difference between resonant and nonresonant evolution. In the machine learning classifier, the two data features, $r_{a,min-max}$ and $r_{a,em}$, related to the semimajor axis (see Table 7.2), encapsulate the information in these histograms.

number of visits to each grid space across the entire 200 grid spaces ($\sigma_n$), the number of empty grid spaces ($n_{empty}$), the average and standard deviation in the visits to each nonempty grid space ($\overline{n}_{nz,avg}, \sigma_{n,nz}$), and the difference between the overall and nonzero standard deviations ($\Delta\sigma_n = \sigma_{n,nz} - \sigma_n$). These data features are calculated for both the short and long simulations.

We consider a few additional data features describing the $a$, $e$, and $i$ time-series. Resonant TNOs often experience large amplitude quasiperiodic variations in their semimajor axes as a result of the resonant perturbations, and they spend more time at the extreme ends of their $a$-range than at the middle. In contrast, completely nonresonant particles tend to have more uniform variations in $a$ within a smaller range; this is illustrated in Fig. 7.5. Particles interacting with a resonance, but not librating, can often have a lop-sided distribution in $a$, where they spend most of their time at one extreme compared to either the middle or the other extreme. We chose a simple binning scheme in $a$ to help parameterize these three kinds of behavior: a bin near the minimum $a$ value spanning $a_{min}$ to $a_{min} + 0.75\sigma_a$, a bin near the mean $a$ value spanning $\overline{a} \pm 0.375\sigma_a$, and a bin near the maximum $a$ value spanning $a_{max} - 0.75\sigma_a$ to $a_{max}$. We then calculate the ratio of output $a$ values in the low-$a$ bin to those in the high-$a$ bin ($r_{a,min-max}$) as well as the average of the low- and high-$a$ bins to the average-$a$ bin ($r_{a,em}$). The exact bin boundaries were chosen empirically based on comparisons of the ratios for typical detached, resonant, and classical TNOs. We note that for scattering TNOs, those chosen bin boundaries sometimes overlap, because their evolution is not well characterized by averages and standard deviations; in those cases we instead define the bins as the lower, middle, and upper 25% of the full $\Delta a$ range. In an attempt to parameterize intermittent resonant behavior, we also examine the maximum change in $r_{a,min-max}$ and $r_{a,em}$ calculated over four nonoverlapping time ranges in both the short and long integrations. For $e$ and $i$, we apply a similar scheme as above to characterize their distributions in both the short and long integrations, though we do not examine the time-dependency of these ratios; these ratios for $e$ and $i$ do not turn out to be as useful to the classifier as their corresponding ratios for $a$, but we include them for completeness. Finally, we calculate correlation coefficients between the $a$ and $e$, $a$ and $i$, and $e$ and $i$ simulation outputs ($C_{ae}$, $C_{ai}$, and $C_{ei}$) to capture whether the variations in these elements are coupled.

Table 7.2 lists all of the data features we calculate to provide to the classifier. In total, we have 227 data features from the short and long integrations.

### 7.2.3 Performance of a gradient boosting classifier for TNO classification

Here we train and test a classifier using the data features described above with different sets of class labels. We chose to use the scikit-learn (Pedregosa et al., 2011) GradientBoostingClassifier.[1] Smullen and Volk (2020) found that this was the best-performing classifier within scikit-learn, and we confirmed that by testing the other classifiers within sklearn.ensemble[2] that support multilabel datasets. For all tests described below, we set the following parameters, which were found to optimize the performance of GradientBoostingClassifier for the standard set of Gladman et al. (2008) classes based on an initial round of training and testing using just the real TNOs and an initial subset of our final data features: max_leaf_nodes = None, min_impurity_decrease = 0.0, min_weight_fraction_leaf = 0.0, min_samples_leaf = 1, min_samples_split = 3, criterion = "friedman_mse," subsample = 0.9, learning_rate = 0.15, max_depth=8, max_features = "log2," n_estimators = 300. We found this set of parameter values by starting with the default values and making changes to one parameter at

[1] https://scikit-learn.org/stable/modules/generated/sklearn.ensemble.GradientBoostingClassifier.html
[2] https://scikit-learn.org/stable/modules/classes.html#module-sklearn.ensemble

**Table 7.2** List of the data features calculated from the integrations.

| feature name | brief description |
|---|---|
| | *basic time-series data features* |
| $\overline{X}$ | average of $X$ for $X = a, e, i, \omega, q, T_N$ |
| $\sigma_X$ | standard deviation in $X$ for $X = a, e, i, \omega, q, T_N$ |
| $\sigma_{X,normed}$ | $\sigma_X/\overline{X}$ for $X = a, q$ |
| $X_{min}$ | minimum value of $X$ for $X = e, i, \omega, q, T_N$ |
| $X_{max}$ | maximum value of $X$ for $X = e, i, \omega, q, T_N$ |
| $\Delta X$ | $X_{max}$ -$X_{min}$ for $X = a, e, i, \omega, q, T_N$ |
| $\Delta X_{normed}$ | $\Delta X/\overline{X}$ for $X = a, q$ |
| $\overline{\dot{X}}$ | average $dX/dt$ for $X = a, e, i, \omega, \Omega, \varpi, q$ |
| $\dot{X}_{min}$ | minimum value of $\dot{X}$ for $X = a, e, i, \omega, \Omega, \varpi, q$ |
| $\dot{X}_{max}$ | maximum value of $\dot{X}$ for $X = a, e, i, \omega, \Omega, \varpi, q$ |
| $\sigma_{\dot{X}}$ | standard deviation in $\dot{X}$ for $X = a, e, i, \omega, \Omega, \varpi, q$ |
| $\sigma_{\dot{X},normed}$ | $\sigma_{\dot{X}}/\overline{\dot{X}}$ for $X = \omega, \Omega, \varpi$ |
| $\Delta \dot{X}$ | $\dot{X}_{max} - \dot{X}_{min}$ for $X = a, e, i, \omega, \Omega, \varpi, q$ |
| $\Delta \dot{X}_{normed}$ | $\Delta \dot{X}/\overline{\dot{X}}$ for $X = \omega, \Omega, \varpi$ |
| | *rotating frame data features based on binning barycentric distance $r_b$ and position angle relative to Neptune $\theta_N$; "near perihelion" and "near aphelion" indicate the minimum and maximum $r_b$ bins* |
| $n_{empty,peri}$ | number of empty bins in $\theta_N$ near perihelion |
| $n_{empty,N}$ | number of empty bins surrounding Neptune in $\theta_N$ near perihelion |
| $\sigma_{n,peri}$ | standard deviation in the number of visits per $\theta_N$ bin near perihelion |
| $\Delta n_{peri}$ | maximum difference in number of visits between $\theta_N$ bins near perihelion |
| $n_{empty,apo}$ | number of empty bins in $\theta_N$ near aphelion |
| $\sigma_{n,apo}$ | standard deviation in the number of visits per $\theta_N$ bin near aphelion |
| $\Delta n_{apo}$ | maximum difference in number of visits between $\theta_N$ bins near aphelion |
| $R_{max,peri}$ | maximum of Eq. (7.4) for $q = 1$ through $q = 10$ near perihelion |
| $R_{max,apo}$ | maximum of Eq. (7.4) for $q = 1$ through $q = 10$ near aphelion |
| $n_{empty}$ | number of empty bins in the $r_b$-$\theta_N$ grid |
| $\sigma_n$ | standard deviation in the number of visits across all bins |
| $\overline{n}_{nz}$ | average number of visits per bin across all nonempty bins |
| $\sigma_{n,nz}$ | standard deviation in the number of visits across all nonempty bins |
| $\Delta \sigma_n$ | $\sigma_{n,nz} - \sigma_n$ |

*continued on next page*

**Table 7.2** (*continued*)

| feature name | brief description |
| --- | --- |
| | *FFT features for $X = a$, $(e \sin \varpi)$, $(\sin i \sin \Omega)$, AMD* |
| $X_{sf}$ | spectral fraction of $X$ |
| $f_{X, i=1,2,3}$ | The three most powerful frequencies from an FFT of parameter $X$ |
| $P_{X,max}$ | Power associated with the peak frequency from an FFT of parameter $X$ |
| $P_{X,max3}$ | Sum of the power associated with the 3 most powerful frequencies from an FFT of parameter $X$ |
| | *other time-evolution features* |
| $r_{X,em}$ | ratio of time spent near $X_{min}$ and $X_{max}$ to $\overline{X}$ for $X = a, e, i$ |
| $r_{X,min-max}$ | ratio of time spent near $X_{min}$ to $X_{max}$ for $X = a, e, i$ |
| $\Delta r_{a,em}$ | maximum change in $r_{a,em}$ over 4 time windows in the simulation |
| $\Delta r_{a,min-max}$ | maximum change in $r_{a,min-max}$ over 4 time windows in the simulation |
| $C_{ae}$ | correlation coefficient between $a$ and $e$ |
| $C_{ai}$ | correlation coefficient between $a$ and $i$ |
| $C_{ei}$ | correlation coefficient between $e$ and $i$ |

a time in moderate increments away from the default (increments of 0.05 for float parameters ranging from 0 to 1, increments of 1 for most integer parameters, and increments of 50 for n_estimators), stopping at a local maximum in the classifier's accuracy. A repeat of this optimization process using the final set of data features and our entire training and testing set yields identical results for all parameters, except min_samples_split. We find that increasing min_samples_split to 4 for the final classifier would yield a 0.05% improvement in performance; we judged this improvement too small to include in the analysis below. For each version of the classifier described below, we use the dataset described in subsection 7.2.1 with all of the data features listed in Table 7.2, training the classifier on 67% of the dataset (10,156 particles), and testing on the remaining 33% (5071 particles). We held the division of the training and testing sets fixed when testing the accuracy of the classifier in predicting different sets of dynamical classes.

We begin by training and testing the classifier based on the Gladman et al. (2008) classes described in subsection 7.1.1: resonant, classical, scattering, and detached. We discuss the performance of the classifier in terms of its accuracy, defined as the percentage of all predictions that are correct. We note that for this classifier, other common performance metrics, such as precision and recall, yield nearly identical percentages (to within $\sim 0.02\%$), so we do not list them. The accuracy of the classifier in assigning the same classes as the human classifier is 97.3%. If we consider the subset of the classifier's predictions that are given at $> 99\%$ confidence (4853 of the 5071 predictions), the accuracy increases to 98.7%. Given the fuzzy nature of some of the dynamical class boundaries (see discussion in subsections 7.1.1 and 7.1.2), it is useful to divide the misclassified particles into four categories:

- not actually incorrect classifications: These are cases where, upon further inspection, the classifier was correct and the human-assigned class was incorrect (almost always high-order resonances that were missed during visual inspection).
- trivial misclassifications: These are cases where the classifier places a particle on the wrong side of either the classical/detached boundary or the scattering/detached boundary. As discussed in subsection 7.1.1, the eccentricity boundary between classical and detached is based more on cosmogonic arguments than on present-day dynamics, and the classifier is not aware of the exact $e$ cut. Similarly, the $\Delta a = 1.5$ au threshold is somewhat arbitrary, and the classifier is not given this rule. Both of these types misclassifications could be trivially swapped during a post-processing check of the classifier's predictions. (Note that some of the swapped classifications would still be incorrect if, for example, the particle is truly resonant.)
- marginally wrong classifications: These are cases where the assigned class describes *part* of the particle's evolution, but doesn't match the human-classifier's assessment of the majority of the evolution. Examples include labeling a particle resonant when it librates some of the time, but does not meet the 50% threshold; labeling an intermittently resonant particle that only just meets the 50% threshold as classical or detached.
- completely incorrect classifications: These are cases where the class predicted for the particle is unrelated to its evolution.

For the Gladman et al. (2008) dynamical classes, the classifier made predictions that differ from the human-assigned classes for 135 particles out of 5071. Of these, 15 particles were trivially misclassified (5 at the classical/detached boundary, 10 at the scattering/detached boundary) and there were 4 cases where the human-assigned labels were wrong (human error!) and the classifier was correct. This brings the classifier's accuracy up to 97.7%. Of the remaining 116 incorrectly classified particles, 50 were predicted to be classical or detached when they were actually resonant. Most of those particles (42 out of 50) display only intermittent resonant libration, so they fall into the marginally wrong category in that they do display significant nonresonant behavior. Only 8 cleanly librating resonant particles were fully incorrectly classified as nonresonant; of these, 7 are in very high-order resonances ($q \geq 10$; including 3 mixed-$e$-$i$ type) in the classical belt region, and 1 is in the distant 54:5 resonance with Neptune. The classifier misclassified 51 classical or detached particles and 1 scattering particle as resonant. Of these 52 particles, 44 show intermittent resonant libration, so only 8 are fully incorrectly classified as resonant while displaying no resonant behavior. Finally, the classifier incorrectly classified 10 resonant particles and 1 detached particle as scattering. Eight of the resonant particles librate in resonances for more than half the simulation (and thus are "correctly" labeled as resonant) before weakly scattering out of resonance; so the classifier is only marginally wrong as they do scatter. The classifier was only entirely wrong about scattering behavior in 3 cases; all 3 cases involve particles librating in or interacting with the wide, symmetric libration zones of N:1 resonances (see, e.g., Lan and Malhotra, 2019), which can mimic the large semimajor axis variations associated with the scattering population.

Overall, the classifier only assigned completely irrelevant labels in 19 cases, or 0.4% of the time; an additional 1.9% of the predictions were only marginally incorrect. Nearly half (7) of the fully incorrectly labeled particles are in very high-order ($q \geq 10$), weak resonances in the classical belt region discussed in subsection 7.1.2. The orbital evolution of these particles are so weakly affected by the libration in those resonances that it is unsurprising that the classifier did not identify them. When we restrict ourselves to the subset of classifier predictions made at > 99% confidence, only 0.14% of the predictions were completely incorrect, with an additional 1.1% of the predictions being marginally incorrect. The classifier performs remarkably well at identifying relevant dynamical behavior.

Given that the boundary between the classical and detached population is somewhat arbitrary, and that the common feature between the two populations is nonresonant, nonscattering behavior, we also tested a simplified version of the Gladman et al. (2008) scheme, which combines them, giving us the classes: resonant, scattering, and classical/detached. With this change, the classifier matched the human classifications in 97.7% of all predictions and 98.8% of the subset of predictions made at > 99% confidence (4876 out of 5071 particles). This represents a small improvement in both the accuracy and the number of high-confidence predictions. Of the

113 incorrectly classified particles, 11 are trivially misclassified at the scattering/detached boundary, and 1 was incorrectly classified by the human, bringing the overall accuracy of the classifier up to 98%. Of the remaining 101 misclassifications, 50 are resonant particles that were incorrectly predicted to be classical/detached. Only 6 of these particles experience clean libration for the entire 10 Myr timespan and are thus fully incorrectly labeled; all 6 are in high-order resonances with $q \geq 10$, including 2 mixed-$e$-$i$ resonant particles. Eleven resonant particles were incorrectly predicted to be scattering; 9 of these are only marginal misclassifications, because the particles do weakly scatter after spending most of the simulation in resonance. There were 36 classical/detached particles incorrectly predicted to be resonant; all but 5 of these particles experience intermittent resonant libration and are thus only marginally misclassified. Overall, the classifier provided only 13 completely incorrect classifications for this simplified classification scheme, an error rate of 0.25%; an additional 1.7% of the predictions were marginally incorrect. Of the predictions made at $> 99\%$ confidence, only 6 are completely incorrect (half of which are in $q \geq 10$ resonances), an error rate of 0.1% with marginally incorrect predictions an additional 0.9% of the time. Though it is plausible that there are dynamical differences between the lower-$e$ nonresonant orbits in the classical belt region and the higher-$e$ nonresonant orbits of the more distant detached population, it seems slightly advantageous to combine them in the classification process (especially as they can be trivially separated after machine learning classification).

The above results are based on our entire training and testing dataset, which includes synthetic TNOs generated to provide more examples of under-populated observational classes. If we exclude these synthetic TNOs from the training and testing set, the classifier doesn't perform as well. Though the classifier still matches the human-assigned classes for the simplified Gladman et al. (2008) scheme 97.5% of the time (compared to 97.7% of the time with the full dataset), it assigns completely incorrect classifications at a higher rate; 1.1% of predictions are dynamically irrelevant compared to only 0.25% above. A significant number of the completely incorrect classifications when the synthetic TNOs are excluded are cases where cleanly librating high-order or mixed-$e$-$i$-type resonant TNOs are incorrectly predicted to be classical/detached. Including more examples of these orbits in the training and testing set clearly improves the classifier's ability to identify them as resonant.

We can examine which of the data features described in subsection 7.2.2 were relied upon the most by the classifier. We note that the exact ranking of the 227 features can vary significantly, depending not just on which set of the above classifications were used, but also on the random seed used to divide the training and testing set or to initialize the classifier itself; there is clearly significant stochasticity to how the classifier uses the data features. Thus we do not provide a full ranking list of the features, but instead note a subset of the features that tended to fall toward the bottom or top of the rankings. Fig. 7.6 shows the distribution of four key features for the simplified Gladman et al. (2008) classification scheme; shown are two features describing the semimajor axis evolution and two describing the particle's distribution in the rotating frame with Neptune. Features that consistently were ranked highly by the classifier include most of the features describing the particle distribution in the rotating frame, most of the data features based on $a$ and many based on $\dot{a}$, some features based on $e$ and $\varpi$, all the data features based on the FFT analysis of $a$ (peak frequencies and spectral fraction), and the $a - e$ and $a - i$ correlation coefficients. Features that consistently ranked very low included all features based on $\dot{e}$ and $\dot{\Omega}$, the features based on FFT analysis of $e$, $i$, and AMD, and the features based on $\omega$ (though those based on $\dot{\omega}$ fell in the middle of the rankings). The 10 Myr maximum timescale of the analyzed time series is likely responsible for the less useful nature of the FFT analysis of $e$ and $i$. As discussed in subsection 7.1.1, the secular timescales for the larger-$a$ TNOs are too long for the frequencies to be captured in a 10 Myr integration. Similarly, 10 Myr is not always long enough to capture the $\omega$ libration that occurs inside some of Neptune's resonances, possibly explaining the relative unimportance of those data features. The heavy emphasis on semimajor-axis based data features and those derived from the rotating frame is unsurprising given how strongly diagnostic those can be for resonant behavior.

For some of the consistently highly ranked data features above, we ran tests of the classifier with those features removed to see how the accuracy changes for the simplified Gladman et al. (2008) scheme. If we exclude all the simple time-series data features based on $a$ and $\dot{a}$, the classifier's accuracy drops slightly to 97.3% (compared to 97.7% with those features included), but the rate of dynamically irrelevant classifications rises to 0.9% (compared to 0.25%). Removing the FFT data features for $a$ reduces the accuracy to 96.9% and increases the rate of dynamically irrelevant classifications to 0.9%. Removing both sets of $a$ features reduces accuracy to 96.6%, while maintaining 0.9% dynamically irrelevant classifications. Removing the fea-

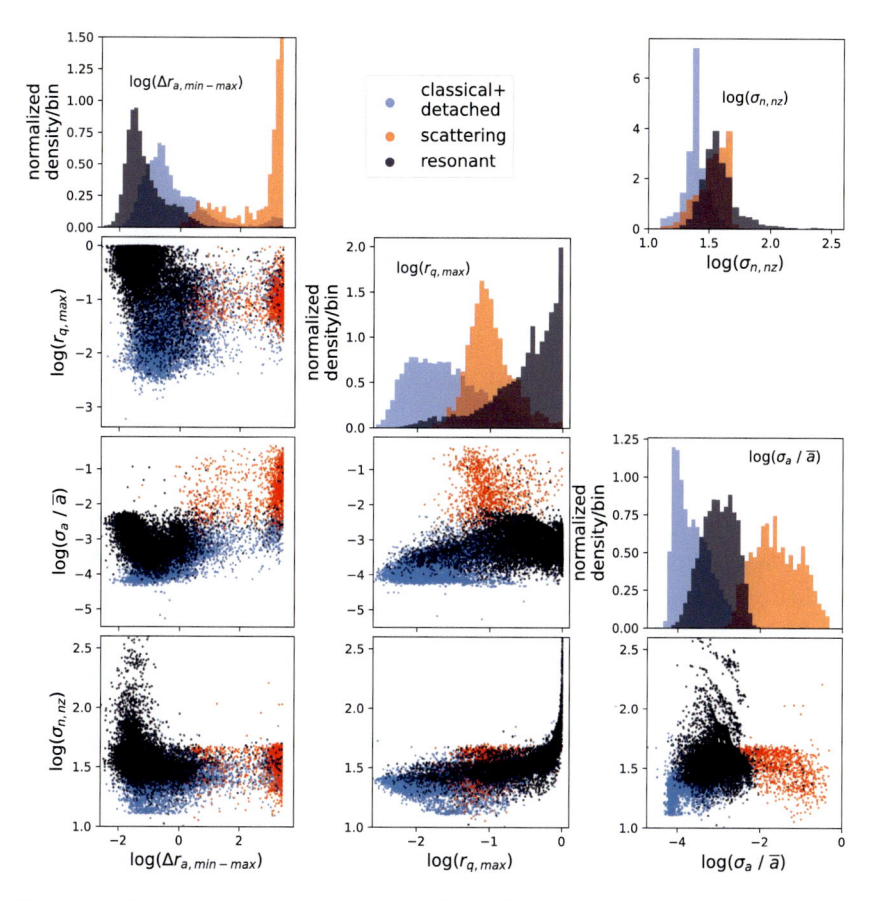

**Figure 7.6** Scatter plots and histograms for four of the consistently highly ranked data features across our entire training and testing dataset for resonant, scattering, and classical/detached TNOs.

tures calculated for the rotating frame resulted in an accuracy of 97.4%, and increased the rate of dynamically irrelevant classifications to 0.7%. As we saw with the inclusion/exclusion of the synthetic TNO training and testing data, the rate of dynamically irrelevant predictions can be more sensitive to the selection of data features than the overall accuracy. This can also be the case when swapping two sets of features that are dynamically equivalent but different in scale (see recent discussion in Smirnov 2024). If we use mean motion in place of $a$ for the simple time-series features, the classifier's accuracy drops slightly to 97.5%, and the rate of dynamically irrelevant classifications rises to 0.5%; including mean motion features in addition to $a$ features actually slightly further decreases the accuracy to

97.4%, though does not further increase the rate of irrelevant classifications. We performed further testing of our classifier with different groups of features excluded (the broad categories of features separated in Table 7.2 as well as the basic time-series features for each individual element) and with each individual feature excluded. Though some exclusions did not significantly decrease the accuracy of the classifier, we also did not find any improvements. We conclude that our classifier performs best when all 227 of our data features are included.

In addition to the classifiers described above, which performed very well, we tested a few additional classification schemes. Starting with the simplified Gladman et al. (2008) scheme above, we split the resonant TNOs into two classes based on $e$-type and mixed-$e$-$i$-type resonances, yielding four classes: scattering, $e$-resonant, $e$-$i$-resonant, and classical/detached. After accounting for trivial misclassifications, this classifier predicted the same classes as the human 96.8% of the time (98.4% of the time for the 4835 predictions made at $> 99\%$ confidence). Of the 159 misclassified particles, 37 were particles assigned to a different resonant class than assigned by the human classifier. However, 33 of these 37 actually show libration (often intermittent) of both the $e$- and $e$-$i$-type resonant arguments for their $p{:}q$ resonance; so these are not really incorrect classifications on the part of the machine learning or the human classifier because either label could be accurately applied. Overall, the machine learning classifier only predicted completely incorrect classifications 0.4% of the time, with marginally incorrect classifications another 2% of the time. For the $> 99\%$ confidence predictions, only 0.15% were completely incorrect (half of these are particles in $q > 10$ resonances) and 1% were marginally incorrect. This performance is very similar to the simplified scheme without two resonant classes, though the utility of adding the $e$-$i$-resonant class is less clear given that many of the $e$-$i$-resonant TNOs also show libration in the $e$-type resonance.

We also tested a scheme to better separate the fully nonresonant classical/detached objects from their resonant-interacting counterparts. In this scheme we have the standard single resonant class, the scattering class, and then split the combined classical/detached population into classes of non-resonant and resonant-interacting. The accuracy of this classifier was 95.7% (97.9% for the 4797 predictions made at $> 99\%$ confidence), the lowest amongst those tested. It also had the highest rate of completely incorrect predictions at 2.3% of all predictions and 1.2% of the high-confidence predictions.

For completeness, we also tried a very simple classification scheme, wherein we combined any particle in or interacting with a resonance into one class, anything stably nonresonant into a "nonres" class, and everything else as scattering. This classifier predicted the same classes as the human 97.2% of the time (98.3% of the time for the 4921 > 99% confidence predictions). But as in the scheme above, this scheme resulted in higher rates of completely incorrect classifications (1.5% of the time across all predictions and 0.8% of the time for the high-confidence predictions). We conclude that using either the full or simplified Gladman et al. (2008) dynamical classes results in the best-performing machine learning classifier. We provide an example Jupyter notebook demonstrating the simplified Gladman et al. (2008) classifier in the Github repository for this book.

## 7.3. Looking forward to future applications and improvements

We have shown that machine learning can provide very accurate TNO classifications when there is a sufficiently large and diverse training set and when we provide the classifier with data features tailored to help identify resonant dynamics. Our best classifier above, using a simplified version of the Gladman et al. 2008 dynamical classes to divide TNOs into resonant, scattering, and classical/detached TNOs, returned correct classifications 98% of the time and dynamically relevant classifications (i.e., the particle displayed properties of the assigned dynamical class for at least part of its evolution) 99.7% of the time. The classifier made high-confidence predictions 96% of the time, with 99% of these high-confidence predictions being correct and 99.9% of them being dynamically relevant.

For the expected 40,000 TNOs from LSST, our classifier would only disagree with a human classifier for $\sim 800$ TNOs, and only 100 of those would be assigned completely irrelevant dynamical classes. For the expected $\sim 38,400$ TNOs classified at > 99% confidence, these numbers would drop to $\sim 400$ and $\sim 40$, respectively; and the number of lower-confidence predictions would be $\sim 1600$, a number small enough for manual classification. These are promising results that show machine learning is a viable replacement for manual classification for the LSST era.

We also anticipate more robust classifications of individual TNOs by making it possible to classify large numbers of clones sampling an individual TNO's orbit-fit uncertainties. We show an example in Fig. 7.7, where we use our classifier to conclude that there is a 91% chance that TNO

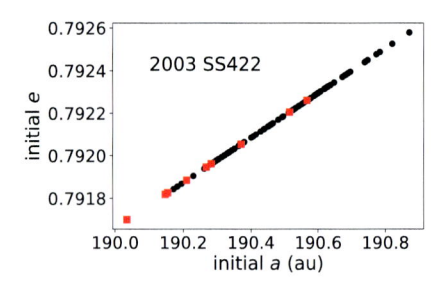

**Figure 7.7** Distribution of initial $a$ and $e$ for 100 orbits sampled from the JPL orbit fit and covariance matrix for TNO 2003 SS422. Black points show orbits classified as resonant by our machine learning classifier, and red squares show orbits classified as scattering. The classifier predicts that 91% of the clones are resonant, which we confirmed by visual inspection of the resonant angle $\phi = 16\lambda - \lambda_N - 15\varpi$. We can thus assign this TNO a 91% probability of being in Neptune's 16:1 resonance.

2003 SS422 is currently in Neptune's 16:1 resonance based on 100 clones sampled from JPL's best-fit orbit and covariance matrix.[3] The ability to assign a probability to this TNO's resonant classification is a significant improvement over noting only a "secure" or "insecure" status using the previous 3-clone approach.

Our classifier can also be used to classify synthetic TNOs, such as those produced by computer simulations of models of the early evolution of the outer solar system (see, e.g., Kaib and Sheppard 2016; Pike et al. 2017; Lawler et al. 2019; Huang et al. 2022; Nesvorný et al. 2023 for recent examples of such simulations) in a way that is fully consistent with the classification of real TNOs. This will enable much more robust comparisons between models and observations.

We note that the rate of fully "incorrect" classifications given by the classifier might be even lower than stated above, depending on how dynamically important very high-order resonances really are. About half of the completely incorrect classifications are of particles that were predicted to be classical or detached but are actually librating in very high-order ($q > 10$) resonances. However, as discussed in subsection 7.1.2 and Fig. 7.3, it is not clear that the perturbations from these very high-order resonances meaningfully affect the long-term orbital evolution of these orbits. Additional investigations are needed to explore which high-order resonant angles reflect substantial perturbations and which might represent mere coincidental

---

[3] Taken from https://ssd.jpl.nasa.gov/tools/sbdb_lookup.html#/?sstr=2003%20SS422 for their 2022-Aug-17 21:39:48 orbit solution.

libration. Machine learning might be a promising approach for finding the limits of Neptune's resonances as well as exploring the fuzzy boundaries between the scattering and detached populations and amongst the resonant/near-resonant/nonresonant populations.

While we anticipate that our existing classifier is robust enough for TNOs in the LSST era, there are still improvements that could be made. As noted in Section 7.2.3, some data features are more influential than others in determining how well the classifier performs compared to a human as well as how often the classifier provides the undesirable dynamically irrelevant predictions. We were able to determine that excluding any of our 227 data features did not improve our classifier, however we have not robustly measured the trade off between accuracy and all possible combinations of those features. The computational cost of calculating all 227 features is relatively small but not trivial for the very large sets of particles we anticipate classifying in the LSST era. A more extensive analysis to determine a potentially smaller subset of those features that yields acceptably accurate classifications could improve the efficiency of the classifier.

Machine learning approaches will be critical to classifying the expected order of magnitude larger number of TNO discoveries over the next decade, and they are well-suited to tasks such as dynamical classification. However, we note that supervised classifiers such as ours are only as good as their training set and data features. If we discover TNOs whose dynamics are not already represented in the training set and/or whose orbital evolution is not well-characterized by the chosen data features, they will not be reliably classified. One might expect that any such novel dynamics would most likely be found for TNOs discovered in presently-unobserved regions of $a$-$e$-$i$ space. Such unusual discoveries will hopefully thus be easily spotted within the large dataset and flagged for human-driven dynamical analyses. As we discover new TNOs, care will need to be taken to continue updating machine learning training sets and/or approaches to ensure accurate classifications. Future unsupervised machine learning tools could also be fruitful for guiding improved classification schemes as the population of known TNOs increases and we are able to more cleanly divide them into additional categories.

## Acknowledgments

This work was supported by NASA grant 80NSSC23K0886. KV acknowledges additional support from NASA (grants 80NSSC22K0512, 80NSSC21K0376, and 80NSSC19K0785) and the Preparing for Astrophysics with LSST Program funded by the Heising Simons Foundation (grant 2021-2975).

## Appendix 7.4. Synthetic TNO orbit generation

To enhance the scattering population training set, we randomly sampled semimajor axes from 52–200 au, perihelion distances from 25–43 au, inclinations from 0–45°, and randomized the other orbital angles; we also generated a set of scattering TNOs with isotropic inclinations to double the small sample of retrograde orbits in the training set. These orbits were integrated in the same manner as the real TNO orbits and examined by eye to label them as scattering, detached, or resonant. From this set, we selected 383 scattering particles with a range of scattering behavior from weak scattering to very rapid scattering. This dataset also included resonant and detached particles, contributing 126 additional high-$a$ resonant particles and 76 detached particles.

To increase the variety of resonant particles in the training set, we started with a list of every $p{:}q$ MMRs with $q \geq 5$ already identified from the observed TNO clones and generated new orbits within a few hundredths of an au of each resonance center over a range of eccentricities similar to the overall observed population (ranging from $e \approx 0$–0.35 for resonances in the classical belt region and $q \approx 30$–40 in the more distant populations). For eccentricity-type resonances, we sampled inclinations up to $\sim 45°$, chose a random longitude of ascending node and mean anomaly, then set the longitude of perihelion such that the initial value of the resonant angle would be in the range 140–220°, near the expected libration center, though this did not guarantee libration. When examining the evolution of these particles, we plotted both eccentricity and mixed-type resonance angles and labeled the resulting resonant particles appropriately. This set of integrations produced 4529 resonant particles, of which 626 were in mixed argument resonances. From the non-resonant particles, we added 275 scattering particles, 299 detached particles, and 28 classical particles (all in the outer classical belt population; we did not include any additional main belt classical particles as those require more time-intensive investigation to be sure they are non-resonant and the observed population already provides a wealth of training set data).

Finally, to increase the number of mixed-argument resonant particles, we ran a set of orbits near every resonance with observed mixed-argument libration targeting inclination ranges seen to librate and choosing initial orbital angles to place the mixed-$e$-$i$ resonant argument near the expected center of libration. These simulations added 451 particles exhibiting mixed argument libration to the training set.

These training set additions are not meant to be exhaustive, but rather to fill out obvious weak spots in the observed dataset without an overwhelming amount of manual classification. The most difficult task still left undone would be further expanding the detached population training set. When populating the high-$a$, high-$q$ orbital space expected for detached TNOs, it is actually very difficult to avoid resonances as Neptune's MMRs become stronger at high-$q$/low-$e$ (see Volk and Malhotra 2022). The resonant librations of the high-order distant resonances can be very subtle and difficult to detect even by eye, so the only way to be sure to have a non-resonant detached object is to plot many, many possible resonance angles to check. This is very time consuming, which is why our additions to the detached training set were more limited than for the targeted resonant populations and scattering populations.

## References

Adams, E.R., Gulbis, A.A.S., Elliot, J.L., et al., 2014. De-biased populations of Kuiper Belt objects from the deep ecliptic survey. Astronomical Journal 148, 55.

Bannister, M.T., Shankman, C., Volk, K., et al., 2017. OSSOS. V. Diffusion in the orbit of a high-perihelion distant solar system object. Astronomical Journal 153, 262.

Bannister, M.T., Gladman, B.J., Kavelaars, J.J., et al., 2018. OSSOS. VII. 800+ trans-Neptunian objects—the complete data release. Astrophysical Journal Supplements 236, 18.

Bernardinelli, P.H., Bernstein, G.M., Sako, M., et al., 2022. A search of the full six years of the dark energy survey for outer solar system objects. Astrophysical Journal Supplements 258, 41.

Chen, Y.-T., Gladman, B., Volk, K., et al., 2019. OSSOS. XVIII. Constraining migration models with the 2:1 resonance using the outer solar system origins survey. Astronomical Journal 158, 214.

Crompvoets, B.L., Lawler, S.M., Volk, K., et al., 2022. OSSOS XXV: large populations and scattering-sticking in the distant trans-Neptunian resonances. Planetary Science Journal 3, 113.

Doner, A., Horányi, M., Bagenal, F., et al., 2024. New Horizons Venetia Burney Student Dust Counter observes higher than expected fluxes approaching 60 au. The Astrophysical Journal Letters 961, L38.

Dones, L., Brasser, R., Kaib, N., Rickman, H., 2015. Origin and evolution of the cometary reservoirs. Space Science Reviews 197, 191.

Elliot, J.L., Kern, S.D., Clancy, K.B., et al., 2005. The deep ecliptic survey: a search for Kuiper Belt objects and centaurs. II. Dynamical classification, the Kuiper Belt plane, and the core population. Astronomical Journal 129, 1117.

Fisher, N.I., Lewis, T., Embleton, B.J.J., 1993. Statistical Analysis of Spherical Data.

Gallardo, T., 2006. Atlas of the mean motion resonances in the Solar System. Icarus 184, 29.

Gladman, B., Marsden, B.G., Vanlaerhoven, C., 2008. In: Barucci, M.A., Boehnhardt, H., Cruikshank, D.P., Morbidelli, A., Dotson, R. (Eds.), The Solar System Beyond Neptune. Univ. Arizona Press, pp. 43–57.

Gladman, B., Volk, K., 2021. Transneptunian space. Annual Reviews of Astronomy and Astrophysics 59, 203.

Gladman, B., Lawler, S.M., Petit, J.M., et al., 2012. The resonant trans-Neptunian populations. Astronomical Journal 144, 23.

Gomes, R.S., Fernández, J.A., Gallardo, T., Brunini, A., 2008. In: Barucci, M.A., Boehnhardt, H., Cruikshank, D.P., Morbidelli, A., Dotson, R. (Eds.), The Solar System Beyond Neptune. Univ. Arizona Press, p. 259.

Gomes, R.S., Gallardo, T., Fernández, J.A., Brunini, A., 2005. On the origin of the high-perihelion scattered disk: the role of the Kozai mechanism and mean motion resonances. Celestial Mechanics & Dynamical Astronomy 91, 109.

Huang, Y., Gladman, B., Beaudoin, M., Zhang, K., 2022. A rogue planet helps to populate the distant Kuiper Belt. The Astrophysical Journal Letters 938, L23.

Ivezić, Ž., et al., 2019. LSST: from science drivers to reference design and anticipated data products. The Astrophysical Journal 873, 111.

JeongAhn, Y., Malhotra, R., 2014. On the non-uniform distribution of the angular elements of near-Earth objects. Icarus 229, 236.

Kaib, N.A., Sheppard, S.S., 2016. Tracking Neptune's migration history through high-perihelion resonant trans-Neptunian objects. Astronomical Journal 152, 133.

Kavelaars, J.J., Jones, R.L., Gladman, B.J., et al., 2009. The Canada-France ecliptic plane survey—L3 data release: the orbital structure of the Kuiper Belt. Astronomical Journal 137, 4917.

Lan, L., Malhotra, R., 2019. Neptune's resonances in the scattered disk. Celestial Mechanics & Dynamical Astronomy 131, 39.

Lawler, S.M., Pike, R.E., Kaib, N., et al., 2019. OSSOS. XIII. Fossilized resonant dropouts tentatively confirm Neptune's migration was grainy and slow. Astronomical Journal 157, 253.

Malhotra, R., 1995. The origin of Pluto's orbit: implications for the solar system beyond Neptune. Astronomical Journal 110, 420.

Malhotra, R., 1996. The phase space structure near Neptune resonances in the Kuiper Belt. Astronomical Journal 111, 504.

Malhotra, R., 2019. Resonant Kuiper Belt objects: a review. Geoscience Letters 6, 12.

Morbidelli, A., Nesvorný, D., 2020. In: Prialnik, D., Barucci, M.A., Young, L. (Eds.), The Trans-Neptunian Solar System. Elsevier, pp. 25–59.

Murray, C.D., Dermott, S.F., 1999. Solar System Dynamics. Cambridge University Press.

Nesvorný, D., Bernardinelli, P., Vokrouhlický, D., Batygin, K., 2023. Radial distribution of distant trans-Neptunian objects points to Sun's formation in a stellar cluster. Icarus 406, 115738.

Pan, M., Sari, R., 2004. A generalization of the Lagrangian points: studies of resonance for highly eccentric orbits. Astronomical Journal 128, 1418.

Pedregosa, F., Varoquaux, G., Gramfort, A., et al., 2011. Scikit-learn: machine learning in Python. Journal of Machine Learning Research 12, 2825.

Pike, R.E., Lawler, S., Brasser, R., et al., 2017. The structure of the distant Kuiper Belt in a nice model scenario. Astronomical Journal 153, 127.

Schwamb, M.E., Hsieh, H., Bannister, M.T., et al., 2019. A software roadmap for solar system science with the large synoptic survey telescope. Research Notes of the American Astronomical Society 3, 51.

Shankman, C., Kavelaars, J.J., Bannister, M.T., et al., 2017. OSSOS. VI. Striking biases in the detection of large semimajor axis trans-Neptunian objects. Astronomical Journal 154, 50.

Smirnov, E., 2024. A comparative analysis of machine learning classifiers in the classification of resonant asteroids. Icarus 415, 116058.

Smotherman, H., Bernardinelli, P.H., Portillo, S.K.N., et al., 2023. The DECam Ecliptic Exploration Project (DEEP) VI: first multiyear observations of trans-Neptunian objects. arXiv:2310.03678.

Smullen, R.A., Volk, K., 2020. Machine learning classification of Kuiper Belt populations. Monthly Notices of the Royal Astronomical Society 497, 1391.

Volk, K., Malhotra, R., 2020. Dynamical instabilities in systems of multiple short-period planets are likely driven by secular chaos: a case study of Kepler-102. Astronomical Journal 160, 98.

Volk, K., Malhotra, R., 2022. Orbital dynamics landscape near the most distant known trans-Neptunian objects. The Astrophysical Journal 937, 119.

Volk, K., Van Laerhoven, C., 2024. Dynamical classifications of multi-opposition TNOs as of 2023 December. Research Notes of the American Astronomical Society 8, 36.

Volk, K., Murray-Clay, R., Gladman, B., et al., 2016. OSSOS III—resonant trans-Neptunian populations: constraints from the first quarter of the Outer Solar System Origins Survey. Astronomical Journal 152, 23.

# CHAPTER EIGHT

# Identification and localization of cometary activity in Solar System objects with machine learning

**Bryce T. Bolin**[a] **and Michael W. Coughlin**[b]

[a]Goddard Space Flight Center, Greenbelt, MD, United States
[b]School of Physics and Astronomy, University of Minnesota, Twin Cities, MN, United States

## 8.1. Introduction to the identification of cometary activity in Solar System objects

Current and next-generation all-sky surveys provide unprecedented opportunities for the discovery of Solar System asteroids and comets (Jedicke et al., 2016; Bauer et al., 2022). Contemporary asteroid surveys employ a variety of algorithms and pipelines to detect and identify moving objects in their image data, such as the association of two or more detections taken in a sequence called "tracklets" (e.g. Jedicke et al., 2015) or through the shifting and stacking of multiple images together (e.g. Whidden et al., 2019; Bolin et al., 2020). This involves taking multiple images to detect an object more than once within a short enough time for the detections to be linked. The actual linkage of detection can include decision trees or the connection of several detections together within similar co-moving velocities (Kubica et al., 2005; Masci et al., 2019).

Multiple detection linking algorithms can link the detections of both point-source, asteroidal detections, and those that are extended, as for comets. Even for comets, whose detections have a large, extended appearance, multiple detections can be linked as for point sources if the measurement of the comet's position relative to the moving frame of the object is consistent from detection to detection (Denneau et al., 2013). However, in some survey imaging pipelines, extended objects, defined as having a width more expansive than the measured width for known point sources, can be flagged as potential outliers and removed from further processing (Duev et al., 2019). The similarities between extended objects and common telescope/detector artifacts can cause the source detection pipeline to fail; therefore the flagging and removal of extended objects can be motivated by the need to keep the false positives at manageable levels for vetting

*Machine Learning for Small Bodies in the Solar System*
https://doi.org/10.1016/B978-0-44-324770-5.00013-1

by human eyes. As a result, the detection and discovery of comets by all-sky surveys can depend not only on the ability of the algorithms to link the detection of comets from multiple images but also on the base level of extraction of detections on a per-image basis.

The ability to detect comets can vary from survey to survey due to the ability to detect extended objects. For example, the image processing pipeline for the ZTF survey rejects extended objects due to the contamination of artifacts (Masci et al., 2019). Therefore the detection of comets is significantly limited in surveys such as ZTF using the standard data pipeline. Furthermore, the ability to link comet detections can be substantially different than the detection and linking of asteroidal or point-source-like detections. The detection and discovery of comets may require their detection to be nonextended to avoid being flagged as an artifact. Examples of such cases are where the appearance of a comet is nonextended, either by its extent not being detectable given the seeing conditions at the observing location of the survey (typically 1–2 arcseconds Jedicke et al., 2015) or by being discovered as a bare nucleus before their activity starts (e.g., Cheng and Wu, 2022). In this review chapter, we will focus on identifying comets as truly extended sources, i.e., that have morphological features that are recognized as extended at the resolution of the survey. We will describe classical methods of identifying comets in astronomical survey data as well as by follow-up data. We will also describe recent advances in machine learning techniques applied to observations of comets that assist in their recognition as extended cometary sources versus being asteroidal, focusing on advancements in deep learning. Lastly, we will describe techniques that are being mastered for near-future use for the detection and discovery of comets in astronomical surveys, such as the forthcoming Vera C. Rubin Observatory's Legacy Survey of Space and Time (LSST, Schwamb et al., 2023).

## 8.2. Review of classical comet detection methods and strategies

We will start this review with an overview of contemporary methods for finding comets in astronomical surveys and follow-up data. We will start this section with a brief discussion of "classical" comet detection methods, i.e., those not based on machine learning or deep learning methods. We refer to these methods as classical since they were, in principle, accessible before the wide-scale adoption of computational methods in astronomy, i.e., when the discovery of comets occurred by recognition of an extended object by the human eye (Festou et al., 1993). However, for this review, we set a lower bound to the time covered by classical comet identification methods to the early to late 1990s when consumer-grade charge-coupled devices (CCDs) became widely accessible for use in astronomical surveys. We will transition from classical methods to methods based on characterizing individual detections to data science and machine learning-driven methods, which became more prevalent in the 2010s.

### 8.2.1 Serendipitous identification of cometary activity of unknown objects

The first discovery of a comet as an unknown object with an automated asteroid detection pipeline was made in 1992 of C/1992 J1 (Spacewatch) by the Spacewatch survey (Scotti et al., 1992). Spacewatch was one of the first surveys to use an automated pipeline for the detection of asteroids in CCD observations, while the telescope was used in drift scan mode (Jedicke et al., 2015). Several asteroid surveys using CCD technology began after Spacewatch, such as the Catalina Sky Survey, Lincoln Near-Earth Asteroid Research, Panoramic Survey telescope and rapid response system (Pan-STARRS) (Chambers et al., 2016), the Near-Earth Object Wide-field Infrared Survey explorer (NEOWISE) (Mainzer et al., 2011), asteroid terrestrial-impact last alert system (ATLAS) (Tonry et al., 2018), and the Zwicky Transient Facility (ZTF) (Bellm et al., 2019) (for a review of asteroid surveys, please see Jedicke et al., 2015). These ground and space-based surveys discovered numerous comets, about 1000 by the end of the 2020s (Bauer et al., 2022). The SOHO and STEREO observatories also discovered several thousand comets as they passed near the Sun in the field of view of these spacecraft (for an overview, please see Battams and Knight, 2017). However, for this review, we will focus on comet discoveries made by asteroid-dedicated surveys.

### 8.2.2 Identification of cometary activity in known objects

As described above, the discovery of cometary activity can be decoupled from the initial recognition of a Solar System object, i.e., it will be recognized as a candidate asteroid detection agnostic of its activity. A known candidate orbit is the main product of a series of observations of an asteroid candidate. Even if the detection of a candidate comet's activity is not known, the orbit resulting from the observations can provide clues as to its cometary nature. A clue to the cometary nature of a comet candidate can be deduced from the known orbit, such as having a Tisserand parameter significantly less than 3 (for a review of the Tisserand parameter and its significance to comets, please see Fraser et al., 2022). In addition, having an eccentricity of $\gtrsim 1$ or retrograde inclination may suggest that an asteroidal object has a cometary origin versus an asteroidal one. This is because most observed asteroids, such as those in the Main Belt, have an eccentricity less than 0.3 and an inclination less than 17 degrees (e.g., Bolin et al., 2017, 2018).

Comets which from their orbital properties provide clues of cometary origins may encourage follow-observations, which are deeper in limiting magnitude and sensitivity, which may result in the detection of cometary activity. A recent example is in the discovery of C/2022 E3 (ZTF) (hereafter E3, Bolin et al., 2022b) which, while possessing evident activity in subsequent observations, had an asteroidal appearance in the discovery images taken by ZTF as seen in the top left and right panels of Fig. 8.1. The lower panel of Fig. 8.1 shows an example of a comet, P/2022 P2, with clearly evident activity as detected by ZTF. Subsequent observations of E3 taken by amateur astronomers specializing in the follow-up of comets, as well as in subsequent observations by ZTF, showed evidence of cometary activity in the form of a tail and coma (Sato et al., 2022; Bolin et al., 2024b, Fig. 1). Clues to its cometary origin, which inspired follow-up observations, came from its eccentricity of $\sim 1$ and retrograde inclination of $\sim 109$ degrees, typical for long-period comets and distinct from the population of asteroids (Jedicke et al., 2002; Gladman et al., 2008; Nesvorný et al., 2023).

## 8.3. Review of cometary detections methods based on measurements of the individual detections

Next, we discuss detecting cometary activity by inspecting objects on an individual image basis. Previous techniques relied on identifying comets following the linkage of multiple detections of objects taken in multiple images. The next set of techniques relies on the previous knowledge of the location of a known object in CCD images and inspects each detection individually. Thus the detection of cometary activity can be done with individual images instead of grouped images.

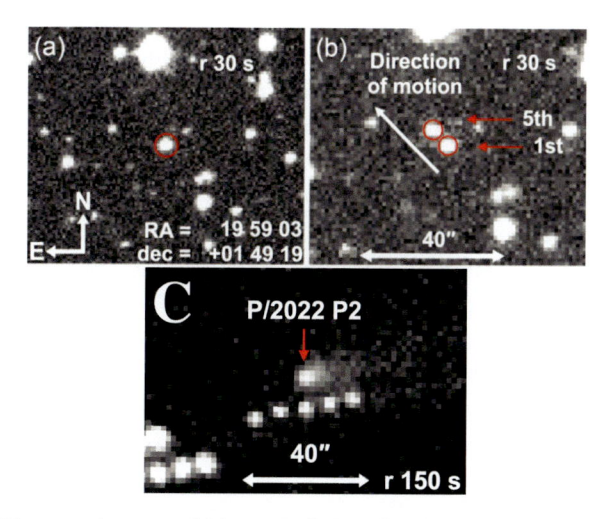

**Figure 8.1** Discovery images of C/2022 E3 (ZTF) taken by ZTF in 2022 March. The top panel shows one of the individual detections of the comet (taken in a series of 5). The top right panel shows a composite stack of the first and fifth images stacked on the background stars. The exposure time, filter, and cardinal directions are indicated. Adapted from Figure 1 of Bolin et al. (2024b) and reproduced with the permission of the authors and MNRAS. The bottom panel shows a co-added composite stack of all five 30 s exposures of images of comet P/2022 P2 (ZTF) taken by ZTF in 2022 August (Bolin et al., 2024a).

### 8.3.1 Identification of comets through spread function parameter analysis

Extended objects in astronomical images, whether in-ground or space-based observations, are defined by their extent relative to the typical width of unresolved detections. In images taken by single-dish telescopes, the width of the point spread function (PSF), the detection of a point or unresolved source, is set by the diffraction limits of the telescope optics, directly proportional to the observed wavelength and inversely proportional to the diameter of the telescope (Born and Wolf, 1999). In CCD detections, the PSF can be super-sampled, that is, when the scale of the individual pixels is less than the critical sample rate, and subsampled if the pixel scale is larger than the critical sampled rate (Howell, 2000).

Detecting extended cometary features, such as a tail or coma, depends on their angular size, which is significantly larger than the imager's resolution. Even when observing with a telescope that enables resolving objects at a diffraction limit smaller than the angular scale of an extended comet, the extendedness may not be resolved if the effective resolution of the images is too coarse due to pixel size. Additionally, the resolution is often limited by the quality of atmospheric seeing when observing from the ground (typically $\sim$1 arcsecond at most professional observing sites) rather than being limited by the inherent diffraction limit of the telescope's optics.

Several attempts have been made to identify comets by quantification of the quality and extendedness of their PSFs. The quality of a PSF is roughly indicated by the reliability of PSF shape measurements made of asteroid detections by imaging pipelines (e.g., as for Pan-STARRS, see Chambers et al., 2016). For the identification of comet candidates, the measured size of the PDF is compared to the expected PSF size for a nonresolved, point-like source. Active comet candidates are more likely to have PSFs that are significantly more extended than compared to stellar-like sources (top panel of Fig. 8.2). Studies of the detection of comets at Palomar observatory and with Pan-STARRS found a significant correlation between parameters describing the extent and quality of PSF for comets when compared to asteroidal detections (see Fig. 8.2) (Waszczak et al., 2013; Hsieh et al., 2015). This method resulted in the discovery of numerous comets with Pan-STARRS, including several Main Belt comets/active asteroids (e.g., Bolin et al., 2013; Hill et al., 2013).

## 8.3.2 Identification of cometary activity by detection of comae and azimuthal variations

Although the PSF extendedness method provides a straightforward and intuitive method for flagging comet candidates by their candidate extendedness when compared to reference point-source PSFs, this comparison is not based on the observed attributes of comets. Comets that can have azimuthal asymmetry due to the presence of tails or a central extended compact coma may not be easily represented by a PSF (e.g., Bolin and Lisse, 2020; Bolin et al., 2021). Improvements to searching for cometary signatures in candidate comet detections can be made by searching for azimuthal asymmetry and the presence of a coma with special aperture functions designed to detect these properties.

Sonnett et al. (2011) were the first to apply the azimuthal asymmetry and coma search technique to $\sim$1000 asteroids detected by the Thousand Asteroid Lightcurve Survey taken with the Canada France Hawaii telescope (Masiero et al., 2009). They used a scheme shown in Fig. 8.3 that divides an annulus outside the extent of the central coma into 18 sections. The section containing part(s) of a candidate comet's tail will be measured to have a higher flux than the other surrounding sections. Sonnett et al. (2011)

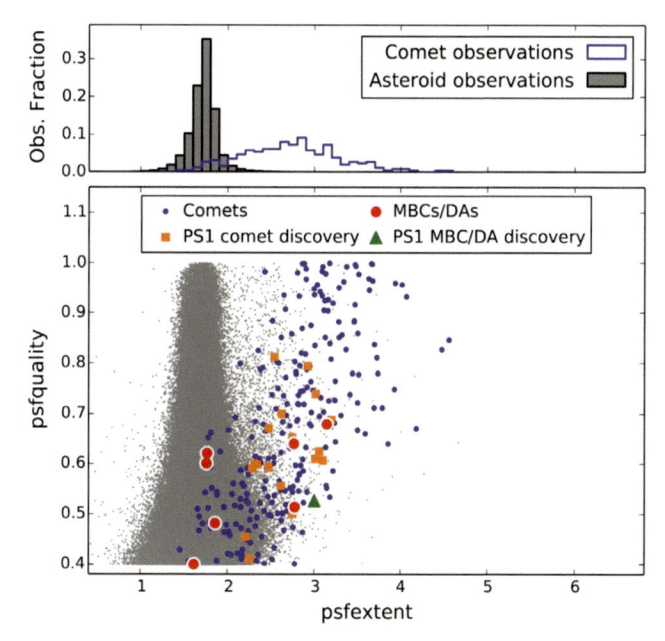

**Figure 8.2** Comparison of the PSF extent, psfextent, and PSF quality, psfquality, properties of asteroid and comet detections from Pan-STARRS taken before 2013 July. psfextent refers to the extendedness of detection PSFs measured as the full width at half maximum of the PSF in arcseconds compared to the expected PSF of point sources. Higher values of psfextent correspond to sources being more extended. psfquality refers to the reliability of PSF measurements with a higher number being more favorable. Top panel: 1-D histograms of PSF extent of asteroid detections (gray) and comet detections (blue). Bottom panel: 2-D distribution of psfextent vs psfquality. The psfextent and psfquality for asteroids is in gray, comets in blue/orange, and Main Belt comets/active asteroids in red and green. Adapted from Figure 5 of Hsieh et al. (2015) and reproduced with the permission of the authors and Icarus. We refer readers to the asteroid and comet detection gallery seen in Fig. 18 of Denneau et al. (2013) for examples of asteroidal and extended comet detections.

used this technique to set a 90% confidence upper limit to the number of Main Belt comets larger than 150 m to be ~400,000.

Chandler et al. (2021) and Ferellec et al. (2023) used similar azimuthal and coma detection techniques applied to comet candidates observed by the Blanco 4.0-m/dark energy camera and the Isaac Newton telescope/wide field camera. Chandler et al. (2021) identified activity in asteroid (248370) 2005 $QN_{173}$ using this technique finding evidence of its recurrent activity throughout its orbit. Ferellec et al. (2023) detected an elevated coma and

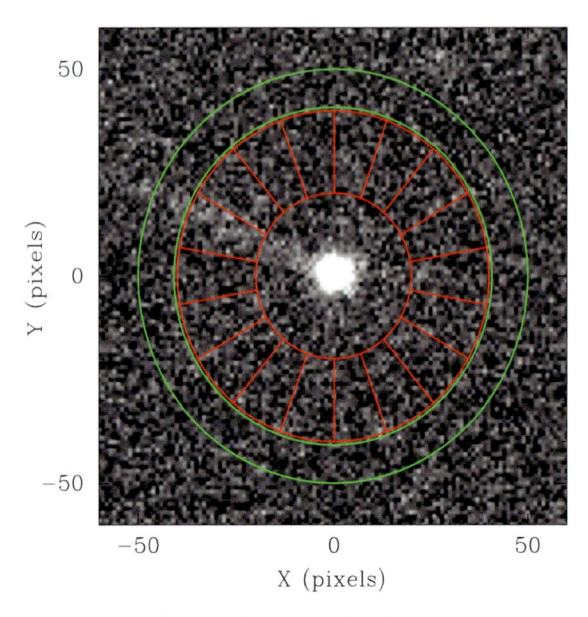

**Figure 8.3** Azimuth scheme for the detection of tails of comets in individual CCD images. The azimuth ring is divided into 18 sections. The section at 10 o'clock encompassing the tail of the detected comet, 133P, measures a higher degree of flux compared to the other sections. The radius of the azimuthal sections was chosen to avoid background stars. Adapted from Figure 2 of Sonnett et al. (2011) and reproduced with the permission of the authors and Icarus.

azimuthal asymmetric flux for the Belt asteroid (279870) 2001 $NL_{19}$ and set an upper limit to the occurrence rate of Main Belt comets of 1:500.

### 8.3.3 Detection of cometary activity through sporadic, significant deviations from phase curve brightness

The brightness of asteroids or bare, inactive comet nuclei follow a steady phase function determined by heliocentric, geocentric, and phase angle parameters (Bowell et al., 1988; Muinonen et al., 2010). The phase function predicts that an asteroid's brightness will increase as it gets closer to the Earth, the Sun, and for a smaller phase angle. In general, this phase function is described by a smooth change in brightness for distant asteroids as their viewing geometry changes due to the motion of their orbit and their evolving view from the Earth.

Monitoring the change in brightness of comet candidates with respect to the secular phase function assuming a bare nucleus provides a test of the presence of sudden outbursts or sporadic activity (Kelley et al., 2019). The

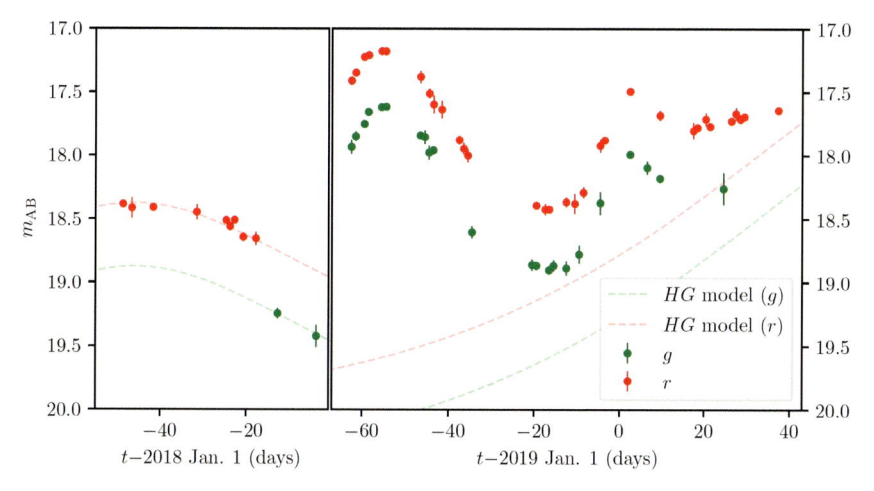

**Figure 8.4** Secular brightness evolution of active asteroid (6478) Gault in 2017 to 2019. The left and right panels show the evolution in the comet's brightness (red and green dots) compared to brightness models for a bare nucleus (red and green dashed lines). The brightness evolution of Gault in 2017 follows a typical pattern for a bare nucleus. The brightness pattern significantly deviates and brightens compared to a bare nucleus in 2018 to 2019. Adapted from Figure 2 of Ye et al. (2019) and reproduced with the permission of the authors and AAS Journals.

brightness of a bare nucleus follows a smoothly varying function describing the increase in an object's brightness as a function of phase angle, heliocentric distance, and geocentric distance (Bowell et al., 1988; Muinonen et al., 2010). This technique has been applied to the detection of activity in active asteroid (6478) Gault, as seen in Fig. 8.4, where the brightness of Gault increased by several magnitudes in late 2018/early 2019 ZTF data compared to its phase function (Ye et al., 2019; Purdum et al., 2021). A similar technique was applied to data from ATLAS for the detection of activity in Centaur comets (Dobson et al., 2023).

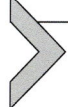

## 8.4. Review of cometary detection methods based on machine learning methods

The next set of methods is based on the growing set of techniques related to machine learning (ML). Carruba et al. have already described many of the core algorithms behind these techniques in Chapter 1 of this volume. We will therefore provide an overview of ML-based comet discovery techniques, highlighting where many of these algorithms have been implemented to find comets. Methods such as crowdsourcing used to find comets, while formidable, are outside the scope of this chapter.

## 8.4.1 Identification of cometary objects with convolutional neural networks

Fully-connected neural networks (NNs) have been used in machine vision applications, such as in image and facial recognition applications (e.g., Bengio, 2009). Applications of NNs adjacent to the recognition of comets include the Morpheus network, a network designed to identify extragalactic sources in *Hubble space telescope* image data (Hausen and Robertson, 2020). The training of NNs usually involves training with a gradient descent-like algorithm to converge on minimum-cost function trained model (e.g., Kolen and Kremer, 2001). Fully connected networks are computationally expensive to train and run since they require weights for every hidden layer. However, convolutional neural networks (CNNs) provide a possible workaround by downscaling the first connected layer through convolution and max pooling (see Section 1.3 of Chapter 1 of this volume for details). This provides an order of magnitude fewer parameters to train and a corresponding considerable speed up in training and reduction in run time, while maintaining accurate results compared to fully connected networks.

### 8.4.1.1 Training of comet identification convolutional neural network models

Previous comet-recognition methods using CNNs relied on a two-stage process combining classical moving object detection methods with CNNs (Chyba Rabeendran and Denneau, 2021). The Tails network by Duev et al. (2021) is the first example to use a CNN trained on comet data to identify comets on a per-image basis. The architecture of the Tails network is provided below in Fig. 8.5. The training set consisted of ∼3,000 image examples of comets with identifiable cometary morphology; an example can be found in Fig. 8.6. Each training example consisted of a science image, a reference image consisting of deep stacks of previous images, and an image consisting of the difference between the two. Approximately 20,000 negative examples of point source-like detections, internal reflection artifacts, hot pixels, and diffraction spikes were used as negative examples. The training set was refined with several instances of active learning. Active learning is an interactive form of machine learning, in which the user manually labels data to improve the accuracy of a classifier (Fang et al., 2017). In the development of Tails, active learning was used in cases near the network's

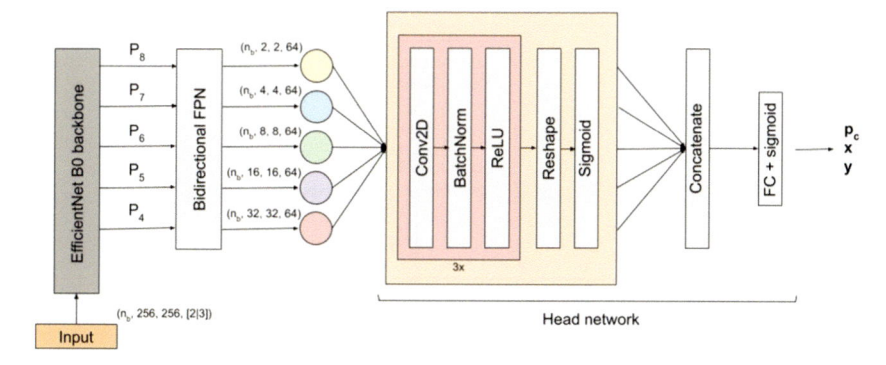

**Figure 8.5** Illustration of the Tails network architecture. Tails uses a combination of bidirectional feature pyramid networks (BiFPN), which are fed by the last five layers of an EfficientDet network. The output of the BiFPN is fed into a head network consisting of several convolutional layers and a fully connected layer and activation function to derive a comet probability score between 0 and 1. Adapted from Fig. 3 of (Duev et al., 2021) and reproduced with the permission of the authors and AAS Journals.

**Figure 8.6** Example of detection of comet C/2019 D1 (Flewelling) used to train the Tails CNN. The left panel shows the science images, the central panel shows the reference panel and the right panel shows the difference between the science and reference image. Adapted from Fig. 2 of (Duev et al., 2021) and reproduced with the permission of the authors and AAS Journals.

decision threshold that were visually inspected, then re-classified and re-added to the revised training set.

### 8.4.1.2 Performance of convolutional neural networks to identify comets

The performance of the Tails network with false positive and false negative rates as a function of the comet probability score is shown in Fig. 8.7; both false positive and false negative rates are below 2%. Tails has been run on the ZTF twilight survey data taken since 2019 September (Bolin et al., 2022a, 2023). The ZTF twilight survey consists of ~50 images taken during both

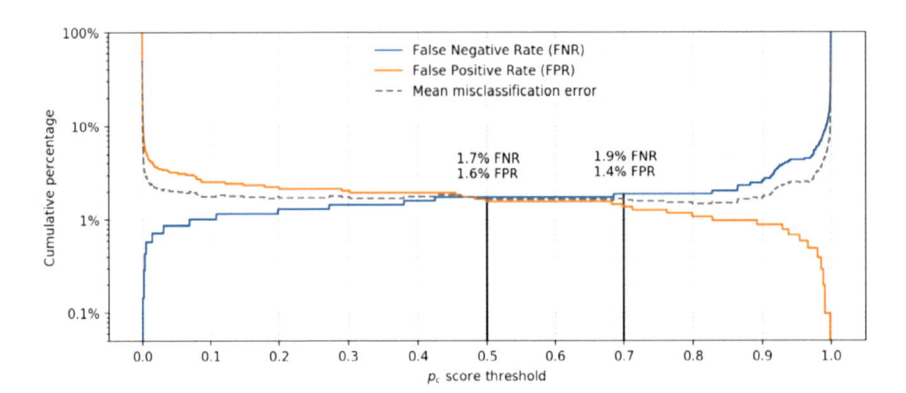

**Figure 8.7** False positive and false negative rates as a function of comet probability score. The false positive and false negative equalize at ∼1.7 percent for a comet probability score of 0.5 and bifurcate into a false positive rate of ∼1.4 percent and a false negative rate of ∼1.9 percent for a score of 0.7. Adapted from Fig. 4 of (Duev et al., 2021) and reproduced with the permission of the authors and AAS Journals.

evening and morning nautical to astronomical twilight on a nightly basis with Sun-centric distances of 30–60 degrees (Bolin et al., 2022a, 2024a). These near-Sun observations enable the detection of asteroids located inside the orbit of the Earth and Venus as well as of comets (Bolin et al., 2024a). It takes several hours of wall time on a 32-core machine to process a nightly data set. The output with comet probability scores, science, reference, and difference images are output to the Fritz user interface (see Fig. 8.8) (Coughlin et al., 2023).

The first discovery of a comet made by AI-assisted methods through Tails was made in 2020 October of comet C/2020 T2 (Palomar) as seen in Fig. 8.9 (Duev et al., 2020). C/2020 T2 exhibits clear cometary features, such as a tail and a coma 3–4 arcseconds wide compared to the seeing of about 2 arcseconds. In addition to T2, Tails contributed to several other comet discoveries from the ZTF twilight survey, including C/2022 E3 (Bolin et al., 2022b), which will be described in a forthcoming publication (Bolin et al., 2022a, 2024a).

## 8.4.2 Other applications of machine learning used to find comets and future developments

Machine learning methods are becoming increasingly used in the detection and discovery of comets that have come out after the development of CNN-based methods, such as Tails (e.g., Rożek et al., 2023; Sedaghat et al.,

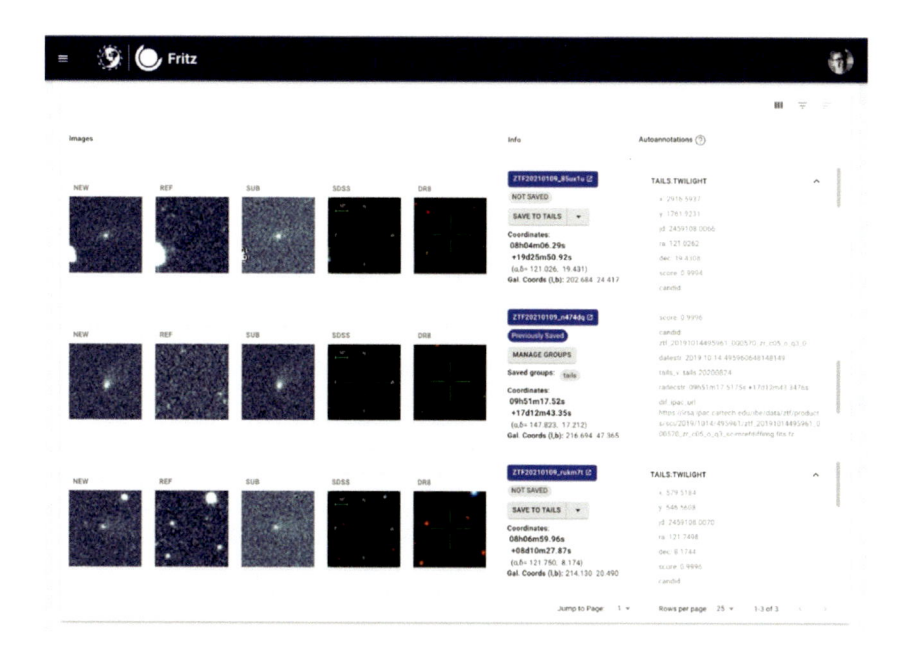

**Figure 8.8** Example of Tails interface for the scanning of comet candidates. Adapted from Fig. 5 of (Duev et al., 2021) and reproduced with the permission of the authors and AAS Journals.

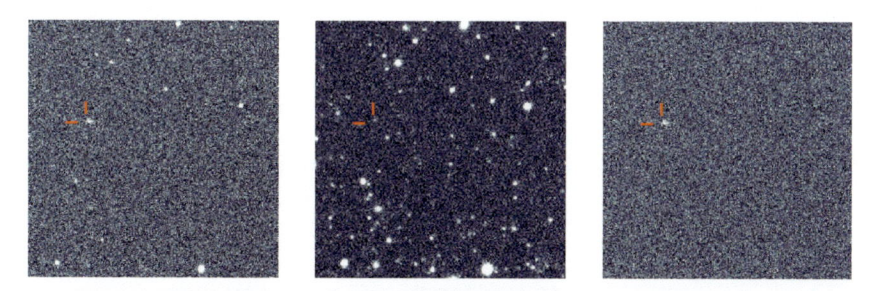

**Figure 8.9** Discovery images of C/2020 T2 (Palomar) taken by ZTF on 2020 October 7. The left panel shows the science image, the center panel shows the reference image, and the right panel shows the difference images. Adapted from Fig. 7 of (Duev et al., 2021) and reproduced with the permission of the authors and AAS Journals.

2024). Future developments and improvements to CCN-based methods, e.g., Tails, include the use of multiple-input CNNs, such as the BTSBot, implemented for identifying supernovae in ZTF images (Rehemtulla et al., 2023). Multiple input methods combine additional features that can be used to enhance the identification of comets, such as orbital information and PSF-based information described in previous sections. By combining both CNN and other feature information, the accuracy of comet identification can be increased, which will be necessary for future surveys, such as LSST.

## Acknowledgments

The authors wish to thank Dr. Valerio Carruba and Dr. Wesley Fraser for providing helpful reviews of this manuscript. Their comments were insightful in the improvement of the manuscript. The authors also wish to thank Ms. Laura-May Abron for her inspiration and artistic insight in the portrayal of ML-methods to identify solar system objects.

## References

Battams, K., Knight, M.M., 2017. SOHO comets: 20 years and 3000 objects later. Philosophical Transactions of the Royal Society of London, Series A 375, 20160257. https://doi.org/10.1098/rsta.2016.0257. arXiv:1611.02279.

Bauer, J.M., Fernández, Y.R., Protopapa, S., Woodney, L.M., 2022. Comet science with ground based and space based surveys in the New Millennium. arXiv e-prints, arXiv:2210.09400.

Bellm, E.C., Kulkarni, S.R., Graham, M.J., Dekany, R., Smith, R.M., Riddle, R., Masci, F.J., Helou, G., Prince, T.A., Adams, S.M., Barbarino, C., Barlow, T., Bauer, J., Beck, R., Belicki, J., Biswas, R., Blagorodnova, N., Bodewits, D., Bolin, B., Brinnel, V., Brooke, T., Bue, B., Bulla, M., Burruss, R., Cenko, S.B., Chang, C.K., Connolly, A., Coughlin, M., Cromer, J., Cunningham, V., De, K., Delacroix, A., Desai, V., Duev, D.A., Eadie, G., Farnham, T.L., Feeney, M., Feindt, U., Flynn, D., Franckowiak, A., Frederick, S., Fremling, C., Gal-Yam, A., Gezari, S., Giomi, M., Goldstein, D.A., Golkhou, V.Z., Goobar, A., Groom, S., Hacopians, E., Hale, D., Henning, J., Ho, A.Y.Q., Hover, D., Howell, J., Hung, T., Huppenkothen, D., Imel, D., Ip, W.H., Ivezić, Ž., Jackson, E., Jones, L., Juric, M., Kasliwal, M.M., Kaspi, S., Kaye, S., Kelley, M.S.P., Kowalski, M., Kramer, E., Kupfer, T., Landry, W., Laher, R.R., Lee, C.D., Lin, H.W., Lin, Z.Y., Lunnan, R., Giomi, M., Mahabal, A., Mao, P., Miller, A.A., Monkewitz, S., Murphy, P., Ngeow, C.C., Nordin, J., Nugent, P., Ofek, E., Patterson, M.T., Penprase, B., Porter, M., Rauch, L., Rebbapragada, U., Reiley, D., Rigault, M., Rodriguez, H., van Roestel, J., Rusholme, B., van Santen, J., Schulze, S., Shupe, D.L., Singer, L.P., Soumagnac, M.T., Stein, R., Surace, J., Sollerman, J., Szkody, P., Taddia, F., Terek, S., Van Sistine, A., van Velzen, S., Vestrand, W.T., Walters, R., Ward, C., Ye, Q.Z., Yu, P.C., Yan, L., Zolkower, J., 2019. The Zwicky Transient Facility: system overview, performance, and first results. Publications of the Astronomical Society of the Pacific 131, 018002. https://doi.org/10.1088/1538-3873/aaecbe.

Bengio, Y., 2009. Learning deep architectures for AI. Foundation and Trends in AI 2, 56.

Bolin, B., Denneau, L., Micheli, M., Wainscoat, R., Tholen, D.J., Lister, T., Williams, G.V., 2013. Comet P/2013 P5 (Panstarrs). Central Bureau Electronic Telegrams 3639.

Bolin, B.T., Ahumada, T., Dokkum, P.v., Fremling, C., Hardegree-Ullman, K.K., Purdum, J.N., Serabyn, E., Southworth, J., 2023. Preliminary estimates of the Zwicky Transient Facility 'Ayló'chaxnim asteroid population completeness. Icarus 394, 115442. https://doi.org/10.1016/j.icarus.2023.115442.

Bolin, B.T., Ahumada, T., van Dokkum, P., Fremling, C., Granvik, M., Hardegree-Ullman, K.K., Harikane, Y., Purdum, J.N., Serabyn, E., Southworth, J., Zhai, C., 2022a. The discovery and characterization of (594913) 'Ayló'chaxnim, a kilometre sized asteroid inside the orbit of Venus. Monthly Notices of the Royal Astronomical Society 517, L49–L54. https://doi.org/10.1093/mnrasl/slac089. arXiv:2208.07253.

Bolin, B.T., Delbo, M., Morbidelli, A., Walsh, K.J., 2017. Yarkovsky V-shape identification of asteroid families. Icarus 282, 290–312. https://doi.org/10.1016/j.icarus.2016.09.029. arXiv:1609.06384.

Bolin, B.T., Fernandez, Y.R., Lisse, C.M., Holt, T.R., Lin, Z.Y., Purdum, J.N., Deshmukh, K.P., Bauer, J.M., Bellm, E.C., Bodewits, D., Burdge, K.B., Carey, S.J., Copperwheat, C.M., Helou, G., Ho, A.Y.Q., Horner, J., van Roestel, J., Bhalerao, V., Chang, C.K., Chen, C., Hsu, C.Y., Ip, W.H., Kasliwal, M.M., Masci, F.J., Ngeow, C.C., Quimby, R., Burruss, R., Coughlin, M., Dekany, R., Delacroix, A., Drake, A., Duev, D.A., Graham, M., Hale, D., Kupfer, T., Laher, R.R., Mahabal, A., Mróz, P.J., Neill, J.D., Riddle, R., Rodriguez, H., Smith, R.M., Soumagnac, M.T., Walters, R., Yan, L., Zolkower, J., 2021. Initial characterization of active transitioning Centaur, P/2019 LD$_2$ (ATLAS), using Hubble, Spitzer, ZTF, Keck, Apache Point Observatory, and GROWTH visible and infrared imaging and spectroscopy. Astronomical Journal 161, 116. https://doi.org/10.3847/1538-3881/abd94b. arXiv:2011.03782.

Bolin, B.T., Fremling, C., Holt, T.R., Hankins, M.J., Ahumada, T., Anand, S., Bhalerao, V., Burdge, K.B., Copperwheat, C.M., Coughlin, M., Deshmukh, K.P., De, K., Kasliwal, M.M., Morbidelli, A., Purdum, J.N., Quimby, R., Bodewits, D., Chang, C.K., Ip, W.H., Hsu, C.Y., Laher, R.R., Lin, Z.Y., Lisse, C.M., Masci, F.J., Ngeow, C.C., Tan, H., Zhai, C., Burruss, R., Dekany, R., Delacroix, A., Duev, D.A., Graham, M., Hale, D., Kulkarni, S.R., Kupfer, T., Mahabal, A., Mróz, P.J., Neill, J.D., Riddle, R., Rodriguez, H., Smith, R.M., Soumagnac, M.T., Walters, R., Yan, L., Zolkower, J., 2020. Characterization of temporarily captured minimoon 2020 CD$_3$ by Keck time-resolved spectrophotometry. The Astrophysical Journal Letters 900, L45. https://doi.org/10.3847/2041-8213/abae69. arXiv:2008.05384.

Bolin, B.T., Lisse, C.M., 2020. Constraints on the spin-pole orientation, jet morphology, and rotation of interstellar comet 2I/Borisov with deep HST imaging. Monthly Notices of the Royal Astronomical Society 497, 4031–4041. https://doi.org/10.1093/mnras/staa2192. arXiv:1912.07386.

Bolin, B.T., Masci, F.J., Coughlin, M.W., Duev, D.W., Stubbs, C.W., 2024a. An artificial-intelligence powered twilight survey of 'Aylô'chaxnim, Atiras, and comets at Palomar Observatory. Icarus. In press.

Bolin, B.T., Masci, F.J., Duev, D.A., Milburn, J.W., Neill, J.D., Purdum, J.N., Avdellidou, C., Saki, M., Cheng, Y.C., Delbo, M., Fremling, C., Ghosal, M., Lin, Z.Y., Lisse, C.M., Mahabal, A., 2024b. Palomar discovery and initial characterization of naked-eye long-period comet C/2022 E3 (ZTF). Monthly Notices of the Royal Astronomical Society 527, L42–L46. https://doi.org/10.1093/mnrasl/slad139. arXiv:2309.14336.

Bolin, B.T., Masci, F.J., Ip, W.H., Helou, G., Kramer, E.A., Lin, Z.Y., Prince, T.A., Sato, H., Paul, N., Yoshimoto, K., Urbanik, M., Denneau, L., Siverd, R., Tonry, J., Weiland, H., Erasmus, N., Fitzsimmons, A., Lawrence, A., Robinson, J., Siverd, R., Tonry, J., Birtwhistle, P., Jacques, C., Hug, G., Korlevic, K., Buzzi, L., Bacci, R., van Buitenen, G., Buczynski, D., Hale, A., Masek, M., Guido, E., Rocchetto, M., Bryssinck, E., Milani, G., Savini, G., Valvasori, A., Ligustri, R., Bacci, P., Maestripieri, M., Tesi, L., Fagioli, G., Lutkenhoner, B., 2022b. Comet C/2022 E3 (ZTF). Minor Planet Electronic Circulars 2022-F13.

Bolin, B.T., Morbidelli, A., Walsh, K.J., 2018. Size-dependent modification of asteroid family Yarkovsky V-shapes. Astronomy & Astrophysics 611, A82. https://doi.org/10.1051/0004-6361/201732079. arXiv:1710.04208.

Born, M., Wolf, E., 1999. Principles of Optics.

Bowell, E., Hapke, B., Domingue, D., Lumme, K., Peltoniemi, J., Harris, A., 1988. Application of photometric models to asteroids. Asteroids II, 399–433.

Chambers, K.C., Magnier, E.A., Metcalfe, N., Flewelling, H.A., Huber, M.E., Waters, C.Z., Denneau, L., Draper, P.W., Farrow, D., Finkbeiner, D.P., Holmberg, C., Koppenhoefer, J., Price, P.A., Saglia, R.P., Schlafly, E.F., Smartt, S.J., Sweeney, W., Wainscoat, R.J., Burgett, W.S., Grav, T., Heasley, J.N., Hodapp, K.W., Jedicke, R., Kaiser, N., Kudritzki, R.P., Luppino, G.A., Lupton, R.H., Monet, D.G., Morgan, J.S., Onaka, P.M., Stubbs, C.W., Tonry, J.L., Banados, E., Bell, E.F., Bender, R., Bernard, E.J., Botticella, M.T., Casertano, S., Chastel, S., Chen, W.P., Chen, X., Cole, S., Deacon, N., Frenk, C., Fitzsimmons, A., Gezari, S., Goessl, C., Goggia, T., Goldman, B., Grebel, E.K., Hambly, N.C., Hasinger, G., Heavens, A.F., Heckman, T.M., Henderson, R., Henning, T., Holman, M., Hopp, U., Ip, W.H., Isani, S., Keyes, C.D., Koekemoer, A., Kotak, R., Long, K.S., Lucey, J.R., Liu, M., Martin, N.F., McLean, B., Morganson, E., Murphy, D.N.A., Nieto-Santisteban, M.A., Norberg, P., Peacock, J.A., Pier, E.A., Postman, M., Primak, N., Rae, C., Rest, A., Riess, A., Riffeser, A., Rix, H.W., Roser, S., Schilbach, E., Schultz, A.S.B., Scolnic, D., Szalay, A., Seitz, S., Shiao, B., Small, E., Smith, K.W., Soderblom, D., Taylor, A.N., Thakar, A.R., Thiel, J., Thilker, D., Urata, Y., Valenti, J., Walter, F., Watters, S.P., Werner, S., White, R., Wood-Vasey, W.M., Wyse, R., 2016. The Pan-STARRS1 surveys. ArXiv e-prints, arXiv:1612.05560.

Chandler, C.O., Trujillo, C.A., Hsieh, H.H., 2021. Recurrent activity from active asteroid (248370) 2005 $QN_{173}$: a Main-belt Comet. The Astrophysical Journal Letters 922, L8. https://doi.org/10.3847/2041-8213/ac365b. arXiv:2111.06405.

Cheng, Y.C., Wu, Y.L., 2022. Cometary activities of the hyperbolic asteroid A/2021 X2 observed at Lulin Observatory. The Astronomer's Telegram 15597, 1.

Chyba Rabeendran, A., Denneau, L., 2021. A two-stage deep learning detection classifier for the ATLAS asteroid survey. Publications of the Astronomical Society of the Pacific 133, 034501. https://doi.org/10.1088/1538-3873/abc900. arXiv:2101.08912.

Coughlin, M.W., Bloom, J.S., Nir, G., Antier, S., du Laz, T.J., van der Walt, S., Crellin-Quick, A., Culino, T., Duev, D.A., Goldstein, D.A., Healy, B.F., Karambelkar, V., Lilleboe, J., Shin, K.M., Singer, L.P., Ahumada, T., Anand, S., Bellm, E.C., Dekany, R., Graham, M.J., Kasliwal, M.M., Kostadinova, I., Kiendrebeogo, R.W., Kulkarni, S.R., Jenkins, S., LeBaron, N., Mahabal, A.A., Neill, J.D., Parazin, B., Peloton, J., Perley, D.A., Riddle, R., Rusholme, B., van Santen, J., Sollerman, J., Stein, R., Turpin, D., Wold, A., Amat, C., Bonnefon, A., Bonnefoy, A., Flament, M., Kerkow, F., Kishore, S., Jani, S., Mahanty, S.K., Liu, C., Llinares, L., Makarison, J., Olliéric, A., Perez, I., Pont, L., Sharma, V., 2023. A data science platform to enable time-domain astronomy. The Astrophysical Journal. Supplement Series 267, 31. https://doi.org/10.3847/1538-4365/acdee1. arXiv:2305.00108, 2023.

Denneau, L., Jedicke, R., Grav, T., Granvik, M., Kubica, J., Milani, A., Vereš, P., Wainscoat, R., Chang, D., Pierfederici, F., Kaiser, N., Chambers, K.C., Heasley, J.N., Magnier, E.A., Price, P.A., Myers, J., Kleyna, J., Hsieh, H., Farnocchia, D., Waters, C., Sweeney, W.H., Green, D., Bolin, B., Burgett, W.S., Morgan, J.S., Tonry, J.L., Hodapp, K.W., Chastel, S., Chesley, S., Fitzsimmons, A., Holman, M., Spahr, T., Tholen, D., Williams, G.V., Abe, S., Armstrong, J.D., Bressi, T.H., Holmes, R., Lister, T., McMillan, R.S., Micheli, M., Ryan, E.V., Ryan, W.H., Scotti, J.V., 2013. The Pan-STARRS moving object processing system. Publications of the Astronomical Society of the Pacific 125, 357–395. https://doi.org/10.1086/670337. arXiv:1302.7281.

Dobson, M.M., Schwamb, M.E., Benecchi, S.D., Verbiscer, A.J., Fitzsimmons, A., Shingles, L.J., Denneau, L., Heinze, A.N., Smith, K.W., Tonry, J.L., Weiland, H., Young, D.R., 2023. Phase curves of Kuiper Belt objects, Centaurs, and Jupiter-family comets from the ATLAS survey. Planetary Science Journal 4, 75. https://doi.org/10.3847/PSJ/acc463. arXiv:2303.08643.

Duev, D.A., Bolin, B.T., Graham, M.J., Kelley, M.S.P., Mahabal, A., Bellm, E.C., Coughlin, M.W., Dekany, R., Helou, G., Kulkarni, S.R., Masci, F.J., Prince, T.A., Riddle, R.,

Soumagnac, M.T., van der Walt, S.J., 2021. Tails: chasing comets with the Zwicky Transient Facility and deep learning. Astronomical Journal 161, 218. https://doi.org/10.3847/1538-3881/abea7b. arXiv:2102.13352.

Duev, D.A., Duev, I.D., Bolin, B.T., 2020. COMET C/2020 T2 (Palomar). Minor Planet Electronic Circulars 2020-U170.

Duev, D.A., Mahabal, A., Ye, Q., Tirumala, K., Belicki, J., Dekany, R., Frederick, S., Graham, M.J., Laher, R.R., Masci, F.J., Prince, T.A., Riddle, R., Rosnet, P., Soumagnac, M.T., 2019. DeepStreaks: identifying fast-moving objects in the Zwicky Transient Facility data with deep learning. Monthly Notices of the Royal Astronomical Society 486, 4158–4165. https://doi.org/10.1093/mnras/stz1096. arXiv:1904.05920.

Fang, M., Li, Y., Cohn, T., 2017. Learning how to active learn: a deep reinforcement learning approach. In: Proceedings of the 2017 Conference on Empirical Methods in Natural Language Processing. Association for Computational Linguistics, Copenhagen, Denmark, pp. 595–605.

Ferellec, L., Snodgrass, C., Fitzsimmons, A., Rożek, A., Gardener, D., Smith, R., Medeiros, H., Opitom, C., Hsieh, H.H., 2023. A targeted search for Main Belt Comets. Monthly Notices of the Royal Astronomical Society 518, 2373–2384. https://doi.org/10.1093/mnras/stac3199. arXiv:2211.01435.

Festou, M.C., Rickman, H., West, R.M., 1993. Comets. The Astronomy and Astrophysics Review 4, 363–447. https://doi.org/10.1007/BF00872944.

Fraser, W.C., Dones, L., Volk, K., Womack, M., Nesvorný, D., 2022. The transition from the Kuiper Belt to the Jupiter-Family (Comets). arXiv e-prints, arXiv:2210.16354.

Gladman, B., Marsden, B.G., Vanlaerhoven, C., 2008. Nomenclature in the Outer Solar System. In: Barucci, M.A., Boehnhardt, H., Cruikshank, D.P., Morbidelli, A., Dotson, R. (Eds.), The Solar System Beyond Neptune, pp. 43–57.

Hausen, R., Robertson, B.E., 2020. Morpheus: a deep learning framework for the pixel-level analysis of astronomical image data. The Astrophysical Journal. Supplement Series 248, 20. https://doi.org/10.3847/1538-4365/ab8868. arXiv:1906.11248.

Hill, R.E., Bolin, B., Kleyna, J., Denneau, L., Wainscoat, R., Micheli, M., Armstrong, J.D., Molina, M., Sato, H., 2013. Comet P/2013 R3 (Catalina-Panstarrs). Central Bureau Electronic Telegrams 3658.

Howell, S.B., 2000. Handbook of CCD Astronomy.

Hsieh, H.H., Denneau, L., Wainscoat, R.J., Schörghofer, N., Bolin, B., Fitzsimmons, A., Jedicke, R., Kleyna, J., Micheli, M., Vereš, P., Kaiser, N., Chambers, K.C., Burgett, W.S., Flewelling, H., Hodapp, K.W., Magnier, E.A., Morgan, J.S., Price, P.A., Tonry, J.L., Waters, C., 2015. The main-belt comets: the Pan-STARRS1 perspective. Icarus 248, 289–312. https://doi.org/10.1016/j.icarus.2014.10.031. arXiv:1410.5084.

Jedicke, R., Bolin, B., Granvik, M., Beshore, E., 2016. A fast method for quantifying observational selection effects in asteroid surveys. Icarus 266, 173–188. https://doi.org/10.1016/j.icarus.2015.10.021.

Jedicke, R., Granvik, M., Micheli, M., Ryan, E., Spahr, T., Yeomans, D.K., 2015. Surveys, astrometric follow-up, and population statistics. Asteroids IV, 795–813.

Jedicke, R., Larsen, J., Spahr, T., 2002. Observational selection effects in asteroid surveys. Asteroids III, 71–87.

Kelley, M.S.P., Bodewits, D., Ye, Q., Laher, R.R., Masci, F.J., Monkewitz, S., Riddle, R., Rusholme, B., Shupe, D.L., Soumagnac, M.T., 2019. ZChecker: finding cometary outbursts with the Zwicky Transient Facility. In: Teuben, P.J., Pound, M.W., Thomas, B.A., Warner, E.M. (Eds.), Astronomical Data Analysis Software and Systems XXVII, p. 471.

Kolen, J.F., Kremer, S.C., 2001. Gradient flow in recurrent nets: the difficulty of learning LongTerm dependencies. https://doi.org/10.1109/9780470544037.ch14. pp. 237–243.

Kubica, J., Moore, A., Connolly, A., Jedicke, R., 2005. Efficiently Identifying Close Track/Observation Pairs in Continuous Timed Data. SPIE.

Mainzer, A., Bauer, J., Grav, T., Masiero, J., Cutri, R.M., Dailey, J., Eisenhardt, P., McMillan, R.S., Wright, E., Walker, R., Jedicke, R., Spahr, T., Tholen, D., Alles, R., Beck, R., Brandenburg, H., Conrow, T., Evans, T., Fowler, J., Jarrett, T., Marsh, K., Masci, F., McCallon, H., Wheelock, S., Wittman, M., Wyatt, P., DeBaun, E., Elliott, G., Elsbury, D., Gautier IV, T., Gomillion, S., Leisawitz, D., Maleszewski, C., Micheli, M., Wilkins, A., 2011. Preliminary results from NEOWISE: an enhancement to the wide-field infrared survey explorer for Solar System science. The Astrophysical Journal 731, 53. https://doi.org/10.1088/0004-637X/731/1/53. arXiv:1102.1996.

Masci, F.J., Laher, R.R., Rusholme, B., Shupe, D.L., Groom, S., Surace, J., Jackson, E., Monkewitz, S., Beck, R., Flynn, D., Terek, S., Landry, W., Hacopians, E., Desai, V., Howell, J., Brooke, T., Imel, D., Wachter, S., Ye, Q.Z., Lin, H.W., Cenko, S.B., Cunningham, V., Rebbapragada, U., Bue, B., Miller, A.A., Mahabal, A., Bellm, E.C., Patterson, M.T., Jurić, M., Golkhou, V.Z., Ofek, E.O., Walters, R., Graham, M., Kasliwal, M.M., Dekany, R.G., Kupfer, T., Burdge, K., Cannella, C.B., Barlow, T., Van Sistine, A., Giomi, M., Fremling, C., Blagorodnova, N., Levitan, D., Riddle, R., Smith, R.M., Helou, G., Prince, T.A., Kulkarni, S.R., 2019. The Zwicky Transient Facility: data processing, products, and archive. Publications of the Astronomical Society of the Pacific 131, 018003. https://doi.org/10.1088/1538-3873/aae8ac. arXiv:1902.01872.

Masiero, J., Jedicke, R., Ďurech, J., Gwyn, S., Denneau, L., Larsen, J., 2009. The Thousand Asteroid Light Curve Survey. Icarus 204, 145–171. https://doi.org/10.1016/j.icarus.2009.06.012. arXiv:0906.3339.

Muinonen, K., Belskaya, I.N., Cellino, A., Delbò, M., Levasseur-Regourd, A.C., Penttilä, A., Tedesco, E.F., 2010. A three-parameter magnitude phase function for asteroids. Icarus 209, 542–555. https://doi.org/10.1016/j.icarus.2010.04.003.

Nesvorný, D., Deienno, R., Bottke, W.F., Jedicke, R., Naidu, S., Chesley, S.R., Chodas, P.W., Granvik, M., Vokrouhlický, D., Brož, M., Morbidelli, A., Christensen, E., Shelly, F.C., Bolin, B.T., 2023. NEOMOD: a new orbital distribution model for near-Earth objects. Astronomical Journal 166, 55. https://doi.org/10.3847/1538-3881/ace040. arXiv:2306.09521.

Purdum, J.N., Lin, Z.Y., Bolin, B.T., Sharma, K., Choi, P.I., Bhalerao, V., Hanuš, J., Kumar, H., Quimby, R., van Roestel, J.C., Zhai, C., Fernandez, Y.R., Lisse, C.M., Bodewits, D., Fremling, C., Ryan Golovich, N., Hsu, C.Y., Ip, W.H., Ngeow, C.C., Saini, N.S., Shao, M., Yao, Y., Ahumada, T., Anand, S., Andreoni, I., Burdge, K.B., Burruss, R., Chang, C.K., Copperwheat, C.M., Coughlin, M., De, K., Dekany, R., Delacroix, A., Drake, A., Duev, D., Graham, M., Hale, D., Kool, E.C., Kasliwal, M.M., Kostadinova, I.S., Kulkarni, S.R., Laher, R.R., Mahabal, A., Masci, F.J., Mróz, P.J., Neill, J.D., Riddle, R., Rodriguez, H., Smith, R.M., Walters, R., Yan, L., Zolkower, J., 2021. Time-series and phase-curve photometry of the episodically active asteroid (6478) Gault in a quiescent state using APO, GROWTH, P200, and ZTF. The Astrophysical Journal Letters 911, L35. https://doi.org/10.3847/2041-8213/abf2ca. arXiv:2102.13017.

Rehemtulla, N., Miller, A.A., Coughlin, M.W., Jegou du Laz, T., 2023. BTSbot: a multi-input convolutional neural network to automate and expedite bright transient identification for the Zwicky Transient Facility. arXiv e-prints, arXiv:2307.07618.

Rożek, A., Snodgrass, C., Jørgensen, U.G., Pravec, P., Bonavita, M., Rabus, M., Khalouei, E., Longa-Peña, P., Burgdorf, M.J., Donaldson, A., Gardener, D., Crake, D., Sajadian, S., Bozza, V., Skottfelt, J., Dominik, M., Fynbo, J., Hinse, T.C., Hundertmark, M., Rahvar, S., Southworth, J., Tregloan-Reed, J., Kretlow, M., Rota, P., Peixinho, N., Andersen, M., Amadio, F., Barrios-López, D., Castillo Baeza, N.S., 2023. Optical monitoring of the Didymos-Dimorphos asteroid system with the Danish telescope around the DART mission impact. Planetary Science Journal 4, 236. https://doi.org/10.3847/PSJ/ad0a64. arXiv:2311.01982.

Sato, H., Yoshimoto, K., Guido, E., Nakano, S., 2022. COMET C/2022 E3. CBET 5111.

Schwamb, M.E., Jones, R.L., Yoachim, P., Volk, K., Dorsey, R.C., Opitom, C., Greenstreet, S., Lister, T., Snodgrass, C., Bolin, B.T., Inno, L., Bannister, M.T., Eggl, S., Solontoi, M., Kelley, M.S.P., Jurić, M., Lin, H.W., Ragozzine, D., Bernardinelli, P.H., Chesley, S.R., Daylan, T., Ďurech, J., Fraser, W.C., Granvik, M., Knight, M.M., Lisse, C.M., Malhotra, R., Oldroyd, W.J., Thirouin, A., Ye, Q., 2023. Tuning the Legacy Survey of Space and Time (LSST) observing strategy for solar system science. The Astrophysical Journal. Supplement Series 266, 22. https://doi.org/10.3847/1538-4365/acc173. arXiv:2303.02355.

Scotti, J.V., Gehrels, T., Rabinowitz, D.L., 1992. Automated detection of asteroids in real-time with the spacewatch telescope. In: Harris, A.W., Bowell, E. (Eds.), Asteroids, Comets, Meteors 1991, p. 541.

Sedaghat, N., Chandler, C.O., Oldroyd, W.J., Trujillo, C.A., Burris, W.A., Hsieh, H.H., Kueny, J.K., Farrell, K.A., DeSpain, J.A., Magbanua, M.J.M., Sheppard, S.S., Mazzucato, M.T., Bosch, M.K.D., Shaw-Diaz, T., Gonano, V., Lamperti, A., da Silva Campos, J.A., Goodwin, B.L., Terentev, I.A., Dukes, C.J.A., 2024. 2016 UU121: an active asteroid discovery via AI-enhanced citizen science. Research Notes of the American Astronomical Society 8, 51. https://doi.org/10.3847/2515-5172/ad2b66.

Sonnett, S., Kleyna, J., Jedicke, R., Masiero, J., 2011. Limits on the size and orbit distribution of main belt comets. Icarus 215, 534–546. https://doi.org/10.1016/j.icarus.2011.08.001. arXiv:1108.3095.

Tonry, J.L., Denneau, L., Heinze, A.N., Stalder, B., Smith, K.W., Smartt, S.J., Stubbs, C.W., Weiland, H.J., Rest, A., 2018. ATLAS: a high-cadence All-sky Survey System. Publications of the Astronomical Society of the Pacific 130, 064505. https://doi.org/10.1088/1538-3873/aabadf. arXiv:1802.00879.

Waszczak, A., Ofek, E.O., Aharonson, O., Kulkarni, S.R., Polishook, D., Bauer, J.M., Levitan, D., Sesar, B., Laher, R., Surace, J., PTF Team, 2013. Main-belt comets in the Palomar Transient Factory survey - I. The search for extendedness. Monthly Notices of the Royal Astronomical Society 433, 3115–3132. https://doi.org/10.1093/mnras/stt951. arXiv:1305.7176.

Whidden, P.J., Bryce Kalmbach, J., Connolly, A.J., Jones, R.L., Smotherman, H., Bektesevic, D., Slater, C., Becker, A.C., Ivezić, Ž., Jurić, M., Bolin, B., Moeyens, J., Förster, F., Golkhou, V.Z., 2019. Fast algorithms for slow moving asteroids: constraints on the distribution of Kuiper Belt objects. Astronomical Journal 157, 119. https://doi.org/10.3847/1538-3881/aafd2d. arXiv:1901.02492.

Ye, Q., Kelley, M.S.P., Bodewits, D., Bolin, B., Jones, L., Lin, Z.Y., Bellm, E.C., Dekany, R., Duev, D.A., Groom, S., Helou, G., Kulkarni, S.R., Kupfer, T., Masci, F.J., Prince, T.A., Soumagnac, M.T., 2019. Multiple outbursts of asteroid (6478) Gault. The Astrophysical Journal Letters 874, L16. https://doi.org/10.3847/2041-8213/ab0f3c. arXiv:1903.05320.

# CHAPTER NINE

# Detecting moving objects with machine learning

**Wesley C. Fraser**

National Research Council of Canada, Herzberg Astronomy and Astrophysics Research Centre, Victoria, BC, Canada

## 9.1. Introduction

The study of natural moving objects, such as the asteroids or comets, has a fruitful history, which has enabled significant insights into the physical and chemical structure of the early protoplanetary disk, the formation processes responsible for the growth of planetesimals and planets, and the delivery of water, organics, and other materials important for the formation of life on the Earth, to name a few topics. The discussion of these topics is beyond the scope of this chapter, but we point the novice reader to review texts such as Asteroids IV (Michel et al., 2015), the Trans-Neptunian Solar System (Prialnik et al., 2019), Protostars and Planets VII (Inutsuka et al., 2023), and the chapters of the upcoming Comets III, which are available on the arxiv.[1] The focus of this chapter is the—often arduous—task of searching for new minor bodies, a requisite first step in the study of these populations, either as a bulk population, or individually. In this chapter, I first summarize various flavors of classic search techniques, which have enabled the current research into minor bodies. I then move on to discussing the nascent field of utilizing machine learning to assist in the search for minor bodies. These techniques are sometimes employed to assist with the most difficult steps in the search process, or to enable new techniques entirely. Hereafter when the distinction matters, I refer to natural Solar System objects as minor bodies to distinguish them from artificial satellites and other spacecraft.

The latter half of this chapter is dedicated to a discussion of the use of a convolutional neural network (CNN) that I developed to perform source classification in a circumstance that classically requires significant human effort to perform. The CNN I introduce was relatively straightforward to

---

[1] https://arxiv.org/

*Machine Learning for Small Bodies in the Solar System*
https://doi.org/10.1016/B978-0-44-324770-5.00014-3

develop, and has been used quite successfully in numerous recent surveys for minor bodies. I use this as an easy to understand example of machine learning (ML), and to discuss some of the pitfalls and best practices in developing an ML tool.

## 9.2. Introduction to the detection of moving objects in astronomical imagery

In this section, I first introduce classic, non–ML-based pipelines for the initial detection of new minor bodies. Generally, the process comes in two stages: detection of a source in astronomical imagery and confirmation of its motion (subsection 9.2.1); and then linking—connecting different detections of a common object—together to form arcs, from which orbital information and future ephemerides can be gleaned (subsection 9.2.2). I summarize common practices for each part in turn in the following sections: In subsection 9.2.3, I consider requisite changes to the general moving object procedure required to implement digital tracking techniques, which I refer to as shift'n'stack. In the section that follow, I only discuss recent or ongoing surveys for moving bodies. These were chosen based on the most modern examples of search efforts, and only discussed as is necessary to understand the most common search processes. There is a long and beautiful history of highly successful surveys prior to these, which are not summarized here. Rather, we point the interested reader to the review texts mentioned above, as well as references in the papers discussed in this section.

### 9.2.1 The image search

Conceptually, the search for moving bodies is quite simple: one must simply find sources in a frame, and identify those that are not stationary. This simple statement hides a long history of search efforts of various scales and difficulties. I make no attempt to fully summarize this topic, which would warrant a dedicated textbook, but rather, just point out some common techniques used in the field.

The simplest search technique is to acquire multiple temporally nearby images of a field and search for sources that move. For example, the Asteroid Terrestrial-impact Last Alert System (ATLAS) is an on-going survey, which typically utilizes four 30 s exposures aimed at a common point on the sky and that span a 1-hour interval (Tonry et al., 2018). This image cadence allows the detection of motion of minor bodies with rates of motion at least $\sim 0.5$ "/hr, or roughly half the width of a point-source in the AT-LAS imagery. Those images undergo image differencing from a background sky map to remove stationary sources. Over the short 1-hour baseline of the imagery, the majority of minor bodies exhibit nearly linear motion, enabling the relatively straightforward step of building up *tracklets*, or groupings of nearby detections that show motion compatible with expectations of bound moving minor bodies.

The above paragraph provides an appreciation of the general steps taken in a search for moving bodies:

- Acquire coincident imagery of a field of interest, with a baseline and cadence that enables detection of motion of the objects of interest. The imagery can span temporal baselines as short as a few minutes for asteroids that exhibit angular rates of motion at opposition of many tens of "/hour, or as long as a few hours to detect the more distant Kuiper Belt population, which exhibit opposition rates of motion of only a few "/hour.[2]

- Sort candidate nonstationary sources from stationary sources. This is most robustly done using image differencing (e.g., Alard and Lupton, 1998) to remove background galaxies and stars, but can also simply be done by spatially associating overlapping detections revealing objects that do not move. This leaves behind *transient* sources that are often, but not always moving bodies.

- Find groups of sources that are compatible with the range of motions that could be exhibited by the minor bodies of interest. As hinted above, this is most commonly linear motion, and often restricted to motions that would only be exhibited by objects with bound orbits, e.g., bound to the Sun for minor bodies, or bound to the Earth for artificial satellites. These groups are commonly referred to as *tracklets* and represent likely real minor bodies.

The detections that are assembled into tracklets can be contaminated by a variety of sources, including cosmic rays, detector and processing defects, subtraction residuals, real astronomical stationary sources (e.g., supernovae), and noise spikes in the data. This is often the stage at which sources need vetting by a human operator, and can represent a huge portion of the human effort required of a project.[3]

---

[2] The opposition angular rate of motion of a minor body on a circular orbit and with zero ecliptic inclination and at heliocentric distance $r$ (geocentric distance $\Delta = r - 1$) in au is approximately given by $\dot{\theta} \sim 148 \left[ \frac{1}{\Delta} - \frac{1}{r^{3/2}} \right]$ "/hr.

[3] Vetting is a right of passage in some groups.

I point the interested reader to a few important recent surveys. ATLAS has been mentioned already, which represents a state of the art in surveys for asteroids, near-Earth asteroids (NEAs) and potentially hazardous asteroids (PHAs). The ATLAS-moving object pipeline is largely based on the code developed for, and lessons learned by Pan-STARRS (A. Fitzsimmons, personal communication). Pan-STARRS, or the Panoramic Survey Telescope and Rapid Response System (Chambers et al., 2016) is an on-going survey with sensitivity to asteroids as well as more distant, and hence fainter bodies, such as Kuiper Belt objects (KBOs) and Centaurs. The notable aspect of this survey is that its pipeline is capable of building tracklets of observations that span a few days, over which the trajectories exhibited by minor bodies are nonlinear (Denneau et al., 2013). The Outer Solar System Origins Survey (OSSOS; Bannister et al., 2018) was specifically designed to observe faint and slow-moving KBOs and other distant bodies. OSSOS utilized triplets of imagery acquired in a single night, which span a few hours with which to find slow-moving objects. Candidate triplets were generated using the routines of Bernstein and Khushalani (2000) and every candidate detection was manually vetted by human operators, thereby forgoing the need for image subtraction. This is just a small selection of recent surveys and by no means fully describes this deep and active field of research.

## 9.2.2 The linking stage

The process of connecting tracklets of a common object is commonly referred to as *linking*. Linking is useful in numerous ways. It naturally extends the temporal length of the arc for the body, thereby making any measure of its orbital parameters more accurate. If the arc-length is long enough, it tends to secure the body against future loss—where the ephemeris uncertainty for the object becomes so large it can no longer be distinguished from other nearby moving bodies, thus necessitating a full rediscovery. Most importantly for the discussion here, linking can provide some measure of good:bad vetting of tracklets. In much the same way that multiple individual detections grouped together into a tracklet provide veracity to the detections being that of a real moving object, if tracklets can be linked together into an orbit that is not unphysical, that link provides some veracity to the detections and tracklets together. As this can be done in a mostly automated sense, linking can in some circumstances reduce the human cost burden.

In concept, the process of linking is quite simple: A set of tracklets or other candidate observations of the same source are passed to an orbit fitting routine, which returns a set of orbital parameters and uncertainties (often referred to as *the orbit*), and a measure of goodness of fit, which is usually a chi-squared, or a root-mean-square (RMS) measure of the astrometric residuals, or likelihood value. It is then up to the user to decide if the orbit fit quality is high enough (RMS is low enough) to be considered "good." Orbit-fitting software packages that are commonly used are BKorbit (Bernstein and Khushalani, 2000), Orbfit,[4] or the more modern HelioLinc (Holman et al., 2018). The exact process varies from survey to survey.

The OSSOS survey mixes detection and linking together, whereby an orbit fit to a tracklet of a triplet of discovery observations is propagated to observations taken within $\sim 5$ days of discovery, and candidate moving sources are searched in a singlet image. A new track and orbit fit spanning 5 days is created, propagated to observations roughly a month from discovery, and then to opposition observation years after discovery. This procedure builds up orbits of exceptional quality and reliability, but comes at high human cost as every stage of linking requires human confirmation.

The Pan-STARRS objects are built up in a conceptually similar manner to OSSOS. Starting from a quad set of images with about 15 minutes of spacing between each image, a tracklet is made. Those tracklets that could not already be associated with known moving bodies are searched for linkages to quad tracklets of other nights. After a certain number of tracklets are linked, the candidate is designated a real object. Additional linkages are searched for in all prior observations taken by the Pan-STARRS survey. The advantage of this technique over the OSSOS method is in the reduced human burden—at no point in the linking process are individual images considered for possible new links, but rather only tracklets are considered (usually four detections in a tracklet), where the likelihood of the tracklet being that of a real object is high. The advantage of the OSSOS pipeline over the Pan-STARRS method is a decrease in the requisite telescope time required to produce reliable orbits of newly detected objects, but comes at increased human cost.

The process of orbit linking and fitting is somewhat of an art, and we encourage the interested reader to explore the references provided.

---

[4] http://adams.dm.unipi.it/orbfit/

### 9.2.3 Digital tracking techniques

Digital tracking, or shift'n'stack, is a process by which temporally nearby images are shifted and stacked to search for fainter sources than would otherwise be visible in a single image. That is, an object's motion through a series of images can be counteracted through an appropriate backwards shift, allowing all of the source's flux contained in those images to be stacked in a single coincident point source (see Fig. 9.1). This process directly enables a much deeper search for moving bodies with limiting magnitudes that often approach the limiting magnitude of a single long exposure of the same time as the image sequence. This search process comes with a high computational burden, as image stacks need to be produced at every possible trajectory, and for every possible region where the objects may reside (normally the entire field of view).

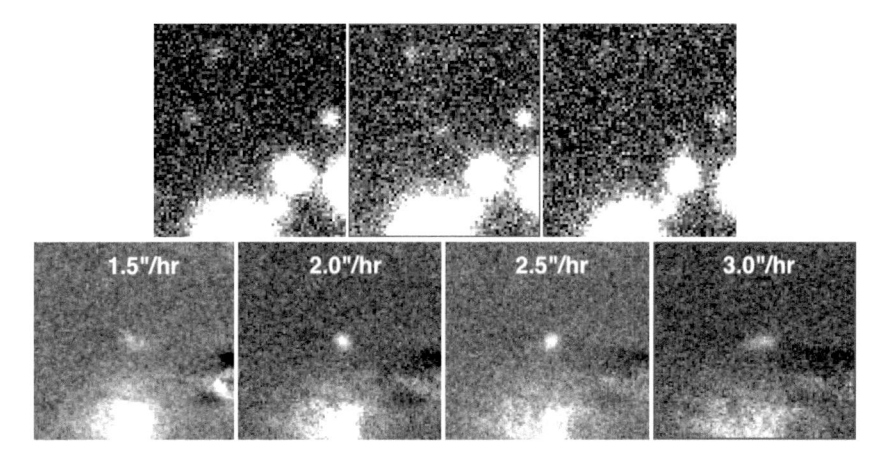

**Figure 9.1 Top:** Three images centered on a real Kuiper Belt object discovered as part of the New Horizons search for distant Kuiper Belt objects (Fraser et al., 2023b). The source is just below the noise floor of the images. **Bottom:** The shift'n'stack sequence of about 100 images, shifted at four different linear rates of motion of the same source shown above. The source's true rate of motion is 2.4 "/hr, and when the shift rate is near that value, the source becomes an easy detection.

Moving object detection has a long history of finding ultra-faint sources. This is especially true in the effort to find Kuiper Belt objects (Luu and Jewitt, 1998; Gladman et al., 2001; Bernstein et al., 2004; Fraser et al., 2008; Fuentes et al., 2009) and other distant, slow-moving bodies (Kavelaars et al., 2004; Holman et al., 2004; Ashton et al., 2020). For these populations, angular rates of motion are highly linear for the duration of a night, and so the shift'n'stack routine can be characterized by only two dimensions: rate of motion in Right Ascension and Declination, or equivalently rate and angle. Applications to more proximate populations are less common due to the higher apparent angular accelerations they exhibit, which make their apparent trajectories nonlinear much more rapidly, though see Heinze et al. (2015). Once the trajectories of objects become nonlinear, either due to proximity, or due to length of image sequence, the computational burden is dramatically increased. For this reason, nonlinear shift'n'stack remains an unsolved problem and an active area of research.

Beyond the massively increased computational expense, shift'n'stack historically has required significant human vetting effort, primarily as a result of a massive ramp up of false sources driven by noise spikes that show up at low signal-to-noise, as well as other types of false positive sources that do not plague nonstacking search techniques. A particularly nefarious example is from faint diffraction spikes associated with bright stars. As an image sequence is acquired, those diffraction spikes rotate around the star. A linear shift of those images can cause the spikes from different images to overlap in the stack, which produces a somewhat round source that can slip past many filters implemented to reject false positives. This and other sources of false positive generally necessitate the manual vetting of each and every candidate, resulting in human cost that is much higher than nonstacking techniques. I discuss the use of ML to tackle this particular problem in subsection 9.3.4.

In general, shift'n'stack routines replace the discovery-to-tracklet steps. A shift'n'stack detection provides both a position, and a rate of motion. This information replaces the tracklet, and is sufficient to search for a set of shift'n'stack discoveries taken at a different epoch for objects that match the orbit predicted from the detection in the first epoch. This is the general search approach that is being used for the shift'n'stack component of the DECam Ecliptic Exploration Project (DEEP; Trilling et al., 2023; Napier et al., 2023), an on-going search for new targets for observations with the New Horizons spacecraft (Fraser et al., 2023b, and also Fraser et al. in prep) and the Classical and Large-A Solar System Survey (CLASSY; Fraser et al., 2023a). Each of the four fields of DEEP will receive observations at opposition and one month after opposition in the discovery year, and then be reobserved one and two years after discovery. From those four visits, observations of common sources will be linked and orbits constructed (Trujillo et al., 2023). CLASSY uses an approach inspired by the OSSOS discovery and linking technique: a field is visited 3 times on three separate nights near opposition and within about 5 nights. Each visit consists of 3 total hours of integration time spanning at least 4 hours. Each of those visits is searched with shift'n'stack techniques and tracklets of candidate sources across all three nights are created. Each of these tracklets is propagated to shift'n'stack search epochs at plus and minus one month of discovery to provide confirmation of real objects and to enable initial orbit fits.

Through the proliferation of graphics processor units (GPUs) and some clever programming, the compute burden to perform the shift'n'stack part of a search has been greatly reduced. By virtue of their design, GPUs can make short work of the median or mean image stacking process reducing the computational time by orders of magnitude. The kernel-based moving object detection (KBMOD Whidden et al., 2019; Smotherman et al., 2021) software package provides generalized linear shift'n'stack search capabilities. As well Burdanov et al. (2023) present a similar tool for digital tracking.

There are other less utilized search techniques that hold some resemblance to recent ML-based search efforts. I highlight a clever effort by Fuentes et al. (2010), who searched for trailed sources in long-exposure observations from the Hubble space telescope (HST). Due to the motion around the Earth, most minor bodies exhibit noticeable parallax as witnessed by the HST resulting in nonlinear trailed arcs for Solar System bodies arcs with characteristic motions that reflect the orbit of the HST. In their search, they generated candidate templates of possible trails that might be exhibited by distant KBOs. By template matching, they found a handful of faint and particularly distant objects, which were previously undetected.

## 9.3. Applications of machine learning

In this section, I summarize a select set of past and current ML efforts designed to find moving objects.[5] Given that the majority of moving sources are discovered in imagery, most modern ML-based applications of moving object detection are based on CNNs of some flavor or other. However, we start out with a likely more familiar ML technique in subsection 9.3.1.

### 9.3.1 Clustering in moving object detection

Some of the earliest uses of ML in moving object detection are not with neural networks or decision trees, which are what is commonly thought

---

[5] While I try to be thorough and complete, inevitably some papers will be missed.

of when the term *machine learning* is mentioned. Early search efforts have made use of clustering algorithms to assist in moving object detection. For example, HelioLinc, mentioned above, requires clustering of detections to sort real moving objects from the chaffe. In the original implementation, a KD-tree (Kubica et al., 2007) with euclidean distance on velocity and distance parameters is used. The KBMOD routine also uses the density-based spatial clustering of applications with noise (DBSCAN) algorithm in position and velocity to remove the candidate shift'n'stack sources that are most likely to be false from the final returned list.

Clustering algorithms in general require selection of parameters that determine the output. To achieve best results, these *hyperparameters* are tuned to maximize performance according to some metric relevant to the circumstance in which the clustering algorithms are used. This is akin to the hyperparameter selection and training that is required of what most readers would consider modern machine learning, albeit usually with simpler training efforts and often more easily understood metrics. Similarly, much like each clustering algorithm is just that—an algorithm, so is any CNN or similarly complex ML technique, and the choices in network architecture are analogous to choice in clustering algorithm.

I bring up clustering as a familiar and comfortable concept that many readers may have already used, that is also essentially a machine learning technique, in hopes to assuage some of the unreasonable skepticism and fear some have over the uses of ML.

### 9.3.2 Moving object detection and classification in direct images

I first discuss the work of Chyba Rabeendran and Denneau (2021), who apply a ResNet-based custom neural network to discovering asteroids in the ALTAS survey (see the introductory chapter for an introduction to the ResNet). Recall that the classic moving object search utilized by ATLAS makes use of a quad set of coincident and temporally close exposures. The authors adopt a novel architecture to perform binary classification—does or does not contain a moving body—in image quads, which were previously identified as containing four separate images of a candidate real moving object (see Fig. 9.2). Specifically, the network makes use of the pretrained ResNet-18 model, which feeds a pair of dense layers. Through this structure, each image of the quad is passed, with the outputs consisting of an 8 element vector with values $\mathbb{R} \in [0, 1)$. Those values represent confidences of the source at the center of the image belonging to one of eight different classes: five bogus classes, including diffraction spikes, detector, and pipeline defects; and three real classes, including fast–moving *streak* objects, point–like slow–moving objects, and comets (see Fig. 9.3).

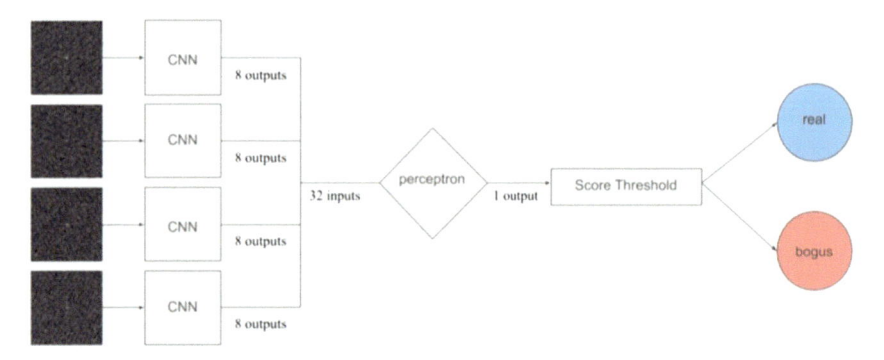

**Figure 9.2** The custom architecture utilized by Chyba Rabeendran and Denneau (2021), where each image of an ATLAS quad is passed through a pretrained ResNet-18 (labeled as "CNN"), and the confidence outputs are concatenated into a single 32-element vector, which is passed through the classifier perceptron. Reproduced from Figure 6 of that work.

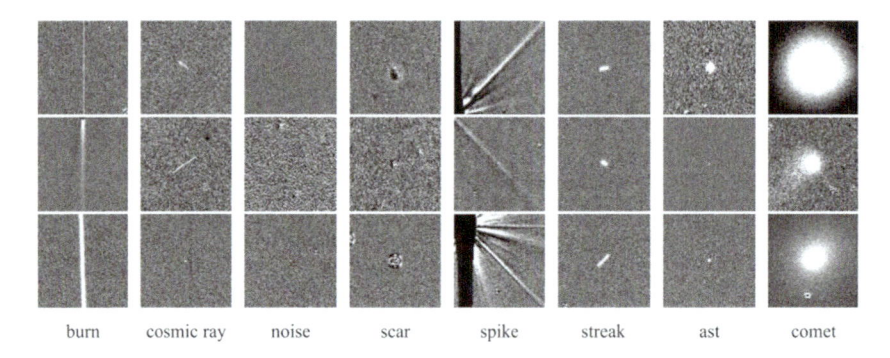

**Figure 9.3** Examples of the 8 different classes utilized in the classification search of Chyba Rabeendran and Denneau (2021). The three real minor body classes are shown at right, with the 5 bogus classes resulting from detector and pipeline artifacts. Reproduced from Figure 3 of that work.

The novelty of their work rests with the perceptron classifier network they develop, which takes as input a 32 element confidence vector, which is the concatenated confidences of the four image outputs of the ResNet–18. The perceptron returns a likelihood that the image quad contains one of the three types of real moving object.

The authors trained only the perceptron stage of their custom network, preserving the weights of the ResNet-18. They focused on maximal discovery. That is, they trained to ensure that real minor bodies were labeled as such as often as possible, without focus on the precision or recall performance, which are more commonly maximized. This resulted in a moderately low 15% false positive rate, and a tiny 0.4% false negative rate. Importantly, the network resulted in a 90% reduction in the final human vetting time required to declare a source as real.

The main weakness of this work was the tiny training sample of which they made use. Data from only one lunation was used in training, with only about 500 sources per each of the eight classes. In some sense, it is surprising that they have such a positive result, though much of this success is likely driven by the perfectly labeled training set they utilized, and the fact that they only trained the weights of the perceptron. Though it is likely that inclusion of more data would enable a significant reduction in the false positive rate, a main takeaway from this work for the aspiring ML user is that good results can be found even with moderately small, but *high quality* training sets.

The results of Chyba Rabeendran and Denneau (2021) reaffirm the common thought that CNNs can provide highly reliable type classifications for sources. For example, that network does a satisfactory job of identifying cometary sources in ATLAS image quads. Another effort to identify comets in astronomical imagery is presented in Duev et al. (2021). In that effort, they utilize a fairly advanced network structure to confirm the presence of comets in image cutouts of Zwicky Transient Facility (ZTF) data. There routine utilizes networks that output segmentation maps to perform binary classification as to whether or not an image contains a comet, as well as perform regression on the location of the candidate comet in those images.

Duev et al. (2021) make use of an EfficientNet (version B0) and bi-directional feature pyramid network (bi-FPN; Tan et al., 2020) to create a multiscale segmentation map. A simple way to think of the segmentation map is a pixel map of the input image that contains large values, where pixels in the original image likely contain a comet, and small values where they likely do not. The bi-FPN can result in a segmentation map that preserves morphological information about a source at multiple scales—one might immediately recognize the importance of this for comets, which show a broad range of morphologies. This segmentation map is fed to a so-called head network, which is essentially a standard shallow CNN+perceptron that outputs a vector $[p_c, x, y]$, where $p_c$ is the probability of an image containing a comet, and x and y are the predicted locations. Their network is presented in Fig. 9.4.

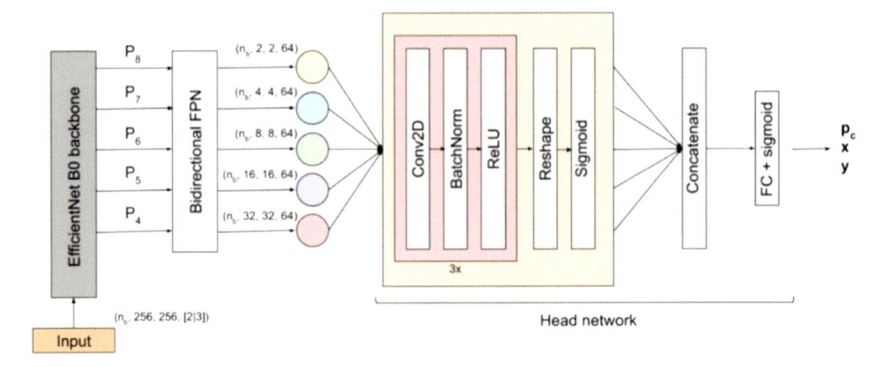

**Figure 9.4** The bi-FPN-based network used by (Duev et al., 2021) to identify comets in ZTF imagery. The network consists of two parts: an EfficientNet-B0 and biFPN section that creates multiscale feature maps from the input imagery, and a CNN+perceptron head network, which performs the binary classification and regression on comet position. Reproduced from Figure 3 of that work.

Their network takes as input a three-layer image. The first layer is a 256x256 pixel cutout. The second layer is a background template of the same region generated from other epochs of ZTF data, and the third layer is the image subtraction of the background template from the first image.

The training was supervised. The training data they utilize reflects the rarity of real comets, with a prelabeled training set that is highly unbalanced, with 5x more cutouts *not* containing comets than cutouts that do. To counteract this, they make use of sample weights to produce a network that can successfully identify images containing a comet. Without sample weights, the network would only have the ability to identify images that *do not* have comets. We discuss the importance of balanced training data further below. Positive labeled training samples were generated using ephemerides of known comets tracked in the Jet Propulsion Laboratory Horizons Service[6] with manual labeling. Other negative labeled imagery was included (though it seems possible that a few of those images may actually contain an unknown comet). The training data also included comets identified with a simple ResNet-based classifier. The resulting training set size was 22,000 samples (compare this with the smaller sample size of Chyba Rabeendran and Denneau (2021) discussed above).

[6] https://ssd.jpl.nasa.gov/

This impressive work resulted in an enviable false negative rate of about 1.7%, with a rough position error on predicted comet location of only a few pixels. The proof is verifiable; they report the discovery of comet C/2020 T2 (Palomar).

### 9.3.3 Trailed moving source detection

The majority of effort by the community towards developing machine learning algorithms for detection of minor bodies has been put into searching for streaks of moving objects within long exposures. These telltale signs are quite common to astronomical surveys that have focus beyond the detection of minor bodies, and so do not take care to avoid circumstances where objects might trail. While not nominal for detection, trailed minor bodies have a unique morphology, that is not produced by any other real astronomical source, thereby affording use of some machine learning techniques that are designed to be sensitive to differences in source morphology.

Jeffries and Acuña (2023) make use of a so-called U-Net (Ronneberger et al., 2015), which they train to produce segmentation maps of regions of imagery likely to contain streaks of moving objects, both natural and artificial. The authors fully simulated an astronomical dataset containing streaks, point sources, and noise, and thus also simulated a perfectly labeled training dataset. They made use of a pretrained U-net encoder stage, and only trained the decoder stage.

The authors found that their streak-detecting U-Net was able to detect and locate streaks with a similar performance to state of the art non-ML-based algorithms, but with a $10 - 20\times$ higher throughput, though their results did come with some other minor weaknesses, which were not present with the other routines.

The work of Jeffries and Acuña (2023) is a clear example of fully manufacturing a training dataset. This is the ultimate form of training data augmentation, and easily allows for a perfectly labeled training set, but comes at the cost that their network will inevitably perform differently (likely more poorly) on real-world data. It may be that the performance is still satisfactory, but extra care needs to be made to ensure that this is the case. Moreover, effort needs to be made to detect any odd behaviors with real-world data that do not occur with the artificial training data.

We next point out Kruk et al. (2022). Like the work of Fuentes et al. (2010), Kruk et al. also searched for streaks of minor bodies in HST data, specifically for data from the Advanced Camera for Surveys (see Ryon and Stark, 2023), and the Wide Field Camera 3 (see Marinelli and Dressel,

2024). For training data, they made use of the Asteroid Hunter Project on the Zooniverse,[7] which presents HST imagery to a small army of volunteers who classify an image as containing a streak, and mark both the beginning and ending of a streak if present. Each image is classed by 10 different users, resulting in high-quality labels, though the sample size itself numbered only a few thousand samples. The training data and labels consisted of those images with asteroids, as well as masquerading bad sources, which include cosmic rays satellite trails, and cosmic rays (appropriately labeled as such).

The authors make use of AutoML (Zoph and Le, 2017) Vision Model.[8] This service aims to automatically explore different network architectures and select the most performant architecture for the problem presented. In some sense, AutoML aims to enable machine learning for nonexperts, and is a good place to start for many budding ML users.

In their dataset, Kruk et al. (2022) identify 1,300 asteroid trails, nearly half of which are from unidentified objects. Their network has a relatively low recall of 61%. It is unclear if this low performance is a result of their training set (size or quality of labels), a failure of AutoML to find a highly performant network, or is simply due to the similarity in morphology between asteroid trails and other nonasteroidal sources in the HST data.

The DeepStreaks network is presented in Duev et al. (2019). Deep-Streaks is a custom network architecture designed to detect streaked fast-moving minor bodies in ZTF imagery. DeepStreaks makes use of three separate classic networks as building blocks: ResNet50, VGG6, and DenseNet121 (see the introductory chapter for a description of each network architecture). A separate incarnation of building block network is trained to tackle binary classification of three separate but similar types of streak defined by their streak length. A logic branch exists for each problem class, which is the ensemble OR output of one of each of the different building block networks. The operation of the architecture is as follows. A given sample is passed through all nine separately trained building block networks. A candidate source is labeled as good if the candidate is labeled good by any one of the building block networks over two or more branches. This is an interesting idea, as it presumably allows each network and each branch to be sensitive to its own particular style of streaked source, and so the ensemble should have better performance than its individual blocks or branches.

[7] https://www.zooniverse.org/projects/sandorkruk/hubble-asteroid-hunter
[8] https://cloud.google.com/vision/automl/object-detection/docs

The training was performed in a supervised fashion, with train data assembled from a dataset made up of images with streaks previously identified by humans, synthetic streaks implanted into otherwise streak-less frames, and a set of imagery containing undesirable streak-like features, such as cosmic rays and image ghost arcs.

The outputs of DeepStreaks are still vetted by humans, but with a $50 - 100\times$ reduction in vetting of sources, which would otherwise be false negative sources. A building block threshold of 0.5 probability for positive classification resulted in an approximately 97% true positive rate. The work of Duev et al. (2019) is an example of how classic networks, which are simple to implement in modern ML frameworks, can be manipulated in a straightforward fashion to produce an excellent outcome with easily understood operations.

The work of Wang et al. (2022) is a direct competitor to the DeepStreaks work. They repurpose the EfficientNet-B0 architecture to search for streaks in two layer cutouts (image of the candidate streak and background template of the same region) of astronomical imagery. They generated a relatively hefty fully simulated training set, weighted 4:1 to those cutouts containing streaks and those that do not. They achieved an excellent low false positive rate of only 0.02%. The EfficiencyNet-B0 model contains 4,161,268 trainable parameters. Though Wang et al. (2022) make use of some techniques to avoid overfitting (see discussion in Section 9.5), such as dropout, they did not present any effort to quantify overfitting. Given the complexity of the EfficientDet-B0 network, it seems likely that overfitting could be an issue. Whether or not this is more relevant to the performance of the network on real astronomical data compared to the fact that training data were fully simulated is unclear.

We would like to point the reader to Lieu et al. (2019) and Varela et al. (2019). Lieu et al. (2019) presents a good example of transfer learning, whereby they only retrain a subset of layers in a large and complex network that was previously trained on nonastronomical data. They consider a large range of available networks, and achieve a relatively good performance on asteroid streak detection, even with a relatively small training set. The budding ML expert may start here, as this paper provides an excellent overview of the many things one needs to consider when embarking on a new ML project. Varela et al. (2019) present a good example of the application of the "You Only Look Once" or YOLO algorithm[9] to the detect streaks

---

[9] YOLO was originally designed for fast facial recognition tasks.

in wide-field ground-based astronomy imagery. YOLO represents a very different approach to the use of CNNs compared to the other techniques discussed here.

The final paper I discuss is the work of Cowan et al. (2023). This tour de force in custom network design aims to detect the streaks left by moving minor bodies. Specifically, they look for the blobby streaks left by bright asteroids in the mean stacks of spatially coincident imagery from the Microlensing Objects in Astrophysics (MOA) survey. The reason these streaks are blobby, as compared to other efforts discussed above, is because MOA uses relatively short exposures, in which asteroids remain nearly point-like, and so mean stacks images of asteroids appear as a nearly linear series of point-like sources.[10]

Cowan et al. (2023) created 5 different custom networks that each outputs a likelihood that an input image contains a streak. Three of those networks are structured like a VGG with varying depths, and two are reminiscent of the ResNet, though these networks use a custom block that performs an addition of three separate convolutional layers of varying depth, and the ResNet skip connection (see Fig. 9.5). Though the performance of each network was tested individually, the best performance arose from the average ensemble of the outputs of all five networks. This is reminiscent of the results of Duev et al. (2019), where the merger of outputs of a set of different network architectures resulted in good performance (though Duev did not discuss the performance of each individual network, only the logic-driven ensemble). The input to the network of Cowan is a set of 128x128 pixel cutouts selected from the full frame mean stacks, with output being a probability of the cutout containing a streak.

Training was done in a supervised fashion. Through a—likely arduous—manual search, a number of real asteroid streaks were identified and labeled. With augmentation (rotation, blurring, flipping, brightening and darkening), a large training set of cutouts was generated, weighted 1:5 with:without streaks. Critically, an independent test dataset was created from data of an entirely different night, and from some chips of the detector mosaic that were excluded from the training set. That is to say, their test set was entirely independent of their training set. This is a crucial feature of ML training, which should be utilized whenever possible so as to preserve the veracity of the test results.

---

[10] They refer to these as asteroid tracklets. We call these streaks to avoid confusion with the astrometric tracklet terminology introduced above.

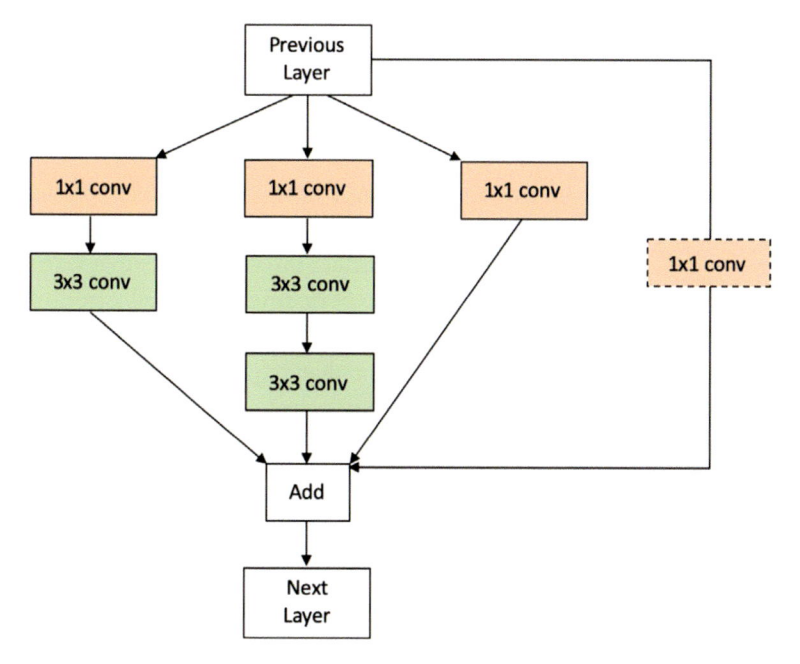

**Figure 9.5** The custom ResNet-style block invented and utilized by Cowan et al., 2023. Note the difference with the standard ResNet block, which sums only one convolutional chain with the skip connection. Reproduced from Figure 9 of that work.

With this ensemble network, Cowan achieved 96% recall and 87% precision, which are extremely good results. It is unclear how the performance of this ensemble will degrade with asteroid brightness, an aspect not discussed in their paper. Though they make it clear from the start that their focus is the detection of bright (by some standards) asteroids with V<21.

### 9.3.4 Detection of moving objects in shift'n'stack imagery

We now turn our attention to the application of CNNs to the good:bad binary classification of candidate sources detected by shift'n'stack-type searches. The first application of this is from Smotherman et al. (2021), who apply a ResNet-50 network to vet the candidate detections produced by KBMOD (see Section 9.2.3). The network accepted as input 21x21 pixel stacks centered on candidate sources. The stacks were produced during the shift'n'stack process. The network output is the probability of a candidate being real. Training of that network was done in a supervised fashion. They chose to create an artificial dataset of positive sources by injecting Gaussian point-sources along predetermined linear trajectories, with real negatives originating from outputs of actual search data. With this process, they created a sample of 40,000 candidate sources, 70% of which went to training. Details of the process are sparse, but they claim an accuracy of 96% on the test set.

A weakness of the effort of Smotherman et al. (2021) is through the use of Gaussian point-spread functions (PSFs) for training. Although these are somewhat point-like, they do not fully reflect the odd shapes that stars may take in astronomical imagery. As such, the performance on real astronomical imagery is inevitably reduced, though they do report about $10\times$ reduction in the number of candidates per detector that require human vetting. We interpret this as about $10\times$ fewer false positives. A better choice would have been a Moffat profile (Moffat, 1969), which more accurately reflects the shape of the core and wings of stellar PSFs. Better still would be to use an actual PSF model.

I now turn my attention to the custom CNN I have developed to perform the same style of vetting as done by Smotherman et al. (2021). That is, real:bogus classification of candidate sources found by shift'n'stack searches.

The impetus for this effort came from the search for Kuiper Belt objects bright enough to be observable by the Long Range Reconnaissance Imager on board the New Horizons spacecraft (Fraser et al., 2023b). The search field was at low galactic latitudes about $10°$, with extremely high stellar density, which resulted in a greater than 1000:1 false positive rate in the KBMOD outputs of those search data, when searching to depths with sources reaching signal- to-noise about 5 in a stack. This false positive rate saw little improvement with robust image subtraction. With thousands of candidates to vet per detector (the Hyper-SuprimeCam mosaic on the Subaru telescope has 103 available detectors), and about 15 epochs, human vetting was practically impossible, and so good:bad classification of the KBMOD stacks with CNNs was explored.

In my experimenting, I had first explored relatively shallow VGG-like CNNs. Generally, the performance matched the reported performance of the deep ResNet trained by Smotherman et al. (2021), which reduced the human vetting workload by a factor of ten. Though to make the work practical, a further $10\times$ reduction was needed. Various custom modifications were explored, including networks that operated on only a finite range of source fluxes, networks that operated on a dual channel input of the mean and median KBMOD stacks, various versions of ensembles, and so on.

The network that I settled on is an ensemble of ResNets. This network has proven extremely useful for the New Horizons search, the CLASSY project, and in a deep search with the James-Webb space telescope that is being performed at time of writing this chapter (Morgan et al., 2023; Eduardo et al., 2023). A network diagram of one instance of the ResNet ensemble is presented in Fig. 9.6. This network takes as input 43x43 pixel mean shift-stacks produced by KBMOD.

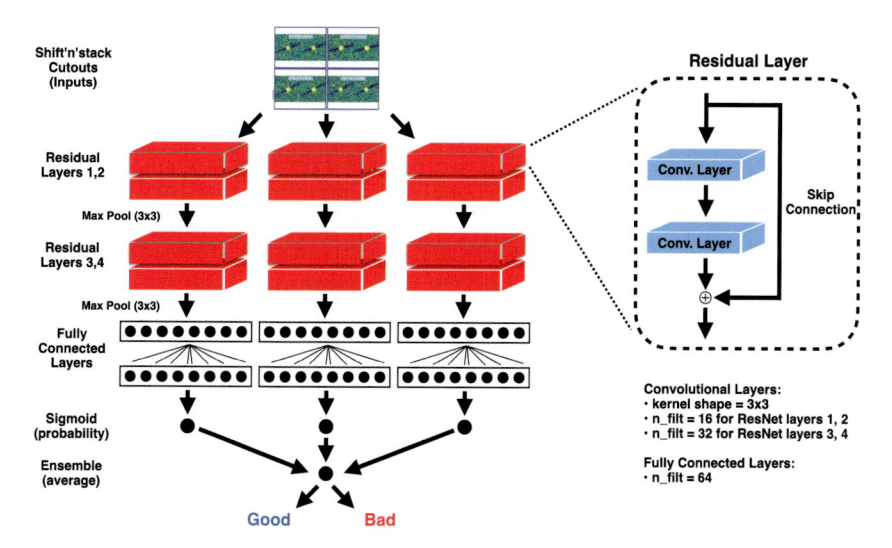

**Figure 9.6** The custom ResNet ensemble I have implemented for the binary classification of KBMOD shift-stacks. The filter parameters shown are those utilized in the New Horizons KBO search (see subsection 9.3.4). Each of the three branches of the ensemble is trained on the same training set, but with a different random initialization of weights. Reproduced from Figure 3 of an upcoming manuscript (Fraser et al. submitted to the Planetary Science Journal). We remind the reader to see Chapter 1 for an introduction to the ResNet and general CNN architectures.

Training is performed in a supervised fashion, and is done independently for each project (New Horizons search, CLASSY, JWST), as the detectors used in each project have vastly different image properties. Creation of a network that can handle data from different telescope facilities is an on-going effort. We present the results of this network applied to a small amount of CLASSY data.

A training dataset is randomly drawn from the KBMOD outputs. A key aspect to ensuring a high-performing network was to train on samples that are morphologically as close to the real minor bodies as possible. To that end, imagery were populated by a large number—roughly 100 per individual detector in a mosaic—of artificial moving objects on varied trajectories and over a range of brightnesses. A swarm of artificial Kuiper Belt object orbits were generated, and propagated to the epochs of each individual image. These *implanted* objects were injected into individual frames using a model PSF generated using the Trailed Imagery in Python package (Fraser et al., 2016), with the PSF modeled from point sources in the frames themselves. As a result, the injected artificial objects possessed morphology and noise properties that accurately reflected real minor bodies in the CLASSY imagery. KBMOD was executed on these implanted frames, from which a randomized training set was generated by labeling as good all implanted sources that were detected by KBMOD. All other KBMOD candidates were labeled as bad.

Bogus sources were selected at random to build up a nearly balanced sample of 1.6:1 bogus:good sources. Effort was made to explore this ratio, as it was felt that increasing the bogus fraction would result in a broader variety of bogus sources to train on. It was found, however, that any further imbalance beyond a factor of about 1.6 in favor of the bogus sources, would result in a network that produced a massive fraction of false negatives, which, needless to say, would be catastrophic. This was true even with the use of sample weights to try to counteract the imbalance.

The above process of generating a training sample inevitably resulted in some real detected minor bodies being labeled as bogus. This mislabeling results in approximately 0.02% mislabeling of bogus sources. Such a small mislabeling of bogus labels did not influence the performance of the outputs. In the circumstance of a higher mislabeling, an iterative training procedure of relabeling between training may help. Rather, I would recommend the training sample be augmented (e.g., with more implanted sources) to avoid the issue. The adage "garbage in garbage out" is true for most machine learning applications, and so the importance of maximizing the quality of the training set cannot be under-emphasized.

By virtue of creating a training set from a KBMOD search of implanted imagery, the distribution of brightness of the sources labeled good inherently followed the distribution of implanted source brightnesses, convolved by the detection efficiency as a function of source brightness. In an attempt to avoid any bias this may introduce, we implemented sample weights for the good labeled sources, which were the inverse frequency of occurrence of an object of that brightness in the detected sample. This was done by fitting a polynomial $f(m)$ to the histogram of brightnesses $m$ of implanted detected sources. The final sample weights of good sources was then $w(m) = \frac{1.6}{f(m)}$, and for bogus sources $w = \frac{1}{1.6}$.

Training of the network was otherwise commonplace, and made use of a cross-categorical loss, and the ADAM (version 2) optimizer, and 20% dropout. The network itself was coded in keras. Augmentation of the training set involved linear shifts by one pixel in the four cardinal directions, rotations by 90, 180, and 270 degrees, and mirror flips vertically and horizontally. In my experience, these augmentations reduced the false positive rate of the network by about 10%. Surprisingly, I found that augmentation by adding noise generally resulted in degradation of the network performance. I attributed this to the somewhat lazy approach I adopted, in adding Gaussian noise with standard deviation equal to that of the pixel values of the unaugmented sample. This created artificial samples with different noise properties than in the real data, and so forced the network to recognize a signal that was not present in real samples. A likely better approach would be to sample noise of the true background noise distribution, though I found that the training dataset was large enough to make use of this extra augmentation method unnecessary.

In practice, an instance of the network consisting of three branches was trained on a single training set, with each branch trained separately using a different random weight initialization. In my experimenting, I discovered that the quality of network output in a training run could be highly variable, a problem that I was unable to solve directly. When a training run succeeded—it did not always succeed—I found that a single branch of the ensemble would better reject certain types of bogus source better than other branches. Generally, the average ensemble outperformed the individual branches that made up the ensemble. I found that three branches was most effective in terms of keeping the number of trainable parameters low, while ensuring a well-performing network, with diminishing returns for higher numbers of branches.

Due to the roughly 100:1 ratio of bogus to good sources, which is the ration output by KBMOD (when searched to a low SNR), a single randomly drawn training set never fully covered the broad range of morphologies exhibited by the KBMOD outputs, at least for the New Horizons KBO search. In that case, we found that an ensemble of three independent instances produced a satisfactory result. Each of three instances of the network shown in Fig. 9.6 was trained on a different randomly generated training set. We trained roughly 10 different unique instances of that network, and selected the three best performers for production. The final classification was done using the average ensemble probabilities of those three instances (each of which itself is an ensemble).

In Figs. 9.7, 9.8, 9.9, and 9.10, I show examples of the diagnostic plots I made most use of when determining the performance of the network. Fig. 9.7 shows the behavior expected of a successful training run: respectively, rapid increase and decrease in model accuracy and loss over the first few epochs, a pattern which stabilizes on low values of loss and accuracy approaching 100%.

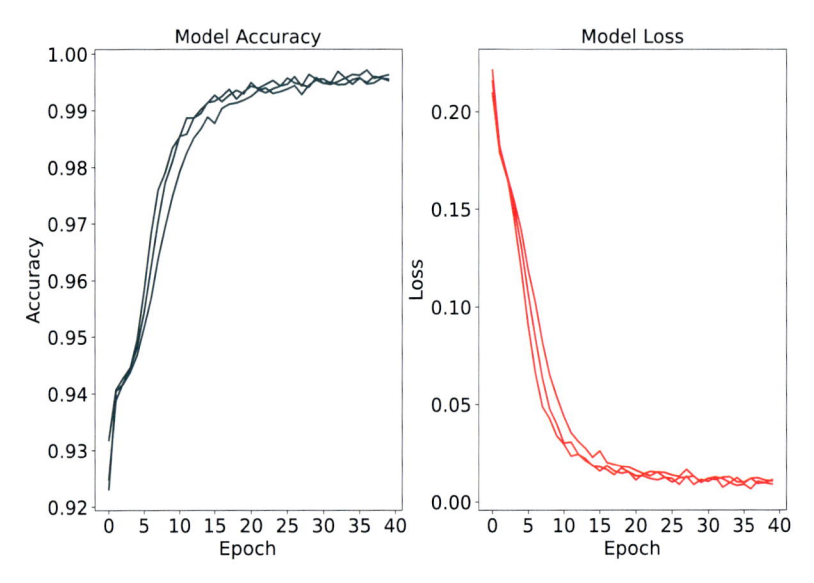

**Figure 9.7** Example of loss and training evolution during the training of one instance of the network depicted in Fig. 9.6. For this run, training data were selected from 6 nights of data from the CLASSY survey, resulting in an augmented training sample size of 317,205.

Fig. 9.8 shows the prediction probability distribution of the training and test datasets. I have found this particularly useful in checking for a failed training, and for significant overfitting. In the case of a failed training, the distribution will not cleanly separate into two classes at some P value as it does here. Moreover, a well-performing network will result in a wide and nearly empty valley between those samples that cluster near P-values of 0 and 1. In the example shown, there is a hint of overfitting evidenced by peaks of samples with middling probabilities at $P \sim 0.3$ and $P \sim 0.7$, which are more prominent in the test data than they are in the train sample.

Fig. 9.9 presents the usual quadrant plots for binary classification for both train and test datasets. The performance is generally good, with precision of 98.7% for the test sample and 99.9% for the train sample, and recall of 99.4% and 99.9% for the test and train samples, respectively. As is

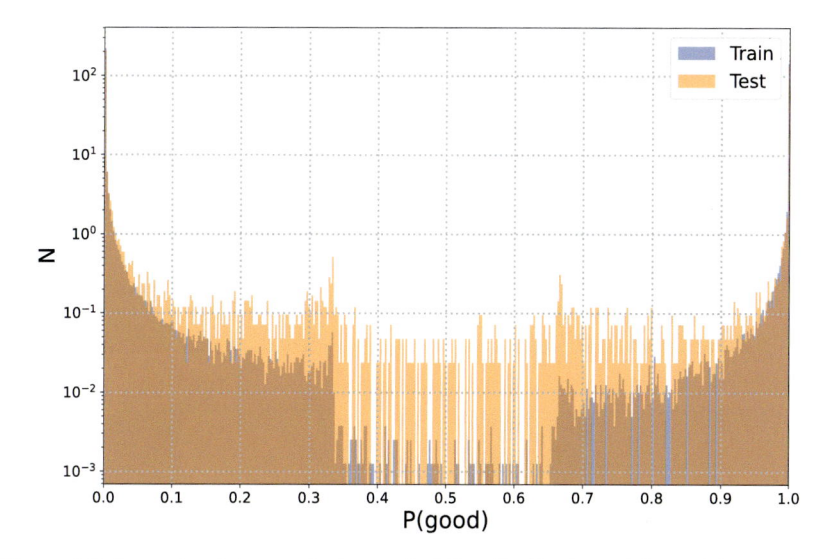

**Figure 9.8** Example of the prediction probability distribution of the training and test distributions of the training run shown in Fig. 9.7. For readability, the training distribution is normalized so that the peaks of the training and test distributions match. The test sample (orange) shows a slightly higher fraction of samples with middling probabilities about 40 to 60% that is not present in the training sample. The general similarity of the training and sample probability distributions however, implies a successful training run resulting in a similar behavior of network outputs for both samples.

common with ML efforts, the test dataset does not perform quite as well as the training dataset, in this case only marginally. Though I point out that a better test of network performance will come from the application of the network to a dataset from a different night than used for training and testing. This is presented in Fig. 9.10, which shows an example of application of the above network ensemble applied to data not used to assemble the training or test datasets. The training and test datasets included data acquired on nights of August 22, 23, and 26, 2022. Data from August 24, 2022 were acquired in similar transparency and image quality as neighboring nights, and are an ideal validation dataset.

Fig. 9.10 shows the typical behavior experienced in the CLASSY search. The KBMOD outputs have the faintest search depths, but report back $94,722$ candidate sources across the 40 detectors of MegaCam, $2,088$ of which are planted and about 70 are real minor bodies. Application of the ensemble reduces the search depth by about $0.1 - 0.2$ magnitudes, and the sample labeled good is a factor of about $25\times$ smaller than the KBMOD detection list, at $3,628$ sources. Final human vetting results in a further

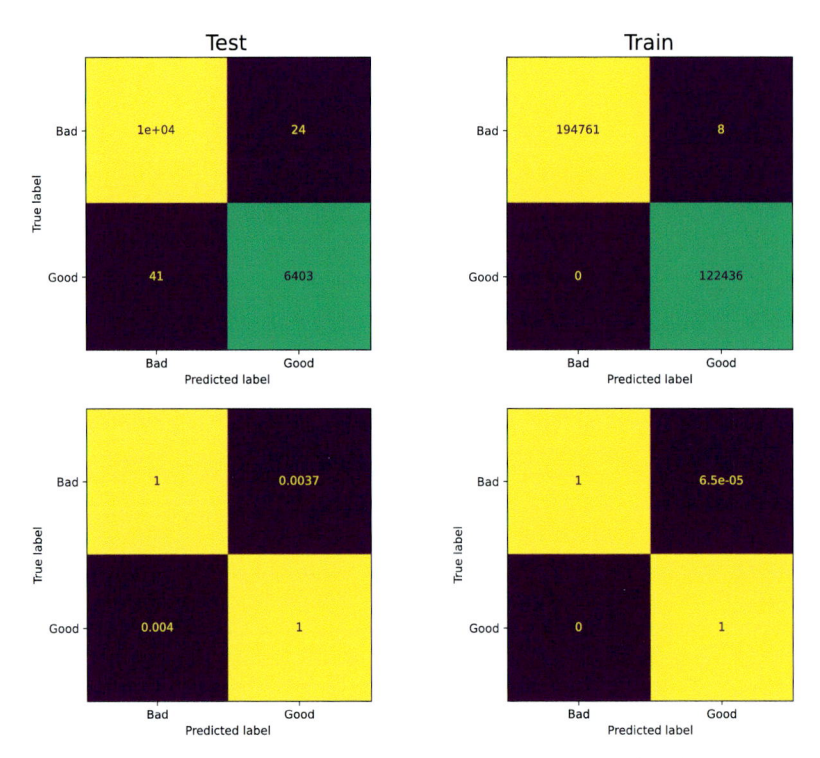

**Figure 9.9** Quadrant plots for the training and test samples used in the training run shown in Fig. 9.7. Absolute and fractional versions are shown. The test sample shows a higher fraction of false positives than occurs for the train sample, implying a small amount of overfitting has occurred.

degradation of the search efficiency, the magnitude of which depends on the performance of the network on the particular night in question, and can sometimes result in approximately 0.2 mags loss in search depth. In this example, of the 3,359 sources labeled good by both human and network, 2016 are implants, and about 70 are real minor bodies. The remaining approximately 1,200 sources are false positives, with the vast majority having SNR approximately 5, or detection efficiencies $\eta < 0.4$, where both machine and human can no longer distinguish real sources from noise peaks. The remaining false positives are ultimately culled at the tracklet creation stage, where three nights of data are grouped together (see subsection 9.2.3). At this point the detection list purity is essentially 100% (Fraser et al., 2023a). The New Horizons search also relied on linking to cull the remaining false positives, though the search had to be limited to a higher SNR due to the significantly higher prevalence of subtraction residuals, which resulted in significantly noisier stacks.

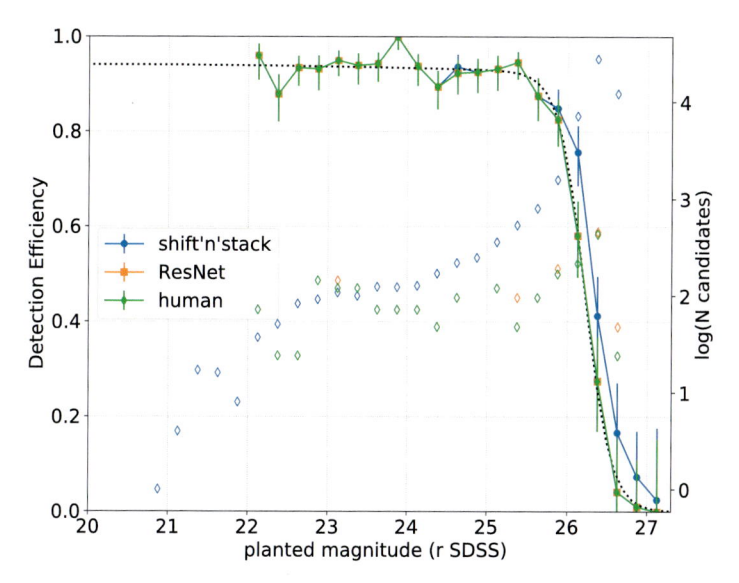

**Figure 9.10** Example detection efficiencies of the KBMOD outputs (blue), those labeled "good" by the ResNet ensemble (orange), and those finally labeled as "good" by a human (green). These are evaluated against the artificial implanted moving sources. Error bars are 1-$\sigma$ poissonian intervals. The black dotted curve is a efficiency curve fit of the form $\eta(r) = \frac{0.94 - 4.1 \times 10^{-4}(r-20.0)^2}{1.0 + \exp((r-26.2)/0.16)}$ to the human vet outputs. The open diamonds plot the number of candidates output at each stage of the vetting, with colors matching the efficiency curve values. Note: the lack of efficiency curve data brightward of r=22 is an artifact of not implanting artificial objects brighter than this value.

## 9.4. Source brightness regression

In this section I discuss the use of a CNN[11] to quickly estimate source brightnesses from the shift'n'stack outputs discussed in the previous section. The network I discuss below can produce photometry estimates orders of magnitude more quickly than PSF fitting techniques with similar precision. The main strength of regression via CNN, however, is that the network can be trained to learn what sources of flux in a difference image stack correspond to the minor body, and what other sources in the stack are superfluous (noise, background, etc.), as long as the sources in the training set

[11] See Chapter 1 for an introduction to the CNN.

contains said superfluous sources. That is to say, through basic training, in most circumstances the CNN will automatically handle effects of contamination in the stacks, which would be extremely difficult to manage through classical PSF fitting methods.

We make use of the methods presented in Bialek et al. (2020), whereby with clever use of an appropriate likelihood function, the network can be trained to predict a brightness value as well as an estimate of the brightness uncertainty, *even when photometric uncertainty is not part of the information available to training*. The network is trained to estimate the probability density function (PDF) of a source's flux from that source's input, and the mean and standard deviation of that probability distribution are returned. As we are measuring fluxes of background limited sources, we are justified in adopting the negative log-likelihood for a Gaussian distribution as the functional representation of that probability density function, which is given by

$$-\log p_\theta\left(y|x\right) = \frac{\log \sigma_\theta^2\left(\boldsymbol{x}\right)}{2} + \frac{\left(y - \mu_\theta(\boldsymbol{x})\right)^2}{2\sigma_\theta^2(\boldsymbol{x})}. \tag{9.1}$$

Here, $y$ is the flux value of stack $\boldsymbol{x}$. The $\mu_\theta(\boldsymbol{x})$ and $\sigma_\theta$ are the network estimates of the mean and standard deviation of the flux PDF, $p_\theta\left(y|x\right)$. $\theta$ are the learned weights of the network.

To accommodate training variances, an ensemble of $M$ predictor networks is trained independently. The final PDF is simply the average of the PDFs of each network:

$$p\left(y|\boldsymbol{x}\right) = \frac{1}{M} \sum_{m=1}^{M} p_{\theta_m}\left(y|x\right), \tag{9.2}$$

and so the ensemble estimate of the flux is just the mean of flux estimates of each of the networks:

$$\mu_*(\boldsymbol{x}) = \frac{1}{M} \sum_{m=1}^{M} \mu_{\theta_m}(\boldsymbol{x}). \tag{9.3}$$

The final variance on the flux estimate is given by

$$\sigma_*^2 = \frac{1}{M} \sum_{m=1}^{M} \left(\sigma_{\theta_m}^2\left(\boldsymbol{x}\right) + \mu_{\theta_m}^2\right) - \mu_*^2(\boldsymbol{x}). \tag{9.4}$$

The above includes errors induced by training variance (aleatoric uncertainty), and variance due to the imagery themselves (epistemic uncertainty).

For this example, we use a shallow—run of the mill VGG-like CNN—with only three convolutional layers that feed two perceptron layers. A CNN as the starting point for this effort seemed obvious given the image nature of the data inputs. I have found that such a simple network is sufficient for the task at hand. As a result, no further effort was expended to search for a more complex architecture, such as a ResNet, though there are a few remaining issues we discuss below, which may be solved with a more complex architecture.

In the network I settled on, the first two convolutional layers have only 8 filters with the third only 4, and the perceptron only 16 neurons. In all cases, max pooling is used after the first CNN layer. The activation on $\sigma$ is a modified *elu* of the form

$$f(x) = \exp(x) + 10^{-16} \text{ for } x < 0 \qquad (9.5)$$
$$= x \text{ for } x \geq 0.$$

Inputs to the network were the same 43x43 pixel stacks used in the ResNet-based binary classifier I discuss above, and the known *implanted* flux values for each stack. For this example, we adopt $M = 5$.

In developing this example, I made use of the advice I give below regarding avoidance of overfitting. In particular, I first experimented to find a CNN that produced satisfactory results, and then experimented with its complexity to find a network that had a low number of trainable parameters, while maintaining satisfactory outputs, and to search for the effects of overfitting. In particular, I explored two additional networks with the same architecture as above, but with double and with half the number of convolutional filters and neurons.

The implanted sources in four nights of CLASSY data from August 1, 22, 23, and 26, 2022 were utilized to make the training set. Implanted sources detected with KBMOD were used, with their known instrumental fluxes—apparent fluxes were inappropriate, because night to night variations in the flux zeropoints would cause massive variations in the mean predictions. The training set was augmented with 1-pixel linear shifts in the four cardinal directions, mirror flips in the horizontal and vertical axes, and rotations at 90°, 180°, and 270°. After augmentation, the training sample was 95% of available sources, or 260,544 samples, and the remaining 13,696 used for test.

Training was done using the ADAM optimizer, batch size of 4096, and sample weights as the inverse of the source flux distribution were used, like with the ResNet binary classifier, though we point out that for this case, regressor performance was not appreciably altered with sample weights.

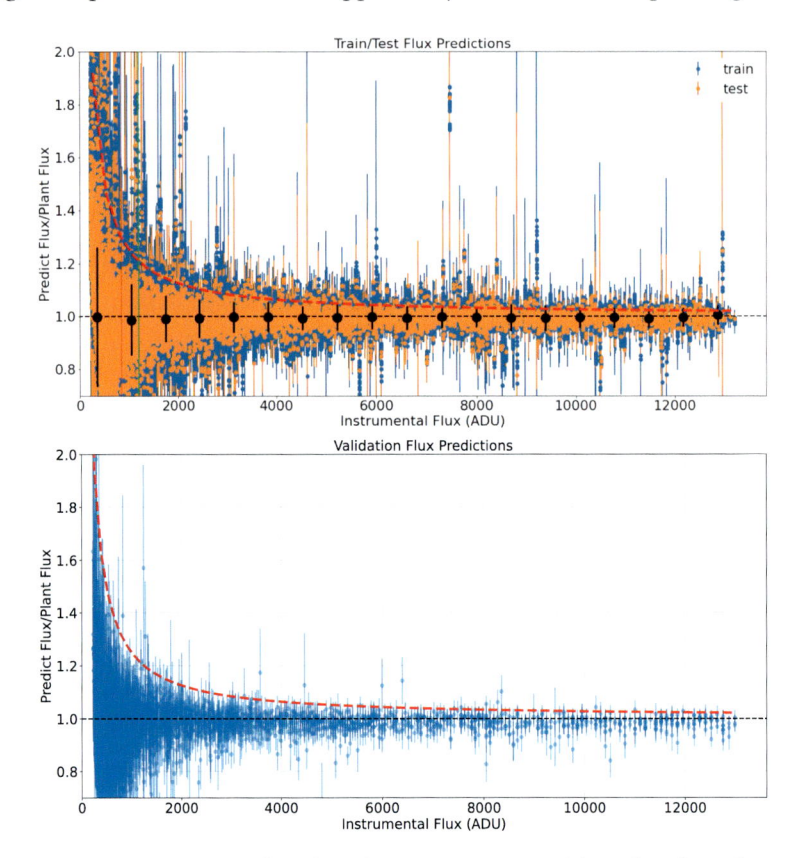

**Figure 9.11  Top:** The ratio of predicted to true instrumental flux of implanted sources as a function of planted source instrumental flux for the train (blue) and test (orange) dataset—the test sample was 5% of the available training sample. Black circles show the mean predicated/planted flux ratios of the test set in 20 bins spanning the range of implanted fluxes. The error bars on the means show the mean predicted uncertainty on the instrumental flux of the samples in each bin. The red curve shows the SNR expectation for background limited aperture photometry. **Bottom:** The ratio of predicted to true instrumental flux of implanted sources in a validation dataset, which is generated from an entirely different epoch of data than the four used to generate the training and test datasets. The red curve is the same as shown in the top panel.

In Fig. 9.11, we present a ratio of the predicted instrumental flux to true instrumental fluxes for implanted sources. We also show an estimate of the SNR that would be achieved with aperture photometry. For that, we consider a circular aperture with radius $1.2\times$ the typical full-width at half maximum for CLASSY imagery, or about 0.6", which corresponds to an aperture area of 68 pixels, and we adopt a background value of 900 ADU. As can be seen, the predictions broadly match the precision one would achieve with well-tuned aperture photometry (Fraser et al., 2016). Furthermore, the output exhibits no appreciable bias down to the level of flux, where Malmquist bias clearly sets in, showing less than 0.5% bias for test, training, and validation datasets, for all three networks.

For validation, we utilized the KBMOD outputs for a night of CLASSY data not utilized in creation of the training or test datasets (see bottom panel of Fig. 9.11). The validation data are the same as those used to generate Fig. 9.10, and show the same performance as for the test and training data, with flux predictions scaling appropriately with expectations from SNR estimates. Though the validation data show lower uncertainties in the flux predictions, measured both by the scatter in true minus predicted fluxes as well as the predicted uncertainties, $\sigma_*$. This is likely due to better observing conditions experienced during the observations used for validation compared to those used for training.

From the black points in Fig. 9.11, the predicted uncertainty $\sigma_*$ is typical of the true scatter of the predicted fluxes around their true values. Moreover, the distribution of scatter in predicted fluxes roughly follow expectations of a normal distribution; 6 of the 2,295 sources have predicted fluxes more than $3\sigma_*$ from their true values, where we would expect 7.

It can also be seen that for a very small fraction of samples, the predicted uncertainty is much higher, and the predicted flux is largely incorrect. These *outliers* were identified by comparing each source's predicted, $\sigma_*$ to those values of the 50 neighboring sources when sorted by implanted flux (not predicted flux). I found that twenty five sources had $\sigma_*$ twice the median value of the 50 neighbors. These are shown in Fig. 9.12. These outliers are flux measurements of those shift-stacks that are contaminated by other astronomical objects in the frame, or in the case of lower than correct predictions, contamination by negative subtraction residuals of nearby bright stars. It is clear however, that those sources with the most incorrect predictions are also those with typically the largest fractional uncertainties.

In a search for overfitting, I compared the outputs of the nominal model to that of the large and small models, which have 1,896, 7,098, and 896 trainable parameters for each of branch of the ensemble, respectively. For all three models, the number of outliers that were chosen as $\sigma_*$ twice the median value of the 50 neighbors, was comparable between models, with 29, 25, and 36 outliers for the small, nominal, and large models, respectively.

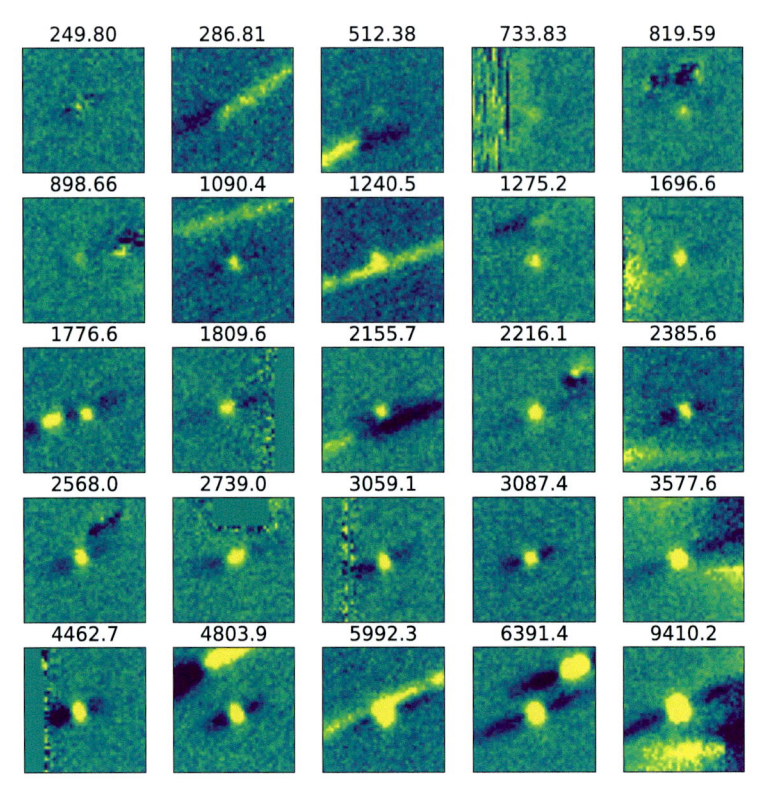

**Figure 9.12** The 25 sources with the most discrepant flux predictions, which represent about 1% of the validation dataset. Background contamination is an issue for the regressor model. The predicted flux for each source is shown above the stack. Most of these sources are relatively bright, though contamination is similarly prevalent for fainter sources, but uncertainties of those fainter sources are appropriately larger, and so the effects of contamination are relatively less influential.

The small model identified the same outliers as the nominal model, but also included four of the brightest sources, which all had underestimated flux predictions. All the brightest sources have significantly different morphologies compared to the fainter sources, with much deeper and longer negative subtraction wells compared to the significantly more numerous fainter sources. It seems the small model possesses too few parameters to maintain the capability of handling the full range of morphologies in the datasets. The large model similarly reports many bright sources as outliers in addition to the same 25 outliers reported by the nominal model. In this case, however, the large model reports appreciably smaller uncertainties for bright sources: for sources with flux $> 6,000$ adu, the median $\sigma_*$ is 0.05,

0.04, and 0.03 mags for the small, nominal, and large models, respectively. From a noise perspective, approximately 0.04 mag uncertainty is expected for this brightness range, and so it seems the largest model is overfitting for the brightest sources, which are relatively rare in the training datasets, resulting in an unreasonably small $\sigma_*$ for those rare sources.

The network I used for this flux regressor is extremely simple compared to typical CNNs seen in the astronomy literature. It is shallow, with very few filters, and as a result, possesses a very small number of trainable parameters. But with an excellent training sample, and appropriate selection of PDF, the performance of the network is admirable.

The result of being able to create a network that can predict both flux and its uncertainty may seem paradoxical given that no information about flux uncertainty was provided at training. The trick is with the use of a likelihood that is appropriate for the measurement in question—in this case a Gaussian for flux—that itself also depends on a parameter that encodes variance, in this case, $\sigma$ of the Gaussian. One can create a network that provides an estimate of the uncertainty on the regressed parameter, seemingly for free.

In summary, it seems the nominal model produces uncertainties that match expectations of photon counting, as measured by values of predicted $\sigma_*$, as well as the number of $3\sigma$ outliers in the validation dataset (once contaminated outliers are removed). This analysis demonstrates that overfitting is not a major issue for the nominal network, as might be expected; the training sample size of about $260,000$ samples (after augmentation) is more than two orders of magnitude larger than the $1,896$ trainable parameters in this simple network. The larger model shows signs of overfitting, resulting in overpredicted precision for bright sources. The comparison of the small, nominal, and large networks demonstrates the effort typically required to reveal any overfitting, which we discuss further in the next section.

This regressor is not without its weaknesses. The main outstanding issue is with management of those sources with incorrect flux predictions as a result of background contamination, which only represent approximately 1% of this validation sample. I have yet to solve this issue. I believe the underlying cause is likely a result of the contamination that largely invalidates the assumption of Gaussian PDFs for these sources. This suggests a possible recourse, whereby these outlier sources are first identified by their unreasonably large uncertainties compared to other similarly bright sources, and their fluxes remeasured with an additional network trained with a PDF likelihood that is more appropriate for contaminated imagery. Future experiments are required to determine the best form of this likelihood.

The three networks discussed above, as well as associated data-files, and an example notebook showing the training and usage of those networks is available at the GitHub repository associated with this textbook (see Section 9.7).

## 9.5. Overfitting

Here we talk about overfitting, and its avoidance. Much of the advice I give in this section comes from practical experience and not necessarily from literature examples, though we point the reader to already discussed papers of Cowan et al. (2023) and Duev et al. (2019) for some discussions on overfitting avoidance. Overfitting remains an active area of research in machine learning.

Overfitting is the circumstance in which a network seemingly returns outputs or demonstrates performance on training data that differs significantly compared to outputs from data not involved in the training. In my opinion, this is one of the most nefarious and damaging potential consequences of the use of machine learning for scientific purposes, which is, unfortunately, not often considered—not all of the papers I summarize in Sections 9.3 even mention the word "overfitting"!

Consider this particularly scary, yet unfortunately true anecdote. A network was being trained to detect moving objects in multiframe inputs of astronomical imagery, whereby any moving objects are not at the same position from frame to frame. Using implanted sources like those that I have discussed in the previous section, the network was trained to great success. During experimenting, a bug was introduced resulting in only the first frame being processed by the network, and yet after training, a magical result occurred in that the network was still able to find the moving sources in just those singular frames. With effort, it was realized that the network was trained too deeply and instead of learning to detect moving sources, it instead learned to detect subtle features unique to the artificial implanted sources that were not found in real point sources, and was not finding real moving sources at all. Such a network behavior would have been catastrophic if gone unnoticed and the network were put into production.

The above anecdote provides a good lesson for ML users both novice and expert and is a stark demonstration of the dangers of overfitting. It is easy to imagine a less fantastical result than being able to detect moving sources in single frames, which would slip past the attention of even an experienced ML user.

Overfitting itself is simply enabled by the huge complexity of many modern neural networks. Consider for example, VGG-15, with 138 million trainable parameters. Such a complex network needs a sufficiently complex training set, or otherwise during training the network could learn to identify individual samples in the training data, rather than meaningful features for the problem at hand (e.g., morphology) that distinguish different classes of source.

Guidance on when a circumstance might allow for overfitting can come from similar lessons learned in classical chi-squared curve fitting. The reduced chi-square considers the degrees of freedom $\nu$ in the problem, where $\nu = N_{samp} - N_{param}$, and $N_{samp}$ and $N_{param}$ are the number of samples used in the fit, and the number of free parameters in the model, respectively. For a meaningful curve fit, $\nu$ should be very large. In a similar way, we can approach neural network training. That is, ensure that the number of samples in a training set is significantly more numerous than the number of trainable parameters. Using $\nu$ as a metric, the ResNet binary classifier I discuss above had $N_{param} = 9.5 \times 10^5$ trainable parameters (316,354 per branch of the ensemble), while the number of samples (after augmentation) used in training was 317,205. Each sample is itself a 43x43 pixel image, and so the total $N_{samp} = 0.56 \times 10^9$, or $\sim 600\times$ larger than $N_{param}$. This is a large difference, which suggests that overfitting in this circumstance should not be an issue, though as I point out above, there are some hints of slight overfitting that has occurred, and so this comparison of $N_{samp}$ to $N_{param}$ should only be used as guidance. In the flux regressor example, the $N_{samp} = 0.5 \times 10^9$, while $N_{param} = 35,490$. With such a huge training sample compared to the trainable parameters, it is unsurprising that I have not been able to find evidence of overfitting for the regressor.

Evidence for overfitting in the ResNet is seen in Figs. 9.8 and in 9.9, with each showing slightly worse behavior for the test dataset than the training dataset. In some circumstances the validation dataset may be useful for identifying overfitting, most likely in the form of worse network performance on the validation data.

Direct tests for overfitting are difficult and circumstantial. For the binary classification problem, a robust test could be to create two *fully* independent training sets, train a network on each training set, and then cross-validate the outputs of the training sets on the networks they were not used to train. To ensure fully independent training sets, one would need to utilize two independent observational datasets, with independently implanted artificial sources, ideally using different PSF models (e.g., TRIPPy PSFs and noiseless Gaussian sources), and which used independent templates for background subtraction. Then one could be confident that the networks trained on each train set would also be independent (see Pöntinen et al., 2023, for an example of this approach). Clearly, the burden for such a check is very high. An equivalent procedure would need to be devised for the flux regression problem, as it would for any different ML application. As of yet I have not seen a robust general method for checking of overfitting in CNN training.

There are many techniques to avoid overtraining. The first and most obvious is to ensure one is using a massive and high quality training set. The creation of a high quality training set is often a large undertaking, and should not be avoided. Steps should be taken to maximize the sample size. Augmentation can greatly help, and there are many techniques that can be implemented, including translations, rotations, mirroring, noise addition, contrast changes, etc., though with the addition of each new augmentation one has to check that network performance is not degraded—a common occurrence with the addition of noise, for example. If the training set is to be artificially generated, effort must be made to ensure that the artificial positives match reality as close as is possible, or one should expect significantly degraded performance in production settings. We direct the reader to Hausen and Robertson (2020) for a short discussion on overfitting avoidance through augmentation.

Effort can be made during training to prevent overfitting. For example, one can halt training before training converges on a minimum loss value (using the keras early stopper callback, for example). This can help prevent the training from reaching a local minimum, which can produce unreasonably confident classification probabilities for the training data. Tuning the stop time is challenging however, because too early and the network performance can be degraded. Tuning early stopping remains very much an art. Furthermore, one should use dropout, which randomly selects some fraction (usually 15–25%) of trainable parameters that are not adjusted at a given training epoch. This approach can greatly help in avoiding local minima in the loss phase space, which are a main cause of overfitting. Batch normalization can also help in overfitting avoidance.

If a pretrained network is used as a starting place, then one can train only certain sections (e.g., the last fully connected layer of a VGG network). This can keep the number of trainable parameters low and reduce the chances of overfitting.

Finally, overfitting can be reduced at the network design level. One is almost always better off to use a less complex network (fewer layers, fewer filters, etc.) than a more complex one, as long as the networks show equal performance. Once one has found a general network architecture that works for an application, an iterative process of reduction in complexity and retraining should be performed to find the minimal network that still retains satisfactory performance. This is the principle technique employed by myself and by Chyba Rabeendran and Denneau (2021) to avoid overfitting.

## 9.6. Applications of machine learning in the era of big data surveys

Here I conclude with some speculative remarks about the applications of machine learning in future large-scale astronomy projects. The future prospects for the observational science of minor bodies is exciting, with many large-scale surveys on the near horizon, including Euclid and the Vera C. Rubin Legacy Survey of Space and Time.

After a successful journey to the L2 Lagrange point, Euclid has begun its visible and near-infrared (NIR) survey (Laureijs et al., 2011), which will cover nearly 15,000 square degrees of the sky in the I, Y, J, and H filters, as well as provide slitless spectroscopy of much of that region. It is expected that Euclid will observe roughly 150,000 minor bodies (Carry, 2018) to a depth unheard of in NIR surveys—Euclid should achieve a limiting magnitude of $m_{AB} \sim 24.8$ in all three NIR filters. Study of how best to detect and extract minor bodies in Euclid imagery is already on-going. Using simulated Euclid observations, Pöntinen et al. (2023) studied the utility of CNNs to identify streaked images of minor bodies in Euclid data. This fairly advanced technique uses CNNs to identify streaks in small cutouts, a recurrent neural network to stitch together streaks spanning neighboring cutouts or images, and then a gradient boost algorithm to identify start and endpoints of the merged streaks. Their procedure outperforms the non-ML version of a competing algorithm in virtually every way: a $2.6\times$ reduction in run time; doubling of purity of the outputs; and an approximately 0.5 magnitude improvement in brightness of detected streaks. This impressive work presents the sort of efforts that are required to effectively search massive datasets, such as that currently being produced by Euclid. Also see the work of Lieu et al. (2019), which we discussed earlier.

The Euclid observing strategy is essentially to acquire four 600 s dithered exposures in 4,400 seconds (Euclid Collaboration et al., 2022). Due to spacecraft pointing constraints, observation geometry is always essentially near quadrature, which naturally provides a circumstance of low apparent velocities for minor bodies in most of the Euclid pointings. For KBOs, for example, the rates of motion should be no more than a few tenths of an arcsecond per hour, resulting in total streaks of no more than a few pixels in length across the entire 4,400 s sequence. At time of writing, I am unaware of any publications discussing the detection of slow-moving— untrailed—minor bodies in Euclid imagery.

It is likely that Euclid could benefit greatly from a technique like that presented by Chyba Rabeendran and Denneau (2021) to look for KBOs and other slow-moving minor bodies. One significant improvement that seems necessary for efficient search of Euclid data is to forgo the use of cutouts, and rather make use of larger image regions, and to make use of a CNN that can manage the actual detection of moving sources within the frames. This would skip the expensive computational step of first producing a list of candidates as is currently done with ATLAS data, and may afford an even deeper search than more classical non-ML methods could provide.

With first light expected in mid-2025, over the following ten years, the Legacy Survey of Space and Time (LSST) will be executed on the Vera C. Rubin telescope, and will observe roughly 6 million minor bodies, up to 400 times each, with observations spanning six optical filters: u, g, r, i, z, Y (Ivezić et al., 2019). The image and database outputs of the LSST will be in the petabytes.

The pipeline that will be used to search for moving bodies in the LSST is largely a classical pipeline, and is developed on top of the decades of experience the ground-based community has in searching for minor bodies using non-ML-based techniques (Heinze et al., 2022). The pipeline will search for objects detectable in single visits of 2 back-to-back 15 s exposures. The anticipated search depth from this pipeline is $m_r \sim 24.0$. Due to the short 15 s exposures, all but the closest near-Earth objects (NEAs) will appear as point sources in the observations made with Rubin.

Though methods to detect trailed sources might be somewhat fruitful for NEA science, the search for minor bodies in LSST observations stand to benefit most from shift'n'stack efforts. The expected typical visit cadence a minor body will receive is two visits in two different filters in one night, spanning about 33 minutes in a night, though some small fraction (about 10 percent) of objects will receive a 3rd or even a 4th visit.[12] Objects will receive another pair of visits again on a few day timescale. Across the annual approximately 2-month observability window, an object should receive about 40 visits, roughly half of which will be in the g, r, and i filters. Of the filters offered by Rubin, these are where most minor bodies are brightest. If imagery across a lunation could be stacked, the search depth would be increased by as much as about 1.5 mags, which would increase the discovery sample by nearly a factor of 3. Even across just one week of observing, shift'n'stack would net an approximate 70% increase in the sample of discovered minor bodies. Needless to say, the potential is enormous.

Across two months, the trajectories of all minor bodies are highly nonlinear—this is true even for KBOs after as little as 4 days at near-opposition geometries. As such, nonlinear shift and stack will be required to maximize the potential search depth available to Rubin observations. The compute complexity of the nonlinear shift'n'stack scales with the time baseline of observations, $t$, as $\mathcal{O}(t^4)$ as the time baseline, determines the number of unique trajectories that need to be stacked. Even across a lunation, the nonlinear shift'n'stack is impossibly intractable without a fundamental algorithm shift. ML may hold the potential for that algorithm shift.

One possible approach to implement a search across multiple images is to develop a deep CNN to take as input multilayer consecutive images of the same field, and train it to recognize sources with brightness below the single-image noise floor. This approach seems plausible through the use of artificial source implantation within the LSST imagery themselves. One of the weaknesses of this approach though is in the fixed size of input imagery, thereby fixing the number of images that can be searched. Not all sources will experience the same observing pattern, and so multiple networks will need to be trained to manage the range of visit circumstances that occur in the dataset.

Another possible approach is through the use of techniques that capture the latent structure of the scene. These techniques include U-nets and image transformers. One can envision an approach that merges recurrent network structures with transformers, whereby a transformer could be trained to produce a weight map marking possible locations of subnoise moving objects, and a recurrent network that cycles over a temporal sequence of transformer outputs. Such a network would not be limited to

---

[12] Some details of the survey cadence are available at http://survey-strategy.lsst.io/.

an image sequence of fixed length. Regardless of the winning solution, it is inevitable that the approach to nonlinear shift'n'stack will involve engineering of advanced networks well beyond the basic styles of CNN that are commonly in use nowadays. If a method is found however, the potential scientific output for Solar System science is astronomical.

## 9.7. Code availability

The trained flux regressor models discussed above and a Jupyter notebook demonstrating their training and usage are available at https://solar-system-ml.github.io/book/chapter9/cnn_mags_production/; associated documentation available is at https://solar-system-ml.github.io/book.

## Acknowledgments

I would like to thank Sebastien Fabbro, Hossen Teimoorinia, and Preeti Cowan for their helpful suggestions regarding overcoming training issues and improvements to the neural networks I have worked with. I would also like to thank JJ Kavelaars and John Spencer for providing the encouragement needed to learn machine learning techniques in the first place. Finally, I would like to thank Sam Lawler for helping to execute the CLASSY program, the data products of which many of these results are derived.

Based in part on observations obtained with MegaPrime/MegaCam, a joint project of CFHT and CEA/DAPNIA, at the Canada-France-Hawaii telescope (CFHT), which is operated by the National Research Council (NRC) of Canada, the Institut National des Science de l'Univers of the Centre National de la Recherche Scientifique (CNRS) of France, and the University of Hawaii. The observations by way of the Canada-France-Hawaii telescope were performed with care and respect from the summit of Maunakea, which is located at a significant cultural and historic site.

This research used the facilities of the Canadian Astronomy Data Centre operated by the National Research Council of Canada with the support of the Canadian Space Agency.

## References

Alard, C., Lupton, R.H., 1998. A method for optimal image subtraction. The Astrophysical Journal 503, 325–331. https://doi.org/10.1086/305984. arXiv:astro-ph/9712287.

Ashton, E., Beaudoin, M., Gladman, B.J., 2020. The population of kilometer-scale retrograde Jovian irregular moons. Planetary Science Journal 1, 52. https://doi.org/10.3847/PSJ/abad95. arXiv:2009.03382.

Bannister, M.T., Gladman, B.J., Kavelaars, J.J., Petit, J.M., Volk, K., Chen, Y.T., Alexandersen, M., Gwyn, S.D.J., Schwamb, M.E., Ashton, E., Benecchi, S.D., Cabral, N., Dawson, R.I., Delsanti, A., Fraser, W.C., Granvik, M., Greenstreet, S., Guilbert-Lepoutre, A., Ip, W.H., Jakubik, M., Jones, R.L., Kaib, N.A., Lacerda, P., Van Laerhoven, C., Lawler, S., Lehner, M.J., Lin, H.W., Lykawka, P.S., Marsset, M., Murray-Clay, R., Pike,

R.E., Rousselot, P., Shankman, C., Thirouin, A., Vernazza, P., Wang, S.Y., 2018. OS-SOS. VII. 800+ trans-Neptunian objects—the complete data release. The Astrophysical Journal. Supplement Series 236, 18. https://doi.org/10.3847/1538-4365/aab77a. arXiv:1805.11740.

Bernstein, G., Khushalani, B., 2000. Orbit fitting and uncertainties for Kuiper Belt objects. Astronomical Journal 120, 3323–3332. https://doi.org/10.1086/316868. arXiv:astro-ph/0008348.

Bernstein, G.M., Trilling, D.E., Allen, R.L., Brown, M.E., Holman, M., Malhotra, R., 2004. The size distribution of trans-Neptunian bodies. Astronomical Journal 128, 1364–1390. https://doi.org/10.1086/422919. arXiv:astro-ph/0308467.

Bialek, S., Fabbro, S., Venn, K.A., Kumar, N., O'Briain, T., Yi, K.M., 2020. Assessing the performance of LTE and NLTE synthetic stellar spectra in a machine learning framework. Monthly Notices of the Royal Astronomical Society 498, 3817–3834. https://doi.org/10.1093/mnras/staa2582. arXiv:1911.02602.

Burdanov, A.Y., Hasler, S.N., de Wit, J., 2023. GPU-based framework for detecting small Solar System bodies in targeted exoplanet surveys. Monthly Notices of the Royal Astronomical Society 521, 4568–4578. https://doi.org/10.1093/mnras/stad808. arXiv:2303.07293.

Carry, B., 2018. Solar system science with ESA Euclid. Astronomy & Astrophysics 609, A113. https://doi.org/10.1051/0004-6361/201730386. arXiv:1711.01342.

Chambers, K.C., Magnier, E.A., Metcalfe, N., Flewelling, H.A., Huber, M.E., Waters, C.Z., Denneau, L., Draper, P.W., Farrow, D., Finkbeiner, D.P., Holmberg, C., Koppenhoefer, J., Price, P.A., Rest, A., Saglia, R.P., Schlafly, E.F., Smartt, S.J., Sweeney, W., Wainscoat, R.J., Burgett, W.S., Chastel, S., Grav, T., Heasley, J.N., Hodapp, K.W., Jedicke, R., Kaiser, N., Kudritzki, R.P., Luppino, G.A., Lupton, R.H., Monet, D.G., Morgan, J.S., Onaka, P.M., Shiao, B., Stubbs, C.W., Tonry, J.L., White, R., Bañados, E., Bell, E.F., Bender, R., Bernard, E.J., Boegner, M., Boffi, F., Botticella, M.T., Calamida, A., Casertano, S., Chen, W.P., Chen, X., Cole, S., Deacon, N., Frenk, C., Fitzsimmons, A., Gezari, S., Gibbs, V., Goessl, C., Goggia, T., Gourgue, R., Goldman, B., Grant, P., Grebel, E.K., Hambly, N.C., Hasinger, G., Heavens, A.F., Heckman, T.M., Henderson, R., Henning, T., Holman, M., Hopp, U., Ip, W.H., Isani, S., Jackson, M., Keyes, C.D., Koekemoer, A.M., Kotak, R., Le, D., Liska, D., Long, K.S., Lucey, J.R., Liu, M., Martin, N.F., Masci, G., McLean, B., Mindel, E., Misra, P., Morganson, E., Murphy, D.N.A., Obaika, A., Narayan, G., Nieto-Santisteban, M.A., Norberg, P., Peacock, J.A., Pier, E.A., Postman, M., Primak, N., Rae, C., Rai, A., Riess, A., Riffeser, A., Rix, H.W., Röser, S., Russel, R., Rutz, L., Schilbach, E., Schultz, A.S.B., Scolnic, D., Strolger, L., Szalay, A., Seitz, S., Small, E., Smith, K.W., Soderblom, D.R., Taylor, P., Thomson, R., Taylor, A.N., Thakar, A.R., Thiel, J., Thilker, D., Unger, D., Urata, Y., Valenti, J., Wagner, J., Walder, T., Walter, F., Watters, S.P., Werner, S., Wood-Vasey, W.M., Wyse, R., 2016. The Pan-STARRS1 surveys. arXiv e-prints, arXiv:1612.05560.

Chyba Rabeendran, A., Denneau, L., 2021. A two-stage deep learning detection classifier for the ATLAS asteroid survey. Publications of the Astronomical Society of the Pacific 133, 034501. https://doi.org/10.1088/1538-3873/abc900. arXiv:2101.08912.

Cowan, P., Bond, I.A., Reyes, N.H., 2023. Towards asteroid detection in microlensing surveys with deep learning. Astronomy and Computing 42, 100693. https://doi.org/10.1016/j.ascom.2023.100693. arXiv:2211.02239.

Denneau, L., Jedicke, R., Grav, T., Granvik, M., Kubica, J., Milani, A., Vereš, P., Wainscoat, R., Chang, D., Pierfederici, F., Kaiser, N., Chambers, K.C., Heasley, J.N., Magnier, E.A., Price, P.A., Myers, J., Kleyna, J., Hsieh, H., Farnocchia, D., Waters, C., Sweeney, W.H., Green, D., Bolin, B., Burgett, W.S., Morgan, J.S., Tonry, J.L., Hodapp, K.W., Chastel, S., Chesley, S., Fitzsimmons, A., Holman, M., Spahr, T., Tholen, D., Williams, G.V., Abe, S., Armstrong, J.D., Bressi, T.H., Holmes, R., Lister, T., McMillan, R.S.,

Micheli, M., Ryan, E.V., Ryan, W.H., Scotti, J.V., 2013. The Pan-STARRS Moving Object Processing System. Publications of the Astronomical Society of the Pacific 125, 357. https://doi.org/10.1086/670337. arXiv:1302.7281.

Duev, D.A., Bolin, B.T., Graham, M.J., Kelley, M.S.P., Mahabal, A., Bellm, E.C., Coughlin, M.W., Dekany, R., Helou, G., Kulkarni, S.R., Masci, F.J., Prince, T.A., Riddle, R., Soumagnac, M.T., van der Walt, S.J., 2021. Tails: chasing comets with the Zwicky Transient Facility and deep learning. Astronomical Journal 161, 218. https://doi.org/10.3847/1538-3881/abea7b. arXiv:2102.13352.

Duev, D.A., Mahabal, A., Ye, Q., Tirumala, K., Belicki, J., Dekany, R., Frederick, S., Graham, M.J., Laher, R.R., Masci, F.J., Prince, T.A., Riddle, R., Rosnet, P., Soumagnac, M.T., 2019. DeepStreaks: identifying fast-moving objects in the Zwicky Transient Facility data with deep learning. Monthly Notices of the Royal Astronomical Society 486, 4158–4165. https://doi.org/10.1093/mnras/stz1096. arXiv:1904.05920.

Eduardo, M., Morgan, A., Fraser, W., Trilling, D., Stansberry, J., Bernstein, G., Hilbert, B., Holman, M., Roman, T., 2023. A pencil beam approach to search for ultra-faint transneptunian objects using jwst. In: Asteroids, Comets, Meteors Conference. Lunar and Planetary Institute, Houston. p. Abstract #2514. https://www.hou.usra.edu/meetings/acm2023/pdf/2514.pdf.

Euclid Collaboration, Scaramella, R., Amiaux, J., Mellier, Y., Burigana, C., Carvalho, C.S., Cuillandre, J.C., Da Silva, A., Derosa, A., Dinis, J., et al., 2022. Euclid preparation. I. The Euclid Wide Survey. Astronomy & Astrophysics 662, A112. https://doi.org/10.1051/0004-6361/202141938. arXiv:2108.01201.

Fraser, W., Alexandersen, M., Schwamb, M.E., Marsset, M., Pike, R.E., Kavelaars, J.J., Bannister, M.T., Benecchi, S., Delsanti, A., 2016. TRIPPy: trailed image photometry in Python. Astronomical Journal 151, 158. https://doi.org/10.3847/0004-6256/151/6/158. arXiv:1604.00031.

Fraser, W.C., Kavelaars, J.J., Holman, M.J., Pritchet, C.J., Gladman, B.J., Grav, T., Jones, R.L., MacWilliams, J., Petit, J.M., 2008. The Kuiper Belt luminosity function from $m_R$=21 to 26. Icarus 195, 827–843. https://doi.org/10.1016/j.icarus.2008.01.014. arXiv:0802.2285.

Fraser, W.C., Lawler, S., Pike, R.E., Kavelaars, J., Ashton, E., Gwyn, S., Chen, Y.T., Huang, Y., Gladman, B., Petit, J.M., Semenchuck, C., Peltier, L., Alexandersen, M., Noyelles, B., Hestoffer, D., Chang, C.K., Connolly, A., Kalmbach, J.B., Wang, S.Y., Eduardo, M., Juric, M., Van Laerhoven, C., Bannister, M., Cowan, P., Tan, N., Volk, K., 2023a. The classical and large — a solar system. In: Asteroids, Comets, Meteors Conference. Lunar and Planetary Institute, Houston. p. Abstract #2346. https://www.hou.usra.edu/meetings/acm2023/pdf/2346.pdf.

Fraser, W.C., Porter, S.B., Lin, H.W., Napier, K., Spencer, R.J., Kavelaars, J., Verbiscer, A.J., Yoshida, F., Terai, T., Ito, T., Gerdes, D., Benecchi, S.D., Stern, S.A., Gwyn, S., Buie, M.W., Peltier, L., Singer, K.N., Brandy, P.C., Team, N.H.L., Team, N.H.G.S., 2023b. Approaches to detecting Kuiper Belt objects for NASA's new horizons extended mission: digging into the noise. In: 54th Lunar and Planetary Science Conference. Lunar and Planetary Institute, Houston. p. Abstract #2361. https://www.hou.usra.edu/meetings/lpsc2023/pdf/2361.pdf.

Fuentes, C.I., George, M.R., Holman, M.J., 2009. A Subaru pencil-beam search for $m_R$ ~27 trans-Neptunian bodies. The Astrophysical Journal 696, 91–95. https://doi.org/10.1088/0004-637X/696/1/91. arXiv:0809.4166.

Fuentes, C.I., Holman, M.J., Trilling, D.E., Protopapas, P., 2010. Trans-Neptunian objects with Hubble Space Telescope ACS/WFC. The Astrophysical Journal 722, 1290–1302. https://doi.org/10.1088/0004-637X/722/2/1290. arXiv:1008.2209.

Gladman, B., Kavelaars, J.J., Petit, J.M., Morbidelli, A., Holman, M.J., Loredo, T., 2001. The structure of the Kuiper Belt: size distribution and radial extent. Astronomical Journal 122, 1051–1066. https://doi.org/10.1086/322080.

Hausen, R., Robertson, B.E., 2020. Morpheus: a deep learning framework for the pixel-level analysis of astronomical image data. The Astrophysical Journal. Supplement Series 248, 20. https://doi.org/10.3847/1538-4365/ab8868. arXiv:1906.11248.

Heinze, A., Eggl, S., Juric, M., Moeyens, J., Jones, L., Sullivan, I., Bellm, E., 2022. Heliolinc3D: enabling asteroid discovery for the Legacy Survey of Space and Time (LSST). In: AAS/Division for Planetary Sciences Meeting Abstracts, p. 504.04.

Heinze, A.N., Metchev, S., Trollo, J., 2015. Digital tracking observations can discover asteroids 10 times fainter than conventional searches. Astronomical Journal 150, 125. https://doi.org/10.1088/0004-6256/150/4/125. arXiv:1508.01599.

Holman, M.J., Kavelaars, J.J., Grav, T., Gladman, B.J., Fraser, W.C., Milisavljevic, D., Nicholson, P.D., Burns, J.A., Carruba, V., Petit, J.M., Rousselot, P., Mousis, O., Marsden, B.G., Jacobson, R.A., 2004. Discovery of five irregular moons of Neptune. Nature 430, 865–867. https://doi.org/10.1038/nature02832.

Holman, M.J., Payne, M.J., Blankley, P., Janssen, R., Kuindersma, S., 2018. HelioLinC: a novel approach to the minor planet linking problem. Astronomical Journal 156, 135. https://doi.org/10.3847/1538-3881/aad69a.

Inutsuka, Shu-ichiro, Aikawa, Yuri, Muto, Takayuki, Tomida, Kengo, Tamura, Motohide (Eds.), 2023. Protostars and Planets VII. Astronomical Society of the Pacific Conference Series, vol. 534.

Ivezić, Ž., Kahn, S.M., Tyson, J.A., Abel, B., Acosta, E., Allsman, R., Alonso, D., AlSayyad, Y., Anderson, S.F., Andrew, J., et al., 2019. LSST: from science drivers to reference design and anticipated data products. The Astrophysical Journal 873, 111. https://doi.org/10.3847/1538-4357/ab042c. arXiv:0805.2366.

Jeffries, C., Acuña, R., 2023. Detection of streaks in astronomical images using machine learning. Journal of Artificial Intelligence and Technology 4, 1–8. https://doi.org/10.37965/jait.2023.0413. https://ojs.istp-press.com/jait/article/view/413.

Kavelaars, J.J., Holman, M.J., Grav, T., Milisavljevic, D., Fraser, W., Gladman, B.J., Petit, J.M., Rousselot, P., Mousis, O., Nicholson, P.D., 2004. The discovery of faint irregular satellites of Uranus. Icarus 169, 474–481. https://doi.org/10.1016/j.icarus.2004.01.009.

Kruk, S., García Martín, P., Popescu, M., Merín, B., Mahlke, M., Carry, B., Thomson, R., Karadağ, S., Durán, J., Racero, E., Giordano, F., Baines, D., de Marchi, G., Laureijs, R., 2022. Hubble Asteroid Hunter. I. Identifying asteroid trails in Hubble Space Telescope images. Astronomy & Astrophysics 661, A85. https://doi.org/10.1051/0004-6361/202142998. arXiv:2202.00246.

Kubica, J., Denneau, L., Grav, T., Heasley, J., Jedicke, R., Masiero, J., Milani, A., Moore, A., Tholen, D., Wainscoat, R.J., 2007. Efficient intra- and inter-night linking of asteroid detections using kd-trees. Icarus 189, 151–168. https://doi.org/10.1016/j.icarus.2007.01.008. arXiv:astro-ph/0703475.

Laureijs, R., Amiaux, J., Arduini, S., Auguères, J.L., Brinchmann, J., Cole, R., Cropper, M., Dabin, C., Duvet, L., Ealet, A., et al., 2011. Euclid definition study report. arXiv e-prints, arXiv:1110.3193.

Lieu, M., Conversi, L., Altieri, B., Carry, B., 2019. Detecting Solar system objects with convolutional neural networks. Monthly Notices of the Royal Astronomical Society 485, 5831–5842. https://doi.org/10.1093/mnras/stz761. arXiv:1807.10912.

Luu, J.X., Jewitt, D.C., 1998. Deep imaging of the Kuiper Belt with the Keck 10 meter telescope. The Astrophysical Journal Letters 502, L91–L94. https://doi.org/10.1086/311490.

Marinelli, M., Dressel, L., 2024. WFC3 instrument handbook for cycle 32 v. 16.0. In: WFC3 Instrument Handbook for Cycle 32 v. 16, vol. 16, p. 16.

Michel, P., DeMeo, F.E., Bottke, W.F., 2015. Asteroids iv. University of Arizona Press.

Moffat, A.F.J., 1969. A theoretical investigation of focal stellar images in the photographic emulsion and application to photographic photometry. Astronomy & Astrophysics 3, 455.

Morgan, A., Eduardo, M., Trilling, D., Stansberry, J., Fraser, W., Bernstein, G., Hilbert, B., Holman, M., Grundy, W., Fuentes, C., Tegler, S., 2023. The search for faint tnos: producing synthetic objects for jwst and hst data. In: Asteroids, Comets, Meteors Conference. Lunar and Planetary Institute, Houston. p. Abstract #2504. https://www.hou.usra.edu/meetings/acm2023/pdf/2504.pdf.

Napier, K.J., Lin, H.W., Gerdes, D.W., Adams, F.C., Simpson, A.M., Porter, M.W., Weber, K.G., Markwardt, L., Gowman, G., Smotherman, H., Bernardinelli, P.H., Jurić, M., Connolly, A.J., Bryce Kalmbach, J., Portillo, S.K.N., Trilling, D.E., Strauss, R., Oldroyd, W.J., Trujillo, C.A., Chandler, C.O., Holman, M.J., Schlichting, H.E., McNeill, A., the DEEP Collaboration, 2023. The DECam Ecliptic Exploration Project (DEEP): V. The absolute magnitude distribution of the cold classical Kuiper Belt. arXiv e-prints, arXiv:2309.09478.

Pöntinen, M., Granvik, M., Nucita, A.A., Conversi, L., Altieri, B., Carry, B., O'Riordan, C.M., Scott, D., Aghanim, N., Amara, A., Amendola, L., Auricchio, N., Baldi, M., Bonino, D., Branchini, E., Brescia, M., Camera, S., Capobianco, V., Carbone, C., Carretero, J., Castellano, M., Cavuoti, S., Cimatti, A., Cledassou, R., Congedo, G., Copin, Y., Corcione, L., Courbin, F., Cropper, M., Da Silva, A., Degaudenzi, H., Dinis, J., Dubath, F., Dupac, X., Dusini, S., Farrens, S., Ferriol, S., Frailis, M., Franceschi, E., Fumana, M., Galeotta, S., Garilli, B., Gillard, W., Gillis, B., Giocoli, C., Grazian, A., Haugan, S.V.H., Holmes, W., Hormuth, F., Hornstrup, A., Jahnke, K., Kümmel, M., Kermiche, S., Kiessling, A., Kitching, T., Kohley, R., Kunz, M., Kurki-Suonio, H., Ligori, S., Lilje, P.B., Lloro, I., Maiorano, E., Mansutti, O., Marggraf, O., Markovic, K., Marulli, F., Massey, R., Medinaceli, E., Mei, S., Melchior, M., Mellier, Y., Meneghetti, M., Meylan, G., Moresco, M., Moscardini, L., Munari, E., Niemi, S.M., Nutma, T., Padilla, C., Paltani, S., Pasian, F., Pedersen, K., Pettorino, V., Pires, S., Polenta, G., Poncet, M., Raison, F., Renzi, A., Rhodes, J., Riccio, G., Romelli, E., Roncarelli, M., Rossetti, E., Saglia, R., Sapone, D., Sartoris, B., Schneider, P., Secroun, A., Seidel, G., Serrano, S., Sirignano, C., Sirri, G., Stanco, L., Tallada-Crespí, P., Taylor, A.N., Tereno, I., Toledo-Moreo, R., Torradeflot, F., Tutusaus, I., Valenziano, L., Vassallo, T., Verdoes Kleijn, G., Wang, Y., Weller, J., Zamorani, G., Zoubian, J., Scottez, V., 2023. Euclid: Identification of asteroid streaks in simulated images using deep learning. Astronomy & Astrophysics 679, A135. https://doi.org/10.1051/0004-6361/202347551. arXiv:2310.03845.

Prialnik, D., Barucci, M.A., Young, L., 2019. The Trans-Neptunian Solar System. Elsevier.

Ronneberger, O., Fischer, P., Brox, T., 2015. U-net: convolutional networks for biomedical image segmentation. In: Medical Image Computing and Computer-Assisted Intervention (MICCAI). Springer, pp. 234–241. http://lmb.informatik.uni-freiburg.de/Publications/2015/RFB15a. available on arXiv:1505.04597 [cs.CV].

Ryon, J.E., Stark, D.V., 2023. ACS instrument handbook for cycle 32 v. 23.0. In: ACS Instrument Handbook for Cycle 32 v. 23.0, vol. 23, p. 23.

Smotherman, H., Connolly, A.J., Kalmbach, J.B., Portillo, S.K.N., Bektesevic, D., Eggl, S., Juric, M., Moeyens, J., Whidden, P.J., 2021. Sifting through the static: moving object detection in difference images. Astronomical Journal 162, 245. https://doi.org/10.3847/1538-3881/ac22ff. arXiv:2109.03296.

Tan, M., Pang, R., Le, Q.V., 2020. Efficientdet: scalable and efficient object detection. arXiv:1911.09070.

Tonry, J.L., Denneau, L., Heinze, A.N., Stalder, B., Smith, K.W., Smartt, S.J., Stubbs, C.W., Weiland, H.J., Rest, A., 2018. ATLAS: a high-cadence all-sky survey system. Publications of the Astronomical Society of the Pacific 130, 064505. https://doi.org/10.1088/1538-3873/aabadf. arXiv:1802.00879.

Trilling, D.E., Gowanlock, M., Kramer, D., McNeill, A., Donnelly, B., Butler, N., Ke-cecioglu, J., 2023. The Solar System Notification Alert Processing System (SNAPS): design, architecture, and first data release (SNAPShot1). Astronomical Journal 165, 111. https://doi.org/10.3847/1538-3881/acac7f. arXiv:2302.01239.

Trujillo, C.A., Fuentes, C., Gerdes, D.W., Markwardt, L., Sheppard, S.S., Strauss, R., Chandler, C.O., Oldroyd, W.J., Trilling, D.E., Lin, H.W., Adams, F.C., Bernardinelli, P.H., Holman, M.J., Juric, M., McNeill, A., Mommert, M., Napier, K.J., Payne, M.J., Ragozzine, D., Rivkin, A.S., Schlichting, H., Smotherman, H., 2023. The DECam Ecliptic Exploration Project (DEEP) II. Observational strategy and design. arXiv e-prints, arXiv:2310.19864.

Varela, L., Boucheron, L., Malone, N., Spurlock, N., 2019. Streak detection in wide field of view images using Convolutional Neural Networks (CNNs). In: Ryan, S. (Ed.), Advanced Maui Optical and Space Surveillance Technologies Conference, p. 89.

Wang, F., Ge, J., Willis, K., 2022. Discovering faint and high apparent motion rate near-Earth asteroids using a deep learning program. Monthly Notices of the Royal Astronomical Society 516, 5785–5798. https://doi.org/10.1093/mnras/stac2347. arXiv:2208.09098.

Whidden, P.J., Bryce Kalmbach, J., Connolly, A.J., Jones, R.L., Smotherman, H., Bektesevic, D., Slater, C., Becker, A.C., Ivezić, Ž., Juric, M., Bolin, B., Moeyens, J., Förster, F., Golkhou, V.Z., 2019. Fast algorithms for slow moving asteroids: constraints on the distribution of Kuiper Belt objects. Astronomical Journal 157, 119. https://doi.org/10.3847/1538-3881/aafd2d. arXiv:1901.02492.

Zoph, B., Le, Q.V., 2017. Neural architecture search with reinforcement learning. arXiv:1611.01578.

CHAPTER TEN

# Chaotic dynamics

**Gabriel Caritá[a], Abreuçon Atanasio Alves[b], and Valerio Carruba[b]**

[a]National Institute for Space and Research (INPE), Division of Graduate Studies, São José dos Campos, SP, Brazil
[b]São Paulo State University (UNESP), Department of Mathematics, Guaratinguetá, SP, Brazil

## 10.1. Introduction

In this chapter, we delve into the exploration of chaotic dynamics within various types of dynamical systems, with a specialized emphasis on its application to chaos in astronomy. The study of chaotic dynamics is inherently captivating and intricate, revolving around the examination of systems that exhibit sensitivity to initial conditions, often demonstrating seemingly random and unpredictable behaviors arising from deterministic rules. This field holds broad implications spanning across physics, engineering, biology, and economics, underlining its relevance in predicting and controlling complex systems and advancing our comprehension of the natural world. Within this context, machine learning (ML) emerges as a formidable tool for unraveling the intricacies of chaos in dynamical systems, capable of discerning patterns within complex, high-dimensional data characteristic of chaotic systems.

Machine learning (ML) has emerged as a powerful tool for studying chaos in dynamical systems. ML techniques can be used to extract patterns and structures from complex and high-dimensional data, which are often characteristic of chaotic systems. In this way, ML can help to identify key features of chaotic systems and provide insights into their behavior. For example, long short-term memory (LSTM) neural networks (NN) can be trained on time series data to predict future trajectories of chaotic systems. LSTM NN are particularly well-suited to this task as they can capture the temporal dependencies in the data, which is important for predicting future behavior (Vlachas et al., 2018).

To further enhance our understanding of chaotic dynamics, we introduce the autocorrelation function (ACF), a statistical method that plays a pivotal role in characterizing unregular systems. Unlike conventional methods, the ACF allows us to stress the dynamics trends in time series analyses, serving as a valuable index for unraveling frequency trends within dynamical systems. This method holds particular relevance as a technique for

*Machine Learning for Small Bodies in the Solar System*
https://doi.org/10.1016/B978-0-44-324770-5.00015-5

pre-processing data, paving the way for integration with ML techniques applicable in astronomy.

In the subsequent sections, we explore the ACF method's potential applications, emphasizing its utility in developing nonlinear dynamics methods and applications. By establishing connections between chaos, ML, and the ACF method, we seek to advance our capabilities in analyzing and understanding complex systems, particularly within the realm of astronomy.

## 10.2. A brief introduction to chaos in dynamical systems

In this section, we are briefly introducing the readers to dynamical systems and chaotic dynamics. This topic is vast and cannot be entirely explained within a few pages. Therefore we encourage readers to search for more specific information in books dedicated to dynamical systems and chaos, for example, Strogatz (2018) and Hilborn (2000). We understand that this is compelling and important to grasp the further parts of this chapter; we are covering the basic information necessary to understand machine learning tools and their applications in chaotic behavior in astronomy.

Dynamical systems have proven over the years to be areas of great relevance in the development of new research and applications in today's world, being applied in different areas, such as hydrodynamics or quantum mechanics. General examples include complex atmospheric systems modeled by modified Lorenz attractors or for galactic dynamics by Henon–Heiles. A very famous system is the circular restricted three-body problem (CR3BP) used in planetary, stellar, and spacecraft dynamics. These models can be analyzed through different techniques and methods depending on their level of complexity or degrees of freedom. We highlight a few examples such as surfaces of sections, stability analyses, and chaotic indicators. To understand the evolution of a dynamical system, typically involving a nonlinear system in the realistic world, it is necessary to employ numerical methods and techniques. As an example in this book, we describe the double pendulum, a classical problem. It is important to note that pendulums are quite important for physics and astronomy because, in general, they can be related through phase space structure, being both harmonic oscillators (Murray and Dermott, 1999; Brunton and Kutz, 2019; Kaheman et al., 2023).

The double pendulum is a nonlinear dynamical system, which we use here to describe and understand the general behavior of such systems. In this book, we don't delve into the intricate origins of the double pendulum equations of motion, because our focus is on machine learning applications. We aim to provide a basic understanding of how the system is described without going into too much detail. For those interested in a more comprehensive example, we recommend checking (Calvão and Penna, 2015).

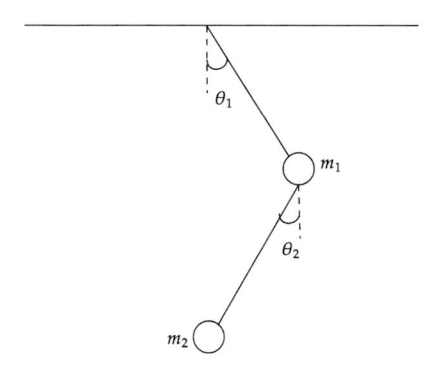

**Figure 10.1** Illustration of a double pendulum.

Following Fig. 10.1, we can easily derive the Lagrangian defined by Eqs. (10.1) and (10.2),

$$\mathcal{L} = T - V, \tag{10.1}$$

where $T$ is the kinetic energy and $V$ is the potential,

$$\mathcal{L} = \frac{1}{2}(l_1^2(m_1 + m_2)\dot{\theta}_1^2 + l_2^2 m_2 \dot{\theta}_2^2 + 2l_1 l_2 m_2 \cos(\theta_1 - \theta_2)\dot{\theta}_1 \dot{\theta}_2) \\ + gl_1(m_1 + m_2)\cos\theta_1 + gl_2 m_2 \cos\theta_2, \tag{10.2}$$

where, $m_1$ and $m_2$ are the respective masses of the primary and secondary pendulum. The lengths of the pendulums are defined by $l_1$ and $l_2$. The angles $\theta_1$ and $\theta_2$ are the respective angles in respect with the horizontal line. Finally, $g$ is the gravitational constant. The equations of motion can be derived from Euler–Lagrange equations.

The sensitivity to initial conditions, often referred to as the "butterfly effect," is a fundamental concept in the study of chaotic dynamical systems. Chaos theory explores the behavior of systems that are highly sensitive to their initial conditions. Let us now assume that our state vector is defined by $X = [\theta_1, \omega_1, \theta_2, \omega_2]^T$, where the angles and velocities corresponding to the positions of the first and second pendulum are represented by the labels $\theta_k$ and $\omega_k$, with $k = 1$ for the first pendulum and $k = 2$ for the second pendulum.

In Figs. 10.2 and 10.3, two orbits with slight differences in initial conditions are presented. These orbits are depicted through their angles and positions as functions of time. In Fig. 10.2, the blue orbit has an initial condition of $X_0 = [\frac{\pi}{2}, 0, \frac{\pi}{2}, 0]^T$, while the orange one has an initial condition similar to $X_0$, with the addition of a small $\delta = 10^{-9}$, making $\bar{X}_0 = X_0 + \delta$. Both orbits were numerically integrated for 150 seconds. It is evident that in this case, the orbits evolve quite differently after about 40 seconds, despite the small difference in initial conditions. This exemplifies the expected sensitivity to initial conditions, indicating that this scenario exhibits chaotic behavior.

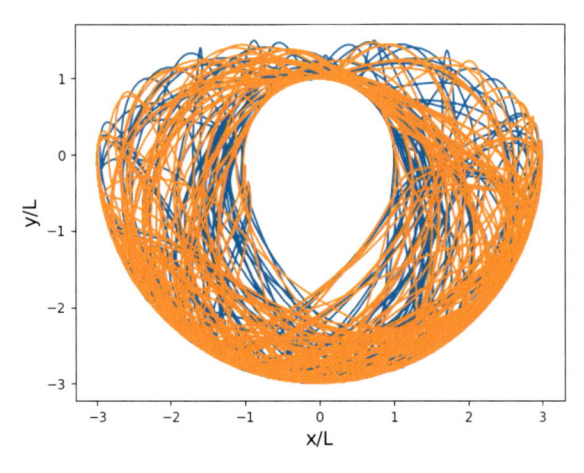

**Figure 10.2** Length normalized planar trajectory of the secondary bulb for the initials conditions $X_0$ in blue and $\bar{X}_0$ in orange.

As another example, we can utilize a complex harmonic oscillator, the double pendulum. In astronomy, the general behavior of celestial bodies may be associated with harmonic oscillators. Delving into the dynamics of a not-so-abstract problem, such as the double pendulum, can be useful to understand the general dynamics of an astronomical system.

A dynamical system in astronomy could encompass various scenarios, such as galaxy dynamics using Hamiltonian mechanics or n-body problems. It could also involve asteroid/comet dynamics, including rotational states, or even an n-body problem to understand the dynamics of exoplanets or asteroid families. Given that these systems are complex and the equations governing their dynamics introduce nonlinearity, the analysis of these problems can be approached using a set of methods and techniques from the theory of dynamical systems. However, as the problems become

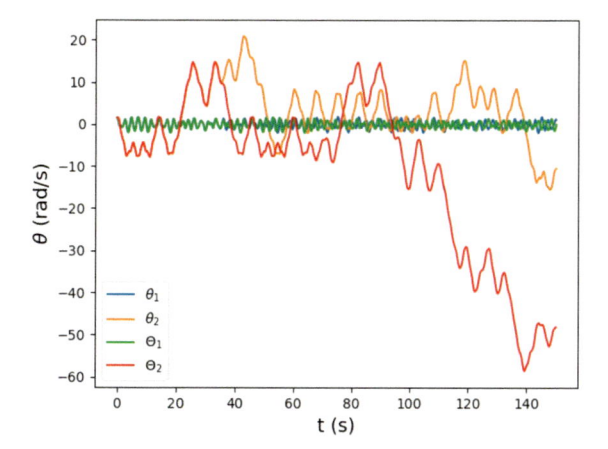

**Figure 10.3** Time evolution of the angles for the trajectories of the initials conditions $X_0$ for $\theta_i$ and $\bar{X}_0$ for $\Theta_i$, assuming $i = 1, 2$.

more complex, classical techniques may no longer suffice. In such cases, we may need to develop or employ nonclassical techniques. Thus we aim to inform and apply statistical and machine learning methods to help describing these intricate systems.

### 10.2.1 Chaos in astronomy

Chaotic effects are phenomena associated with nonlinear dynamics, and most real systems are governed by nonlinear equations, making chaos a potential outcome. For example, in the Solar System, the chaotic motion of small bodies, such as asteroids, can be caused by close encounters, collisions, or even the overlap of resonant effects, whether they are of mean or secular motion. In few words, chaos can be identified through techniques that measure the rate of perturbed and nonperturbed trajectory separation. Similarly, it can also be discerned through the temporal analysis of the number of frequencies present in a given orbit.

For example, Hyperion, one of Saturn's enigmatic moons, stands out in our Solar System due to its intriguing chaotic behavior. What sets Hyperion apart from many other celestial bodies is its irregular shape, resembling a battered and elongated potato. This peculiar shape is a significant contributor to its chaotic orbit and rotation (Wisdom et al., 1984; Wisdom, 1987; Furi et al., 2005; Binzel et al., 1986). As it travels along its elliptical path around Saturn, Hyperion experiences substantial gravitational perturbations not only from the planet itself but also from other nearby moons, particularly the massive Titan. These gravitational interactions, combined with its irregular shape, create a complex dance of forces, which make predicting Hyperion's position and behavior over extended periods a formidable challenge for astronomers. Moreover, the strong tidal forces exerted by Saturn further contribute to the Moon's chaotic dynamics. Hyperion's chaotic rotation is equally captivating. Unlike the orderly, synchronous rotations of many moons and planets, Hyperion tumbles unpredictably as it orbits Saturn. So the chaotic rotational behavior is a result of the irregular shape and variable gravitational forces acting upon it. Essentially, Hyperion doesn't have a stable axis of rotation; instead, it wobbles and tumbles through space, making it a unique and intriguing object of study in our Solar System (Wisdom et al., 1984; Wisdom, 1987; Furi et al., 2005). Its chaotic nature serves as a reminder that even in the seemingly predictable realms of space, celestial objects can exhibit mesmerizing complexity and defy simple models, prompting scientists to delve deeper into the mysteries of our cosmic neighborhood.

Another example for chaos in the Solar System are the asteroids affected by secular effects. Secular effects are in general related with long-term changes in a time series; in astronomy, secular disturbances could result from gravitational interaction among celestial bodies. In celestial mechanics, the gravitational interactions between celestial bodies can lead to secular resonances (see Chapter 2). These resonances can affect the long-term stability and behavior of planetary and asteroid orbits. When the rates of change of certain orbital elements become commensurate, a secular resonance can occur, influencing the evolution of the orbits over extended periods.

Historically, secular resonances have been studied since the 19th century, starting with Lidov and Kozai (Lidov, 1962; Kozai, 1962). As highlighted by Knežević et al. (1991), a modern approach to secular resonances was initiated by Williams (1969). One example of a secular resonance is the Lidov–Kozai resonance. In this resonance, there is a coupling between the inclination and eccentricity of an orbit. The gravitational perturbations from a third body, such as another planet or the Sun, can lead to the periodic exchange of angular momentum between the inclination and eccentricity of the orbit. This results in significant variations in the orbit's shape and orientation over long periods. Secular resonances are important in understanding the dynamics of planetary systems, especially in situations where multiple bodies interact. They can play a crucial role in shaping the architecture and stability of planetary orbits over astronomical timescales. Sometimes frequencies can overlap with other frequencies resulting in chaos. One example of secular variations and chaos is the perturbation caused by the moons in the Saturnian system (Ferraz-Mello and Dvorak, 1987). In this study, the authors pointed out secular variation and the rise of chaotic motion in a presence of a noncircular Dione for a satellite in 2/1 resonance with Dione.

There are numerous asteroid families in the main belt, and they interact with secular resonances raising questions about orbital location (Carruba et al., 2018). The interaction between $\nu_5$, $\nu_6$ secular resonances causes the existence of an extensive chaotic zone for a 2:1 commensurability of mean motion. The perihelion argument resonance causes significant excursions in eccentricity and inclination, and the secular resonance $\nu_{16}$ outlines the Hilda family in the plane (Morbidelli and Moons, 1993). On the other hand, for other commensurabilities, the $\nu_5$, $\nu_6$ interaction is responsible for chaos in almost the entire phase space and causes jumps in the eccentricity of the asteroids (Moons and Morbidelli, 1995).

Chaos is present in the Solar System, and we need to understand them to understand the past, present, and future of the Solar System.

## 10.2.2 Chaotic indicators

Techniques for detecting chaos, referred to as chaotic indicators, have been developed over the years. In this chapter, we will present some of the most relevant ones for astronomy. These indicators include LCE (Lyapunov characteristic exponents) (Sandri, 1996), FLI (fast Lyapunov indicator) (Froeschlé et al., 1997b,a), FAM (frequency amplitude method) (Laskar et al., 1992), MEGNO (mean exponential growth of nearby orbits) (Cincotta and Simó, 2000; Goździewski et al., 2001), GALI (generalized alignment index), SALI (smaller alignment index) (Skokos and Manos, 2016), and spectral analysis (Michtchenko and Ferraz-Mello, 2001; Ferraz-Mello et al., 2005).

One of the most famous in dynamical systems is the Lyapunov exponent, which involves verifying how rapidly nearby trajectories diverge from each other in phase space analyses. The analysis aims to determine if the phase space volume changes for nearby trajectories. The difference between chaotic indicators or frequency analysis lies in how quickly a dynamical system can be identified as chaotic or not.

For example, in galaxy dynamics, identifying chaos quickly is crucial due to the large number of iterations required for the numerical evolution of these dynamics. Therefore GALI and SALI indicators are commonly used, but spectral analysis has also been employed in recent studies in galaxy dynamics (Michtchenko et al., 2018). The GALI and SALI methods serve as robust tools for identifying chaos, distinguishing themselves from other chaotic indicators. Unlike some alternatives, their efficacy in chaotic identification doesn't depend on the entire time evolution history of a system; rather, it is rooted in the assessment of the system's current state. This unique feature makes them notably faster in practice (Skokos and Manos, 2016).

## 10.3. Autocorrelation function indicator (ACFI)

An autocorrelation function indicator (ACFI) method based on a time series' self-similarity can likewise be used to identify chaos. The strength of the association between two time series is measured using the Pearson correlation coefficient $R$ (Pearson, 1895). When there is a strong correlation between the two variables that the series describes, $R$ approaches the maximum value of 1. $R$ will be near the minimal value of $-1$ if they are significantly anticorrelated, and $R \approx 0$ if there is no correlation at all. Although there are other definitions for a correlation coefficient, Pearson's definition is the one that is most frequently applied. Assuming that the $i - th$ term of the series in x and y coordinates is defined by $x_i$ and $y_i$, then

$$R = \frac{cov(X, Y)}{\sigma_x \sigma_y}, \tag{10.3}$$

where $cov(X, Y)$ is the covariance of the two series, defined as

$$cov(X, Y) = \frac{1}{N^2} \sum_{i=1}^{N} \sum_{j=1}^{N} \frac{1}{2}(x_i - x_j)(y_i - y_j), \tag{10.4}$$

where $N$ is the number of terms in the two series, and $\sigma_x$ is the standard deviation of the $x_i$ series, defined by

$$\sigma_x = \sqrt{\frac{1}{N} \sum_{i=1}^{N} (x_i - \mu_x)^2}, \tag{10.5}$$

where $\mu_x = \frac{1}{N} \sum_{i=1}^{N} x_i$ represents the mean value of the series. A similar expression applies to $\sigma_y$. The autocorrelation coefficient of a time series is essentially the correlation function of the series when compared to a lagged copy of itself. Consider creating a time series $y_i = x_{i-1}$ with a lag of one. Applying Eq. (10.3) to $y_i$ allows us to obtain the autocorrelation function. Similarly, coefficients for lags of 2 ($y = x_{i-2}$), or 3 ($y = x_{i-3}$) can be obtained. The autocorrelation function (ACF) of $x_i$ is the spectrum of autocorrelation coefficients for different the time lags. This enables us to evaluate the predictability of a time series' behavior.

Beginning with an example, consider a periodic function, such as a sinusoidal wave with zero initial phase and unitary amplitude, as illustrated on the left side of Fig. 10.4 along with its ACF. The units are defined in terms of time steps. Since we are conducting a time series analysis, it is advisable to use small time steps to avoid frequency aliasing issues. In this instance, we have selected a periodic function, and thus we anticipate that all autocorrelation coefficients in this segment will be 1. Consequently, we can predict the behavior of the sine function based on one period.[1]

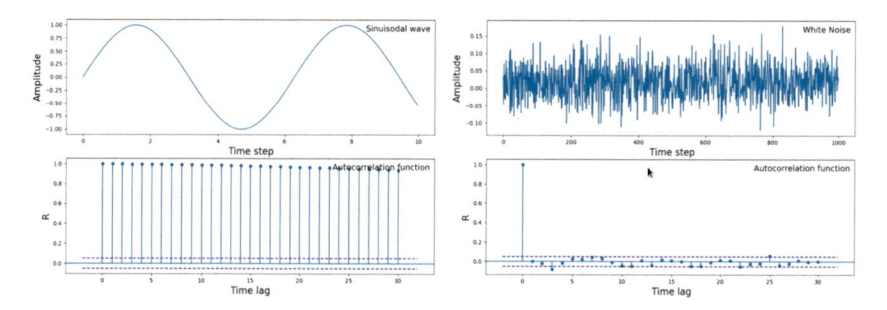

**Figure 10.4** On the left side, we show a plot of the time behavior of a sinusoidal wave (top panel) and of its autocorrelation function (bottom panel). On the right side, we have simulated white noise (top panel) and its ACF (bottom panel). The area between dashed horizontal lines in the ACF plots displays the region for which autocorrelation coefficients are lower than 5%; it also represents negligible autocorrelation. Reproduced from Carruba et al. (2021) with permission from the authors and CMDA (©CMDA).

Now, let's assume a random sample of uncorrelated random variables with an identical distribution, which can be illustrated as a white noise, for example. Given that white noise exhibits no correlation, we are unable to predict future observations based on time series information. In this scenario, the autocorrelation at all lags is zero, except for the first lag, which correlates with the series itself. The right panel of Fig. 10.4 depicts the time behavior of the simulated white noise and its corresponding ACF.

The examples presented so far represent extreme cases, whereas in general, time series often exhibit autocorrelation functions (ACF) that fall be-

---

[1] We observe a decreasing value of coefficients; this is caused by a limited sample of the sine function.

tween the patterns seen with sinusoidal waves and white noise. In Fig. 10.5, we display two ACF plots for a time series of the semimajor axis $a$ of a particle in the Veritas orbital region, differentiating between a regular (left panel) and chaotic (right panel) particle. The regular particle demonstrates significantly larger autocorrelation coefficients, especially with higher time lags, exceeding $\pm 0.05$ and extending beyond 200 time lags.

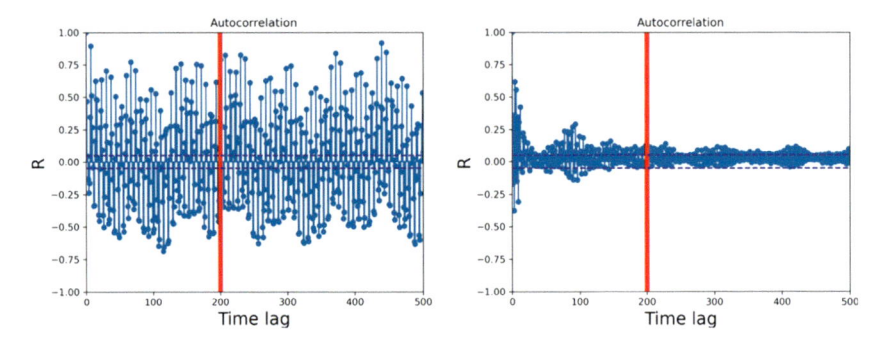

**Figure 10.5** The left panel displays the ACF of a regular particle in the map for the Veritas family region. The right panel shows the ACF of a rather chaotic orbit. The vertical line displays a lag equal to 200; the other symbols are the same as in the first Fig. 10.4. Reproduced from Carruba et al. (2021) with permission from the authors and CMDA (©CMDA).

Predicting the time evolution of the semimajor axis for chaotic particles becomes an impractical task on a short timescale. In light of this, it becomes feasible to establish a correlation with an autocorrelation function index (ACFI), defined by

$$ACFI = \frac{1}{i_{fin} - i_{in}} \sum_{i=i_{fin}}^{i_{fin}} n_i(|R|) > 0.05)_i, \qquad (10.6)$$

where $n_i(|R|) > 0.05)_i$ is the number of autocorrelation coefficients larger, in absolute value, than 5%. In our case $i_{fin} - i_{in} = 500 - 200 = 300$, and we only consider coefficients between time lags of 200 and 500 to avoid including autocorrelation at short timescales. The values of $i_{in}$ and $i_{fin}$ were chosen after experimenting with upper limits in the range from 100 to 300 and 300 to 1000, respectively. This needs to be adapted depending on the problem. Changing either from the chosen values affects the values of ACFI by 0.01 for chaotic particles and 0.04 for regular particles, at most. Mean values of ACFI are also affected by less than 1% by changing either parameter. More details on the dependence of ACFI on these parameters can be found Carruba et al. (2021).

We will apply the ACFI to dynamical systems used in astronomy to assess its effectiveness in detecting chaos.

## 10.3.1 Galaxy dynamics: Henon–Heiles system

The Henon–Heiles system (Hénon and Heiles (1964); Ferrer et al. (1998) is a form to described galactic dynamics assuming axial symmetry to model the potential of galaxies. The Hamiltonian described by Henon and Heiles is

$$H(x, y, \dot{x}, \dot{y}) = \frac{1}{2}(\dot{x}^2 + \dot{y}^2) + \frac{1}{2}(x^2 + y^2) + x^2 y - \frac{1}{3}y^3. \tag{10.7}$$

The equations of motion can be easily obtained following the classical methods. We numerically integrated the equations of motion until the x-axis was crossed at least 1000 times. In Fig. 10.6, the left side displays a (x,y) projection of the ACF for a regular orbit, whereas the right side shows chaotic trajectories for the system. Using the exact method described earlier for the ACFI, we set the free parameters $i_{in}$ and $i_{fin}$ of the ACFI to be equal to 1000 and 2000, respectively. In this case, regular orbits of the ACFI will exhibit higher values of autocorrelation indexes. These indexes play a pivotal role in the ACFI method as they delineate the segment of the autocorrelation function under analysis. Specifically, $i_{in}$ denotes the starting index, whereas $i_{fin}$ signifies the concluding index within the autocorrelation function time series. To test the ACFI, we compare it with an alternative chaos identification method, such as the widely used SALI method. Though we won't delve into the details of the SALI method here, interested readers can find information about it in (Skokos and Manos, 2016).

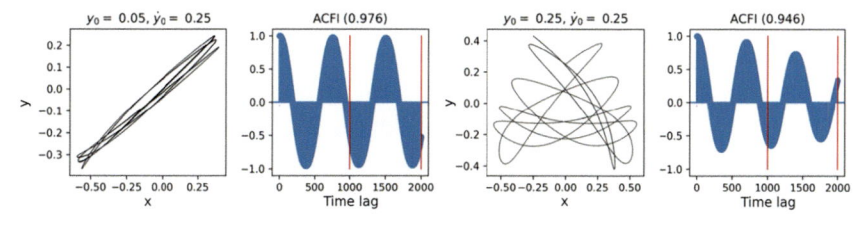

**Figure 10.6** Left panel: a (x, y) projection of a regular orbit (left) with its ACF (right). The vertical red lines display the values of the $i_{in}$ and $i_{fin}$ parameters. Right panel: the same, but for a chaotic orbit. Reproduced from Carruba et al. (2021) with permission from the authors and CMDA (©CMDA).

The SALI method involves calculating the area of a parallelogram formed by two vectors in the tangent space of the orbit. This method allows for a quick identification of chaos, which is particularly useful in galaxy dynamics, where chaos tends to evolve slowly in comparison with other methods.

Additionally, we opted to use another method known as a Poincaré surface of section to plot and understand the overall behavior of the dynamics. In this case, it provides a way to identify periodic and nonperiodic orbits in a Hamiltonian system. This method is versatile and applicable to various dynamical systems, including flows and maps across different physical domains (Strogatz, 2018).

The method relies on utilizing the property of an integral of motion, which, in this case, is related to the Hamiltonian system (Strogatz, 2018). The Poincaré surface of section (PSS) arises in a four-dimensional problem, and we need to reduce degrees of freedom to a two-dimensional problem. To achieve this, we select one of the freedom degrees to be a specific value, typically zero, for example, $\dot{x} = 0$. We determine the initial condition of the orbit by setting initial values for the Hamiltonian. For instance, given an energy $H = \epsilon$ and a initial $x$ position, we obtain the velocity in $y$ direction in function of $\dot{y}(H, x, y)$, and then we choose one section to cross $x > 0$ or $x < 0$. The crossing values can be found using an optimization algorithm or a bisection method to find a value close to zero, considering a threshold $\dot{x} < \alpha$ with $\alpha < 10^{-6}$, for example.

In the PSS, we expect to observe periodic orbits represented by islands, divided by regions with randomly distributed points. In the middle of the islands, there is an elliptical point of the dynamical system being an equilibrium solution or a resonant solution. Each island is surrounded by a periodic orbit, and as we approach the middle of the island, we encounter quasiperiodic orbits until we reach the main periodic orbit. The islands might be associated with resonances in the system. These islands are separated by a chaotic sea, represented by scattered points. Further information on this method for another cases can be found in Strogatz (2018).

Fig. 10.7 illustrates a Poincare surface of section for $H = 0.125$ in the coordinates $(\dot{y}, y)$; in the left side we present the PSS with the points related to the color to the SALI, whereas on the right side we show the very same PSS relating the colors with the ACFI. We chose values lesser than $10^{-6}$ for SALI to be regular motion, and higher values were defined as chaotic motion. It's notable that the regions chaotic regions can be identified in comparison with the SALI method. On the other hand, the ACFI allows us to verify more detailed and intricate dynamics related to the time series

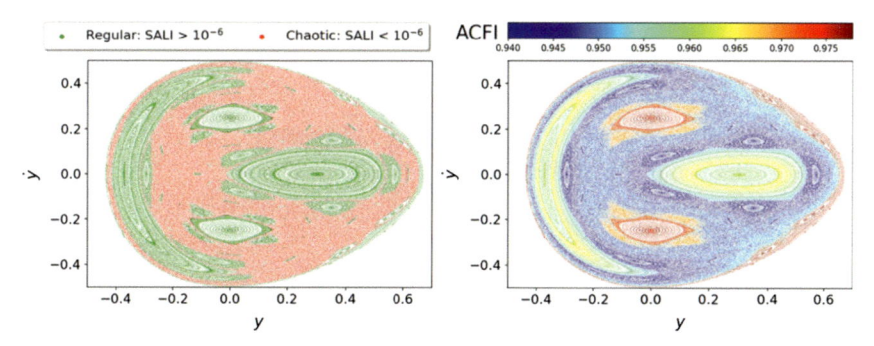

**Figure 10.7** The $H = 0.125$ PSS of 2D Henon–Heiles system. Left panel: classification of orbits using SALI method. Right panel: the values of ACFI indicator on the PSS. Reproduced from Carruba et al. (2021) with permission from the authors and CMDA (©CMDA).

of each initial condition of the Henon–Heiles system; the latter can be also related to Poincare time recurrences, giving enough information about the global dynamical "temperature" of the system (Shevchenko et al., 2020). In depth discussion regarding the ACFI method and its application to this problem can be found in Carruba et al. (2021).

### 10.3.2 Asteroid families

The Veritas asteroid family is subject to the influence of the three-body mean-motion resonance, potentially inducing chaos through the overlapping of resonances Milani et al. (1997).

To discern chaos within these asteroids, it is essential to generate a dynamical map using synthetic proper elements (Knežević and Milani, 2003). In Fig. 10.8, we present an example that has been extensively discussed in Carruba (2010); Carruba et al. (2017). This example involves a 50 by 50 numerically integrated grid in osculating semimajor axis and eccentricity, with elements ranging from 3.15 to 3.20 AU in semimajor axis, and 0.02 to 0.0935 in eccentricity. The computation is based on the theory introduced by Knežević and Milani (2003). The figure depicts blue circles representing the osculating elements of the asteroids and red lines indicating the three-body resonances. In this chapter, we refrain from delving deeply into the specifics of resonance effects and instead focus on the application of chaotic indicators and machine learning for predicting chaos.

Once the synthetic proper elements for the asteroid family are acquired, the identification of chaos becomes feasible. As mentioned earlier, several conventional methods for recognizing chaos exist, and it is imperative to

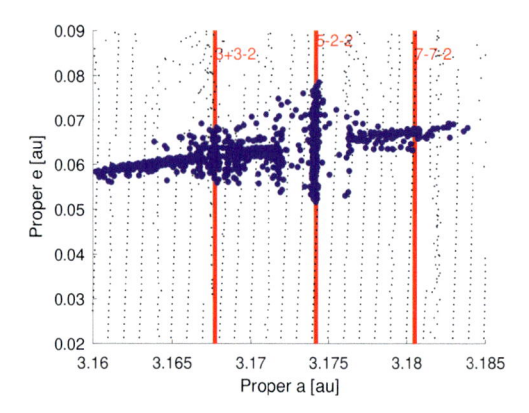

**Figure 10.8** An (a,e) dynamical map of synthetic proper elements for the region of the Veritas asteroid family. Black dots display values of proper elements for particles in the map. Blue full circles show the orbital location of Veritas family members. Vertical lines identify the location of three-body resonances. Reproduced from Carruba et al. (2021) with permission from the authors and CMDA (©CMDA).

validate their alignment with our machine learning model through the auto-correlation function.

In Fig. 10.9, we present four chaotic indicators: FLI, MEGNO, FAM, and ACFI, from the upper left side to the bottom right side. The colorbar denotes chaotic behavior, with red tones indicating chaos and blue colors representing regular behavior. The shades between these regimes in general represents signs of chaotic diffusion and stickiness. It is noteworthy that MEGNO effectively captures the general features of chaotic behavior, whereas ACFI demonstrates comparability with FAM. This alignment is anticipated, given that both ACFI and FAM involve the analysis of frequencies, in contrast to MEGNO, which examines nearby trajectories and their mutual differences in the phase space.

To quantify the similarity among the four approaches, we present the correlation matrix (R) coefficients in Fig. 10.10, reflecting the values of the four indicators used in the presented maps. Weak correlations exist between FLI and MEGNO, whereas FAM and ACFI show no correlation with FLI and MEGNO. However, as anticipated, FAM and ACFI exhibit a high correlation with each other, with a coefficient value of $R = 0.62$. Based on this we can say that the ACFI provides similar ways, in comparison to FAM, to identify chaos produced by mean-motion and secular dynamics.

Further details regarding the applications to other asteroid families, such as Phocaea, Veritas, or trans-neptunian objects (TNOs), are elaborated in Carruba et al. (2021).

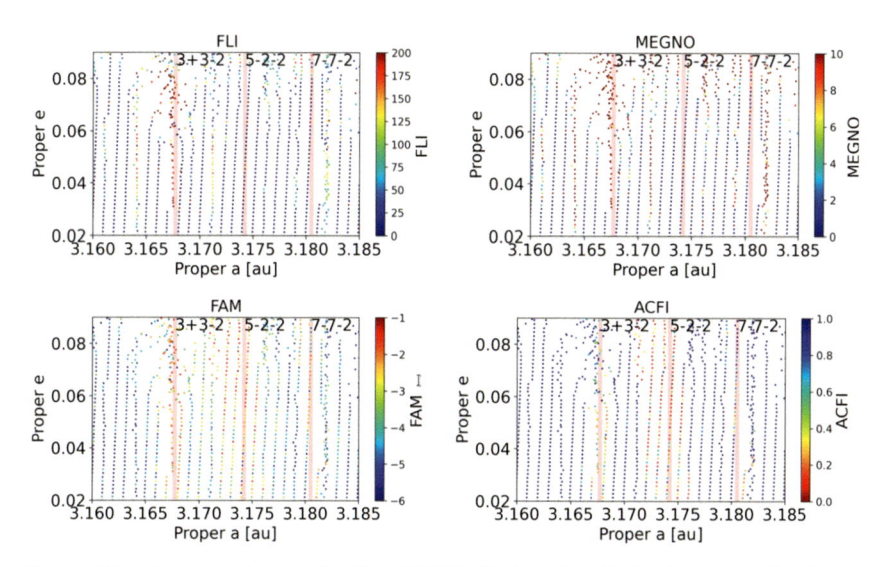

**Figure 10.9** Dynamical maps for FLI, MEGNO, FAM, and ACFI for the Veritas family orbital region. The color legend of each map identifies the range of values of each chaos indicator; other symbols have the same meaning as 10.9. Reproduced from Carruba et al. (2021) with permission from the authors and CMDA (©CMDA).

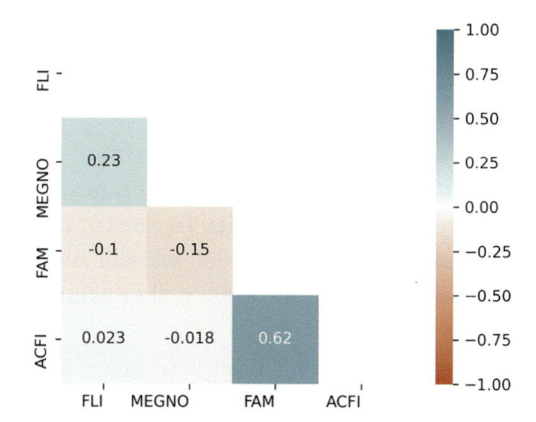

**Figure 10.10** Correlation matrix for the four chaos indicators tested for the Veritas family region. Values of the correlation coefficients are annotated in each of the cells. Reproduced from Carruba et al. (2021) with permission from the authors and CMDA (©CMDA).

## 10.3.3 Circular restricted three-body problem

Let's revisit another common astronomical problem—the restricted three-body problem (R3BP). In this scenario, we consider a three-body system, where the third body has negligible mass (e.g., an asteroid orbiting a planet)

$m_3 = 0$. The masses of the primary and secondary are normalized, being $m_1 = 1 - \mu$ and $m_2 = \mu$ where $\mu$ represents the mass ratio between the binary. The binary system, denoted as $m_1$ and $m_2$, orbits their common barycenter, and the secondary has a circular orbit around the primary. An illustrative example is the Sun-Jupiter system, where the mass ratio is approximately 0.001. Detailed information about this problem is provided in (Murray and Dermott, 1999).

Simulations for this system can be conducted using a simple n-body numerical integrator implemented in the PYTHON programming language called REBOUND (Rein and Liu, 2011). We can perform a similar study as previously done for Galaxy dynamics.

The initial condition of the third body will be planar and prograde ($i = 0$), starting with the argument of pericenter ($\omega$) equal to $\pi$. The equations of motion will be numerically integrated using REBOUND, and we will obtain the chaotic indicator MEGNO, provided by a tool implemented in REBOUND. Additionally, we will test the system with the ACFI indicator.

Simulations will be terminated in the case of collisions with the primary or secondary, assuming the radius of the Sun and Jupiter in normalized units. We will also stop simulations when escape events occur ($R > 10$), where the orbit's radius exceeds 10. For testing purposes, we will consider a short time span of 1000 periods of the binary, with an output density of 10,000 points required for ACFI computation. We will construct a grid of 80 by 80 points over semimajor axes, ranging from 0.5 to 1.5 units of distance and eccentricity between 0 and 0.3, for illustration purposes. ACFI will be calculated from the generated orbits. In cases of collisions or escape events with fewer points than defined earlier, we will assign a low value to ACFI. The time-lag between 900 and 9,000 will be assumed in the indices of the output.[2]

It is crucial to emphasize that the ACFI is an index closely tied to time lag and frequencies within a specific time span of an orbit. The density of points in the time series used for ACFI calculation plays a pivotal role. It should be sufficiently high to discretize and capture all possible frequencies, ensuring the accurate interpretation of the index. In our analyses, we focused on the y-coordinate time series. However, it's noteworthy that

---

[2] The time-lag is a crucial aspect of the ACFI method. It's worth emphasizing that the time series used for ACFI computation should encompass both rapid and slow oscillations. Furthermore, these oscillations need to be accurately considered by the algorithm. This ensures that the ACFI is able to capture and quantify the relevant dynamics of the system over varying timescales, contributing to a more comprehensive understanding of its behavior.

one could also leverage x, x*y, or radius time series for orbit analyses. The threshold set earlier of 5% in the time-series is consistently applied to maintain consistency in the evaluations.

In Fig. 10.11, the left side displays the MEGNO map in a color bar plot, where values of 2 indicate regular orbits and 8 signify chaotic orbits. On the right side, we illustrate the ACFI in a logarithmic scale. As mentioned earlier, the ACFI does not directly indicate whether an orbit is chaotic or regular. Instead, it provides an overview of the global temperature of the dynamical variables in the input.

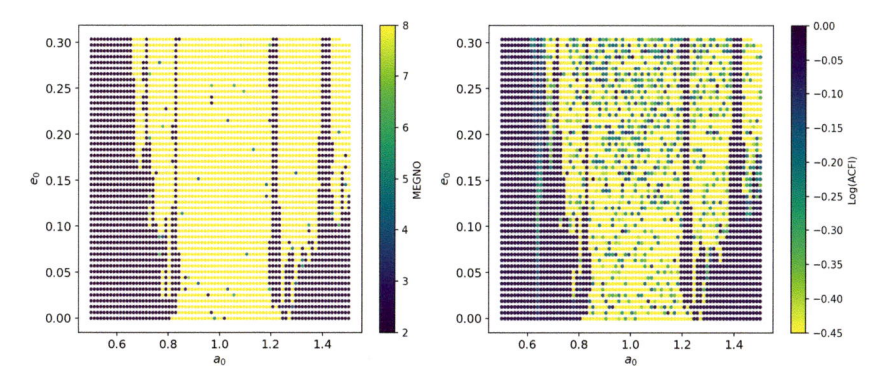

**Figure 10.11** MEGNO and ACFI comparison. Right panel: ACFI indicator. Left panel: MEGNO.

Comparing methods, MEGNO necessitates nearby trajectory information or the evolution of variational equations, making it computationally expensive compared to the more straightforward ACFI. Additionally, evolving only nearby trajectories would require a small $\delta$ applied to the initial conditions, a value that can vary depending on the system, adding complexity to the task. However, since ACFI is a time series method relying on time lag analyses, achieving high precision demands a specific number of points in the dataset, encompassing both rapid and slow oscillations. In both presented cases, the overall dynamical structures are similar, but the ACFI provides more detailed information about the dynamics, showcasing its potential for a richer analysis.

Similar to frequency analysis methods (Laskar et al. (1992); Ferraz-Mello et al. (2005)), the ACFI has the potential to reveal resonances and the general structure of frequencies in the phase space. This characteristic can be explored through the application of a surface of section method, as demonstrated in the first subchapter (refer to the Henon–Heiles–Poincaré surfaces). This method is directly related to chaotic diffusion (Shevchenko et al., 2020; Murray and Dermott, 1999). For instance, the colored line near the 0.63 of the semimajor axis represents the 2/1 mean motion resonance region (Winter and Murray (1997)). By utilizing the ACFI, we can also identify frequencies in this region, as the resonant behavior manifests in the time series. In such cases, chaos may arise from the overlap of resonances or simply from the interaction of various frequencies. These patterns within the chosen time series can be effectively identified through the ACFI method.

Indeed, distinguishing between chaotic and nonchaotic orbits can be a challenging task. Indicators such as MEGNO may require extensive integration time to provide sufficient information about chaotic orbits, particularly depending on the perturbations and the number of bodies involved. On the other hand, the analysis of time-lags using the ACFI method proves to be valuable in comprehending the general dynamics of systems. It aids in understanding various aspects, including resonances such as mean motion and secular resonances, providing insights into the overall stability of the system. The ACFI method thus offers a useful and efficient approach for gaining a deeper understanding of the intricate dynamics of celestial systems.

## 10.4. Conclusions and future developments

The plethora of methods and techniques for chaos identification and dynamical system analyses has expanded over time. As discussed earlier, the methods described in the literature were compared with the ACFI, revealing that the ACFI can stand as a robust tool for dynamical system and time series analyses. What makes ACFI particularly appealing is its simplicity in implementation and application. This simplicity, coupled with its ability to provide meaningful insights into the global dynamics of systems, underscores the potential strength of ACFI as a valuable tool in the field of dynamical systems research.

In contrast to classical techniques, such as MEGNO, the ACFI method doesn't inherently determine whether an orbit is chaotic. Instead, it captures the general system behavior, akin to classical frequency methods analyses. As discussed earlier, whereas classical methods necessitates significant numerically integration time for convergence, the ACFI's analysis of time-lag proves to be valuable for some applications. Coupled with its ability to comprehend various aspects, such as resonances and overall stability, it establishes ACFI as a robust tool for dynamical system analyses.

As discussed earlier, this method could be useful as a pre-processing tool for further machine learning applications, which are still under development or being studied. In the evolving landscape of dynamical systems analysis, the emergence of new methods, for instance, PySindy (a Python library for machine learning and dynamical systems; de Silva et al. (2020); Kaptanoglu et al. (2021)), marks a significant advancement, and future works about chaos should be developed using this kind of tool. Such methods introduces a novel avenue for dynamical system. The expansion of the literature in this direction underscores the ongoing pursuit of more efficient tools for dynamical system analyses.

## 10.5. Code availability

The ACFI code is available at this GitHub repository: https://github.com/valeriocarruba/ACFI-Chaos-identification-through-the-autocorrelation-function-indicator.

## Acknowledgments

This study was financed in part by the Coordenação de Aperfeiçoamento de Pessoal de Nível Superior – Brasil (CAPES) – Finance Code 001 / PRInt. Carruba V. acknowledges the support of the Brazilian National Research Council (CNPq, grant 304168/2021-1). We acknowledge the support grant 2021/08274-9 from São Paulo Research Foundation (FAPESP).

## References

Binzel, R.P., Green, J.R., Opal, C.B., 1986. Chaotic rotation of hyperion? Nature 320, 511.

Brunton, S.L., Kutz, J.N., 2019. Data-Driven Science and Engineering: Machine Learning, Dynamical Systems, and Control. Cambridge University Press.

Calvão, A., Penna, T., 2015. The double pendulum: a numerical study. European Journal of Physics 36, 045018.

Carruba, V., 2010. The stable archipelago in the region of the Pallas and Hansa dynamical families. Monthly Notices of the Royal Astronomical Society 408, 580–600.

Carruba, V., Aljbaae, S., Domingos, R.C., Huaman, M., Barletta, W., 2021. Chaos identification through the autocorrelation function indicator (acfi). Celestial Mechanics & Dynamical Astronomy 133, 38.

Carruba, V., Vokrouhlický, D., Nesvorný, D., 2017. Detection of the Yarkovsky effect for C-type asteroids in the Veritas family. Monthly Notices of the Royal Astronomical Society 469, 4400–4413.

Carruba, V., Vokrouhlický, D., Novaković, B., 2018. Asteroid families interacting with secular resonances. Planetary and Space Science 157, 72–81.

Cincotta, P.M., Simó, C., 2000. Simple tools to study global dynamics in non-axisymmetric galactic potentials–I. Astronomy & Astrophysics. Supplement Series 147, 205–228.

Ferraz-Mello, S., Dvorak, R., 1987. Chaos and secular variations of planar orbits in 2: 1 resonance with Dione. Astronomy & Astrophysics (ISSN 0004-6361) 179 (1–2), 304–310.

Ferraz-Mello, S., Michtchenko, T., Beaugé, C., Callegari, N., 2005. Extrasolar Planetary Systems. Springer.

Ferrer, S., Lara, M., Palacián, J., Juan, J.S., Viartola, A., Yanguas, P., 1998. The Hénon and Heiles problem in three dimensions. I Periodic orbits near the origin. International Journal of Bifurcation and Chaos 8, 1199–1213.

Froeschlé, C., Gonczi, R., Lega, E., 1997a. The fast Lyapunov indicator: a simple tool to detect weak chaos. Application to the structure of the main asteroidal belt. Planetary and Space Science 45, 881–886.

Froeschlé, C., Lega, E., Gonczi, R., 1997b. Fast Lyapunov indicators. Application to asteroidal motion. Celestial Mechanics & Dynamical Astronomy 67, 41–62.

Furi, M., Landsberg, A., Martelli, M., 2005. On the chaotic behavior of the satellite hyperion. Journal of Difference Equations and Applications 11, 635–643.

Goździewski, K., Bois, E., Maciejewski, A., Kiseleva-Eggleton, L., 2001. Global dynamics of planetary systems with the megno criterion. Astronomy & Astrophysics 378, 569–586.

Hénon, M., Heiles, C., 1964. The applicability of the third integral of motion: some numerical experiments. Astronomical Journal 69, 73.

Hilborn, R.C., 2000. Chaos and Nonlinear Dynamics: an Introduction for Scientists and Engineers. Oxford University Press.

Kaheman, K., Bramburger, J.J., Kutz, J.N., Brunton, S.L., 2023. Saddle transport and chaos in the double pendulum. Nonlinear Dynamics 111, 7199–7233.

Kaptanoglu, A.A., de Silva, B.M., Fasel, U., Kaheman, K., Goldschmidt, A.J., Callaham, J.L., Delahunt, C.B., Nicolaou, Z.G., Champion, K., Loiseau, J.C., et al., 2021. Pysindy: a comprehensive python package for robust sparse system identification. arXiv preprint, arXiv:2111.08481.

Knežević, Z., Milani, A., 2003. Proper element catalogs and asteroid families. Astronomy & Astrophysics 403, 1165–1173.

Knežević, Z., Milani, A., Farinella, P., Froeschlé, C., Froeschle, C., 1991. Secular resonances from 2 to 50 au. Icarus 93, 316–330.

Kozai, Y., 1962. Secular perturbations of asteroids with high inclination and eccentricity. The Astronomical Journal 67, 591–598.

Laskar, J., Froeschlé, C., Celletti, A., 1992. The measure of chaos by the numerical analysis of the fundamental frequencies. Application to the standard mapping. Physica D. Nonlinear Phenomena 56, 253–269.

Lidov, M.L., 1962. The evolution of orbits of artificial satellites of planets under the action of gravitational perturbations of external bodies. Planetary and Space Science 9, 719–759.

Michtchenko, T.A., Ferraz-Mello, S., 2001. Resonant structure of the outer solar system in the neighborhood of the planets. The Astronomical Journal 122, 474.

Michtchenko, T.A., Lépine, J.R.D., Barros, D.A., Vieira, R.S., 2018. Combined dynamical effects of the bar and spiral arms in a galaxy model. Application to the solar neighbourhood. Astronomy & Astrophysics 615, A10.

Milani, A., Nobili, A.M., Knežević, Z., 1997. Stable chaos in the asteroid belt. Icarus 125, 13–31.

Moons, M., Morbidelli, A., 1995. Secular resonances in mean motion commensurabilities: the 4/1, 3/1, 5/2, and 7/3 cases. Icarus 114, 33–50.

Morbidelli, A., Moons, M., 1993. Secular resonances in mean motion commensurabilities: the 2/1 and 3/2 cases. Icarus 102, 316–332.

Murray, C.D., Dermott, S.F., 1999. Solar System Dynamics. Cambridge University Press.

Pearson, K., 1895. Correlation coefficient. In: Royal Society Proceedings, p. 214.

Rein, H., Liu, S.F., 2011. Rebound: an open-source multi-purpose n-body code for collisional dynamics. arXiv preprint, arXiv:1110.4876.

Sandri, M., 1996. Numerical calculation of Lyapunov exponents. Mathematica Journal 6, 78–84.

Shevchenko, I.I., Rollin, G., Melnikov, A.V., Lages, J., 2020. Massive evaluation and analysis of Poincare recurrences on grids of initial data: a tool to map chaotic diffusion. Computer Physics Communications 246, 106868.

de Silva, B.M., Champion, K., Quade, M., Loiseau, J.C., Kutz, J.N., Brunton, S.L., 2020. Pysindy: a python package for the sparse identification of nonlinear dynamics from data. arXiv preprint, arXiv:2004.08424.

Skokos, C., Manos, T., 2016. The smaller (sali) and the generalized (gali) alignment indices: efficient methods of chaos detection. Chaos Detection and Predictability, 129–181.

Strogatz, S.H., 2018. Nonlinear Dynamics and Chaos with Student Solutions Manual: With Applications to Physics, Biology, Chemistry, and Engineering. CRC Press.

Vlachas, P.R., Byeon, W., Wan, Z.Y., Sapsis, T.P., Koumoutsakos, P., 2018. Data-driven forecasting of high-dimensional chaotic systems with long short-term memory networks. Proceedings of the Royal Society A. Mathematical, Physical and Engineering Sciences 474, 20170844.

Williams, J.G., 1969. Secular Perturbations in the Solar System. University of California, Los Angeles.

Winter, O., Murray, C., 1997. Resonance and chaos: I. First-order interior resonances. Astronomy & Astrophysics, 290–304.

Wisdom, J., 1987. Chaotic behavior in the solar system. Nuclear Physics. B, Proceedings Supplement 2, 391–414.

Wisdom, J., Peale, S.J., Mignard, F., 1984. The chaotic rotation of hyperion. Icarus 58, 137–152.

# CHAPTER ELEVEN

# Conclusions and future developments

**Valerio Carruba**[a], **Evgeny Smirnov**[b], **and Dagmara Oszkiewicz**[c]

[a]São Paulo State University (UNESP), Department of Mathematics, Guaratinguetá, SP, Brazil
[b]Belgrade Astronomical Observatory, Belgrade, Serbia
[c]Astronomical Observatory Institute, Faculty of Physics and Astronomy, Adam Mickiewicz University, Poznań, Poland

## 11.1. Introduction

In this book, we discussed applications of *AI* and *ML* for several problems concerning Solar System small bodies. Among them, the identification of asteroid family members, asteroids in mean-motion resonances, asteroids interacting with secular resonances, the orbital dynamics around asteroids, the spectro-photometric classification of asteroids, Kuiper belt object, the identification and localization of cometary activity in Solar System objects, the detection and characterization of moving objects, and chaotic dynamics. Here, we will assess the impact of *ML* applications and try to quantify the maturity of this field using an approach introduced by Fluke and Jacobs (2020). We will discuss some possible future trends for new methods and applications of *ML* to Solar System small bodies, and, finally, we will draw our conclusions.

## 11.2. Assessing the impact of *ML* applications to small bodies in the Solar System

To evaluate recently released literature, Fluke and Jacobs (2020) used seven categories, the last two of which had a greater scientific outcome:

1. **Classification**: Characteristics or objects are classified into several groups or labels. The machine learning algorithm uses a training set (labeled or unlabeled) to identify the characteristics that associate an instance with a category. The most likely category label is assigned to a new instance by the algorithm when it is applied.

2. **Regression**: Determining a numerical value (or values) by applying a machine learning algorithm to features it has learned or has otherwise

anticipated. Like with classification, a training set may be used, or the features may be inferred from the data set.

3. **Clustering**: These methods determine whether a feature or item belongs to or is a component of anything. This could be a region inside an N-dimensional parameter space or a physical structure or association, such as an asteroid family.

4. **Forecasting**: The objective of the machine learning algorithm is to foresee or predict the occurrence of a comparable event by using lessons from previous occurrences. There is an intrinsic temporal dependence in the forecast.

5. **Generation and reconstruction**: To ensure that the missing data is consistent with the underlying reality, it is generated and rebuilt. There are a few possible reasons for the lack of information: noise, processing artifacts, or other astronomical phenomena, which work together to mask the needed signal.

6. **Discovery**: Using artificial intelligence or machine learning technologies leads to discovering of new celestial objects, features, or relationships.

7. **Insight**: The application of machine learning yields new scientific insights that extend beyond the discovery of celestial objects. This covers scenarios where data is acquired on the suitability of applying machine learning, dataset selection, hyperparameters, and contrasts with classifications based on human judgment.

A similar human-centered technique is often used to compare classification, regression, and clustering approaches, but with the requirement to "scale up" the size of the dataset to be investigated or the length of time required to finish the work. The outcomes of a classification and regression analysis might serve as the report card for an inquiry or as input for a process that generates, finds, forecasts, or provides insight. These seven categories, which make up a general hierarchy of sophistication, offer an evaluation of how advanced the application of artificial intelligence and machine learning is in an area of astronomy. A common place to start is with a classification, regression, or clustering problem, and a machine learning technique. Once machine learning has proven to be on par with or better than a more conventional approach, it can be used to identify new candidates for rare objects (e.g., (Zhang et al., 2018)) or estimate expected future outcomes (e.g., future solar flares, (Florios et al., 2018)).

In what way might articles written about small bodies in the Solar System be categorized using these criteria? We divide the sample of works provided in this book by chapters (for the sake of brevity, some of the titles have been shortened):

2. **Identification of Asteroid Families' Members**
3. **Asteroids in Mean-Motion Resonances**
4. **Asteroid Families Interacting with Secular Resonances**
5. **Orbital Dynamics Around Asteroids**
6. **Asteroid Spectro-Photometric Classification**
7. **Kuiper Belt Objects**
8. **Identification and Localization of cometary activity in Solar System Objects with Machine Learning**
9. **Detection and Characterization of Moving Objects with Machine Learning**
10. **Chaotic Dynamics**

We then classified the articles presented in these chapters on the basis of the categories presented in this section. Our results are presented in Table 11.1 and in Fig. 11.1.

**Table 11.1** A qualitative summary of the categories of *ML* algorithms and the most common subareas of research for the chapters in this book. In the last three columns, we report the total number of articles published in the area, the fraction of papers in the **Generation, Discovery**, and **Insight** areas (GDI), and the classification of the subareas in terms of Fluke and Jacobs (2020) hierarchy. Em stands for *emerging*. P for *progressing*, and Es for *established*. The last row reports our analysis of the whole area of Solar System small bodies.

| Ch. | Class. | Regr. | Clust. | FC | Gen. | Disc. | Insight | Papers | GDI | Classif. |
|---|---|---|---|---|---|---|---|---|---|---|
| 2 | 2 | 0 | 2 | 0 | 0 | 1 | 0 | 2 | 20.0% | Em |
| 3 | 6 | 0 | 0 | 5 | 0 | 5 | 0 | 6 | 31.3% | P |
| 4 | 7 | 0 | 2 | 7 | 1 | 3 | 1 | 7 | 20.0% | Em |
| 5 | 2 | 4 | 0 | 4 | 2 | 0 | 2 | 4 | 50.0% | Em |
| 6 | 23 | 0 | 8 | 3 | 6 | 11 | 7 | 23 | 41.4% | Es |
| 7 | 2 | 0 | 0 | 1 | 0 | 2 | 1 | 2 | 42.9% | Em |
| 8 | 5 | 0 | 0 | 0 | 0 | 5 | 2 | 5 | 58.3% | P |
| 9 | 2 | 4 | 0 | 4 | 2 | 2 | 2 | 4 | 37.5% | Em |
| 10 | 1 | 0 | 0 | 1 | 0 | 1 | 1 | 1 | 50.0% | Em |
| All | 48 | 8 | 12 | 23 | 13 | 30 | 16 | 50 | 39.3% | Es |

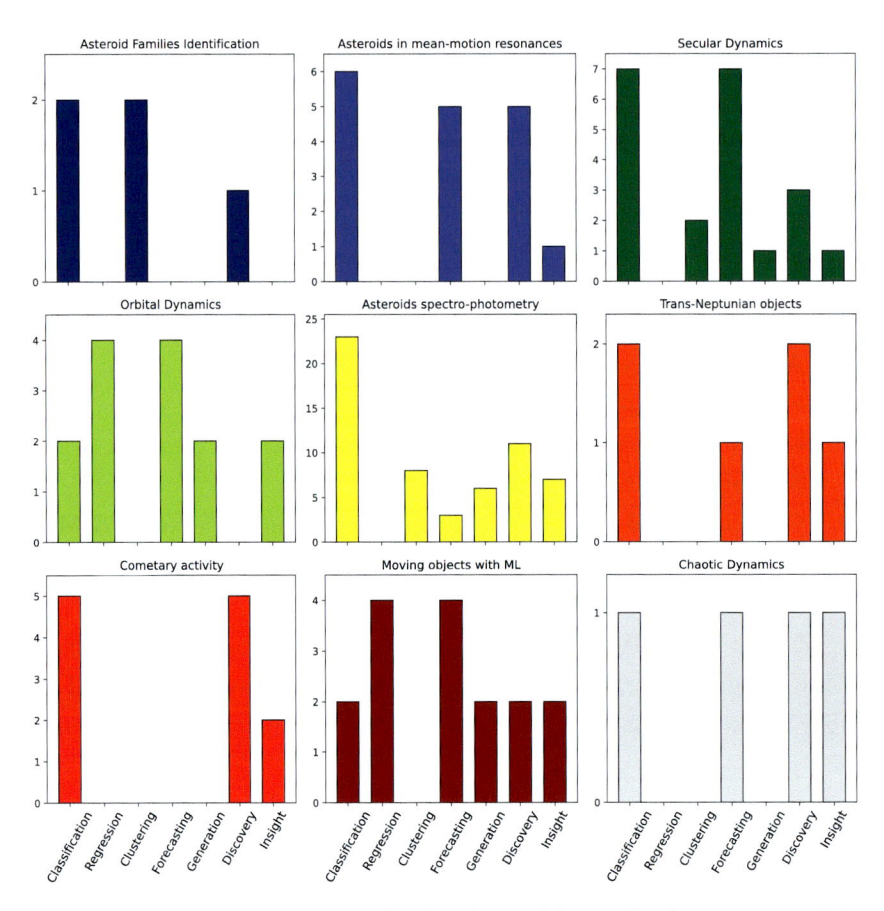

**Figure 11.1** Bar plots summarizing the data from Table 11.1 for the categories of *ML* algorithms applied to the nine chapters of this book.

What insights can this analysis provide? Fluke and Jacobs (2020) introduced a hierarchy of three categories—*emerging, progressing*, and *established*—to assess the *ML* maturity within an astronomical subfield.

The *emerging* stage is used for subfields of astronomy and astrophysics, which are starting to investigate the application of artificial intelligence and machine learning, usually by focusing on easier-to-solve problems. This can be a problem that calls for a regression or classification plan, or it might require contrasting machine learning with another, established technique. The *ML* approaches that are most appropriate for a particular problem could not have been encountered yet, or the community might be small, in the case of *emerging* communities. In our context, we consider an area to be *emerging* if either less than five papers have been published in recent years, and/or the fraction of advanced categories, such as **Generation, Discovery**, and **Insight** (GDI), is less than 30%.

A *progressing* stage is characterized by using a wider variety of approaches, the application of a single technique multiple times, or an abrupt shift to the forecasting, discovery, or insight phases. We consider an area to have reached a *progressing* stage if i) more than five papers have been published in the last five years, and ii) the GDI fraction is higher than 30%.

Ultimately, a subfield is said to have reached the *established* stage if there is a significant amount of research in the area. The emphasis is mostly on prediction, discovery, or insight, and the use of machine learning has become indispensable in established subfields. Because machine learning is already widely used, evaluating its applicability is no longer necessary. *Established* stages are reached for areas with i) more than 10 articles published in the last five years, and ii) a GDI fraction larger than 30%.

The study of *Solar System small bodies* objects was categorized by Fluke and Jacobs (2020) as having reached a *progressing* stage. Carruba et al. (2022) also confirmed this analysis, while finding that the smaller *asteroid dynamics* field was still in the *emerging* stage. The results of our analysis of the literature reviewed in this book are summarized in Table 11.1. Most subareas are still in the *emerging* stage, but two of them, asteroids interacting with mean-motion resonances and identification of cometary activity with *ML*, are in the *progressing* stage, while asteroid spectro-photometric classification can be considered to have reached the *established* stage. Fig. 11.2 displays the results for our analysis of the whole book. Mostly thanks to the asteroid spectro-photometric classification subarea, *ML* for Solar System small bodies can be thought to have reached an *established* stage.

## 11.3. Future trends

The recent Kaggle contests have been a driving force behind advancements in the field of machine learning (*ML*).[1] Machine learning challenges created by Kaggle or other businesses, such as Google or WHO, are used in Kaggle contests. Competitions vary in difficulty and problem kinds, although they are usually accessible to beginners. Typically, the best problem solution is given a cash prize. In the subject of asteroid dynamics, a recent example of a Kaggle-like competition is the challenge centered

[1] See also https://www.kaggle.com/competitions.

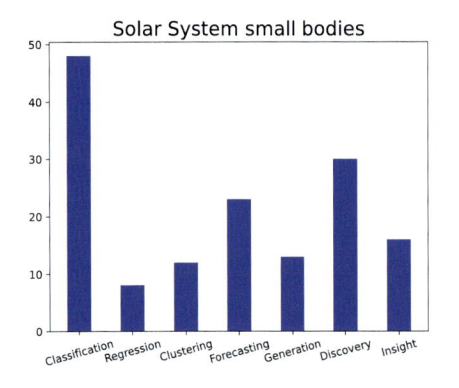
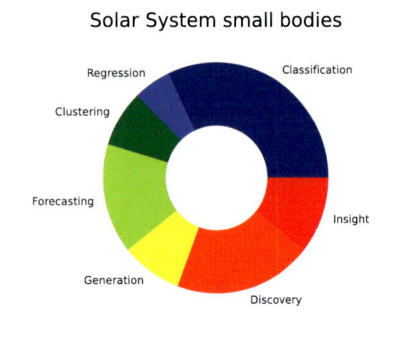

**Figure 11.2** Bar and pie plots summarizing the data from Table 11.1 for the categories of *ML* algorithms applied to the whole book.

around predicting the impact's effects, while deflecting asteroids within the framework of the HERA mission.[2] A dataset was offered to investigate how this task could be completed entirely using *ML* and *AI*. A number of highly accurate models of a kinetic impactor mission over a binary system were created and concealed from the participants. Each of the simulated scenarios' lightcurves as seen from Earth before and after the impact were made open to the public. The problem was to determine whether the given lightcurves could be used to determine the impact geometries and orbits of asteroids. Redrock was victorious, obtaining a final score of 0.453.

Another area that is likely to produce significant progress is big data in astronomy, such as the influx of new data expected to be produced by the Vera C. Rubin observatory. Presently under construction in Chile, the 8.4-meter three-mirror design telescope is anticipated to begin operations in 2025. According to Jones et al. (2015), the LSST catalogs are expected to produce ≃ 1 million new asteroid discoveries in the first year of operations, increasing the known number of small bodies in the Solar System by a factor of 10–100 across all populations. Other massive data releases are expected or are already available from mission such as Gaia (Brown et al., 2016), Euclid (Laureijs et al., 2010), SPHEREx (Doré et al., 2019), NEOSurveyor (Mainzer et al., 2023), PANSTARRS (Chambers et al., 2016), ATLAS (Heinze et al., 2020), etc. New techniques, such as the application of the Lomb–Scargle algorithm on graphics processing units, have already been developed in response to the necessity to handle such a large amount of

[2] https://kelvins.esa.int/planetary-defence/

data (Gowanlock et al., 2021). The upcoming years should see an increase in software development focused on *ML*.

Improvements in the field of computer vision could involve the use of multiple-input CNNs, such as the BTSBot, implemented for identifying supernovae in ZTF images, as discussed in Chapter 8 (Rehemtulla et al., 2023). Another possible improvement in computer vision applied to dynamical astronomy could be the application of vision transformer architectures (*ViT*). In 2020, Google researchers introduced the vision transformer (ViT, (Touvron et al., 2021)) architecture for image categorization problems. Its foundation is the Transformer, a potent architecture that was first created for applications involving natural language processing (Vaswani et al., 2017). At its core, the Transformer architecture relies on a mechanism called self-attention or scaled dot-product attention. Self-attention allows the model to weigh the importance of different words in a sentence when generating representations. This mechanism enables the model to capture dependencies and relationships between words in a more flexible way compared to traditional recurrent neural networks (RNNs) or convolutional neural networks (CNNs). *ViT*'s primary concept is to use the Transformer to transform images by treating them like a series of patches. Each patch in the grid-like representation of the image is regarded as a token, much like a word in natural language processing. After that, a conventional Transformer encoder is supplied with the patches in a sequence that has been flattened. The ultimate output is then generated by passing the output of the last Transformer layer through a classification head.

*ViT*'s ability to perform at the cutting edge of image classification tasks with a far smaller computing footprint than conventional convolutional neural network (*CNN*) designs is one of its main advantages. This is because the Transformer's self-attention mechanism makes it possible to analyze long-range relationships in images more effectively, which is crucial for handling large-scale picture datasets. *ViT* has been used to classify images for a variety of purposes, such as medical imaging, object recognition, and scene classification, and it has frequently shown outstanding results.

The ImageNet dataset, a popular benchmark for image classification tasks, is one instance of an image classification test, where ViT has demonstrated superior performance over conventional CNN architectures. With a top-1 accuracy of 90.7% in the 2021 ImageNet Challenge, the ViT architecture outperformed the top-performing CNN architecture by a small margin. Another example is the COCO object detection dataset, where ViT has outperformed conventional CNN-based object detectors in specific scenarios, achieving state-of-the-art results. With a mean average precision (mAP) of 53.7% on the COCO dataset, the ViT-based object detector, known as DETR-ViT, outperformed the prior state-of-the-art.

It is, however, important to remember that ViT is not always the best option for every picture classification task and that the size, complexity, and particular hyperparameters utilized during training can have a significant impact on the model's performance. ViT may not always perform better than conventional CNN designs on smaller datasets. This is because ViT needs a lot of training data to efficiently learn the representations required for image classification tasks. Caution and a case-by-case approach should be used when applying *ViT* models to new databases.

A promising tool in dynamical astronomy is the sparse identification of nonlinear dynamics (*SINDy*, Brunton et al. (2016)) method. SINDy is a data-driven modeling technique used to discover governing equations from observational data. It is particularly used in the field of dynamical systems and data science to uncover the underlying dynamics of complex systems. Instead of relying on explicit physical models or equations, SINDy aims to learn the equations directly from data. It has been successfully applied in various domains, including physics, biology, engineering, and finance.

The SINDy method uses a sparse regression approach to identify the most relevant terms in a set of candidate functions that could describe the system dynamics. It starts by collecting data from the system of interest, typically in the form of time series measurements. Then, it constructs a library of candidate functions, which are combinations of the system variables and their derivatives. The library can include polynomial terms, trigonometric functions, and other relevant mathematical operations.

The next step is to solve a sparse regression problem to identify the coefficients of the candidate functions that best match the observed data. The sparsity constraint encourages the identification of the fewest number of terms necessary to capture the underlying dynamics accurately. The resulting sparse regression problem can be solved using various optimization techniques, such as LASSO (least absolute shrinkage and selection operator) or iterative thresholding.

Once the sparse regression problem is solved, the identified terms and their corresponding coefficients provide an approximation of the underlying dynamical system. These terms can be interpreted as physical laws or equations that govern the system's behavior. By using SINDy, it is possible to discover the underlying equations without prior knowledge of the system's dynamics, making it a valuable tool for system identification and discovery.

Finally, the recent development of large language models (LLMs) outlines new possibilities for a wide range of problems. These models are special types of artificial neural networks that are designed to perform natural language processing (NLP) tasks, such as to generate human-like text. They are often based on the transformer architecture, which has already been introduced for the VITs models (Vaswani et al., 2017).

LLMs can have different architectures. The two most common are GPT (generative pretrained transformer; Radford et al. (2018)) and BERT (bidirectional encoder representations from transformers; Devlin et al. (2019)), which have demonstrated very good performance on a wide range of NLP tasks, including text generation, language translation, and, what is more important, question answering. These models are trained on large amounts of text data and learn to generate human-like text based on the patterns and structures present in the data.

It is reasonable to assume that these tools are often used in social sciences, such as psychology, linguistics, or philosophy, because they work heavily with texts. However, LLMs can be applied to astronomy as well. Though astronomy does not explicitly deal with texts, there are multiple tasks requiring human agents to perform them. This is exactly what LLMs were designed to do. They can behave as human agents and resolve tasks that require human-like reasoning.

These tasks can be applied to different astronomical problems. One can use OpenAI's GPT-4 for automatic classification of astronomical images based on the algorithm specified in natural language. For example, Smirnov (2024) used `gpt4-vision-preview` model to identify whether or not the resonant angle librates. The reported metrics outperform significantly the traditional methods and requires no coding, only prompting (requesting) an LLM through the API.

A significant advantage of these tools is that they can be used by non-experts in machine learning, do not require a priori knowledge or training dataset, and can be used for a wide range of tasks. The only requirement is an ability to formulate a task in natural language. For the example mentioned above, the actual time required to create from scratch a production-ready workflow that automatically analyses astronomical time series and identifies patterns in the data is reduced to a few minutes. Furthermore, it is possible to send multiple API requests simultaneously, which allows for parallel processing of multiple tasks and significantly reduce the overall computational time.

However, this approach has some disadvantages as well. Firstly, the usage of `gpt4-vision-preview` is proprietary, and hence not cost-effective, because each request costs some money. It can be avoided by the usage of open-source LLMs, such as `llama` or `mistral`. However, it requires further investigation.

Secondly, LLMs can "hallucinate," which means that they can provide wrong outputs like real human beings (Rawte et al., 2023). There are a number of techniques that can mitigate such a risk. Some of them are mentioned by Smirnov (2024).

Overall, this approach is promising and can be used in a wide range of astronomical tasks. It reduces the time necessary to develop a software solution required to resolve astronomical problems. Moreover, the overall results can outperform not only classical machine-learning methods, but also human beings. It is expected that the usage of LLMs in astronomy will increase in the upcoming years.

## 11.4. Conclusions

In this chapter, we summarized the main results presented in this book, and analyzed them in terms of the metrics and approaches used by Fluke and Jacobs (2020). Based on our analysis, though subareas of applications of *ML* to Solar System small bodies have reached various development stage, the whole area can be considered as having reached an *established* stage, mostly thanks to the developments in asteroid spectro-photometric classifications.

Although it is very difficult to predict future trends in a very fluid research field, such as *AI* and *ML* applied to Solar System small bodies, some possible trends were noticed by the authors of this book. Kaggle *ML* competitions have been shown to stimulate scientific progress, as in the case of the HERA mission. In the age of big data in astronomy, just as what is currently seen for the Zwicky Observatory, and soon to be produced by the Vera C. Rubin project, will require new methods to handle large volumes of small bodies' discoveries, some of which are already under development (Gowanlock et al., 2021).

New algorithms, such as the vision transformer (ViT, (Touvron et al., 2021)) architecture for image categorization problems, the sparse identification of nonlinear dynamics (*SINDy*, Brunton et al. (2016)) method for discovering governing equations from observational data, or the use of LLM models to identify the resonant behavior of asteroids in three-body resonances Smirnov (2024), have all been suggested as interesting new tools for scientific advancement. By reducing the time and effort required to complete tasks that typically necessitate the expertise of human specialists, LLM models can automate complex tasks, thereby enabling astronomers to process and interpret vast amounts of data more efficiently and cost-effectively. Many new fascinating discoveries in this sector are to be expected in the upcoming years.

## Acknowledgments

VC acknowledges support from the Brazilian National Research Council (CNPq, grant 304168/2021-1).

## References

Brown, A.G.A., Vallenari, A., Prusti, T., de Bruijne, J.H.J., et al., 2016. Gaia Data Release 1. Summary of the astrometric, photometric, and survey properties. Astronomy & Astrophysics 595, A2. https://doi.org/10.1051/0004-6361/201629512.

Brunton, S.L., Proctor, J.L., Kutz, J.N., 2016. Discovering governing equations from data by sparse identification of nonlinear dynamical systems. Proceedings of the National Academy of Sciences 113, 3932–3937.

Carruba, V., Aljbaae, S., Domingos, R.C., Huaman, M., Barletta, W., 2022. Machine learning applied to asteroid dynamics. Celestial Mechanics & Dynamical Astronomy 134, 36. https://doi.org/10.1007/s10569-022-10088-2. arXiv:2110.06611.

Chambers, K.C., Magnier, E.A., Metcalfe, N., et al., 2016. The Pan-STARRS1 surveys. arXiv e-prints, arXiv:1612.05560.

Devlin, J., Chang, M.W., Lee, K., Toutanova, K., 2019. BERT: pre-training of deep bidirectional transformers for language understanding. arXiv:1810.04805.

Doré, O., Green, J., Hirata, C., et al., 2019. The Spectro-Photometer for the History of the Universe, Epoch of Reionization, and ices Explorer (SPHEREx): report on mission concept. arXiv e-prints, arXiv:1903.09208.

Florios, K., Kontogiannis, I., Park, S.H., Guerra, J.A., Benvenuto, F., Bloomfield, D.S., Georgoulis, M.K., 2018. Forecasting solar flares using magnetogram-based predictors and machine learning. Solar Physics 293, 28. https://doi.org/10.1007/s11207-018-1250-4. arXiv:1801.05744.

Fluke, C.J., Jacobs, C., 2020. Surveying the reach and maturity of machine learning and artificial intelligence in astronomy. WIREs Data Mining and Knowledge Discovery 10, e1349. https://doi.org/10.1002/widm.1349. arXiv:1912.02934.

Gowanlock, M.G., Kramer, D.A., Trilling, D.E., Butler, N.R., Donnelly, B., 2021. Fast period searches using the Lomb-Scargle algorithm on graphics processing units for large datasets and real-time applications. Astronomy and Computing 36, 100472.

Heinze, A.N., Tonry, J.L., Denneau, L., et al., 2020. The Asteroid Terrestrial-impact Last Alert System (ATLAS). The Astronomical Journal 160, 120. https://doi.org/10.3847/1538-3881/aba633.

Jones, R.L., Jurić, M., Ivezić, v., 2015. Asteroid discovery and characterization with the large synoptic survey telescope. Proceedings of the International Astronomical Union 10, 282–292. https://doi.org/10.1017/s1743921315008510.

Laureijs, R., Amiaux, J., Arduini, S., et al., 2010. Euclid: ESA's mission to map the geometry of the dark universe. Proceedings of the SPIE 7731, 77310E. https://doi.org/10.1117/12.857222.

Mainzer, A.K., Masiero, J.R., Abell, P.A., Bauer, J.M., Bottke, W., Buratti, B.J., Carey, S.J., Cotto-Figueroa, D., Cutri, R.M., Dahlen, D., Eisenhardt, P.R.M., Fernandez, Y.R., Furfaro, R., Grav, T., Hoffman, T.L., Kelley, M.S., Kim, Y., Kirkpatrick, J.D., Lawler, C.R., Lilly, E., Liu, X., Marocco, F., Marsh, K.A., Masci, F.J., McMurtry, C.W., Pourrahmani, M., Reinhart, L., Ressler, M.E., Satpathy, A., Schambeau, C.A., Sonnett, S., Spahr, T.B., Surace, J.A., Vaquero, M., Wright, E.L., Zengilowski, G.R., Team, N.S.M., 2023. The near-earth object surveyor mission. arXiv:2310.12918.

Radford, A., Narasimhan, K., Salimans, T., Sutskever, I., 2018. Improving Language Understanding by Generative Pre-Training.

Rawte, V., Sheth, A., Das, A., 2023. A survey of hallucination in large foundation models. arXiv:2309.05922.

Rehemtulla, N., Miller, A., Coughlin, M., Jegou du Laz, T., 2023. BTSbot: a multi-input convolutional neural network to automate and expedite bright transient identification for the Zwicky Transient Facility. arXiv e-prints, https://doi.org/10.48550/arXiv.2307.07618, arXiv:2307.07618, 2023.

Smirnov, E., 2024. Fast, simple, and accurate time series analysis with Large Language Models: an example of mean-motion resonances identification. The Astrophysical Journal.

Touvron, H., Vedaldi, A., Douze, M., Jégou, H., 2021. Training data-efficient image transformers & distillation through attention. In: Proceedings of the IEEE/CVF Conference on Computer Vision and Pattern Recognition, pp. 11926–11935.

Vaswani, A., Shazeer, N., Parmar, N., Uszkoreit, J., Jones, L., Gomez, A.N., Kaiser, L., Polosukhin, I., 2017. Attention is all you need. arXiv:1706.03762.

Zhang, J., Zhang, Y., Zhao, Y., 2018. Imbalanced learning for RR Lyrae stars based on SDSS and GALEX databases. The Astronomical Journal 155, 108. https://doi.org/10.3847/1538-3881/aaa5b1.

# Index

## U

## V

## X

## Z

Printed in the United States
by Baker & Taylor Publisher Services